Recent Advances in Epilepsy Research

ADVANCES IN EXPERIMENTAL MEDICINE AND BIOLOGY

A Continuation Order Plan is available for this series. A continuation order will bring delivery of each new volume immediately upon publication. Volumes are billed only upon actual shipment. For further information please contact the publisher.

Recent Advances
in Epilepsy Research

Edited by

Devin K. Binder

*Department of Neurological Surgery, University of California at San Francisco,
Moffitt Hospital, San Francisco, California, U.S.A.*

Helen E. Scharfman

*Center for Neural Recovery and Rehabilitation Research, Helen Hayes Hospital,
New York State Department of Health, West Haverstraw, New York, U.S.A.*
*Departments of Pharmacology and Neurology, Columbia University, New York,
New York, U.S.A.*

Kluwer Academic / Plenum Publishers
New York, Boston, Dordrecht, London, Moscow
Landes Bioscience / Eurekah.com
Georgetown, Texas U.S.A.

Kluwer Academic / Plenum Publishers
Eurekah.com / Landes Bioscience

Copyright '2004 Eurekah.com and Kluwer Academic / Plenum Publishers

Printed in the U.S.A.

Kluwer Academic / Plenum Publishers, 233 Spring Street, New York, New York, U.S.A. 10013
http://www.wkap.nl/

Please address all inquiries to the Publishers:
Eurekah.com / Landes Bioscience, 810 South Church Street
Georgetown, Texas, U.S.A. 78626
Phone: 512/ 863 7762; FAX: 512/ 863 0081
www.Eurekah.com
www.landesbioscience.com

Recent Advances in Epilepsy Research edited by Devin K. Binder and Helen E. Scharfman, Landes /
Kluwer dual imprint / Kluwer series: Advances in Experimental Medicine and Biology

ISBN: 0-306-47860-9

Library of Congress Cataloging-in-Publication Data

Recent advances in epilepsy research / edited by Devin K. Binder, Helen
E. Scharfman.
    p. ; cm. -- (Advances in experimental medicine and biology ; v.
548)
Includes bibliographical references and index.
 ISBN 0-306-47860-9
 1. Epilepsy--Genetic aspects. 2. Epilepsy--Research. 3.
Neuromuscular diseases--Research.
 [DNLM: 1. Epilepsy--physiopathology. 2. Epilepsy--genetics. WL 385
R294 2003] I. Binder, Devin K. II. Scharfman, Helen E. III. Series.
RC372.R38 2003
616.8'53--dc22

                    2003022445

INTRODUCTION

Epilepsy research has entered an exciting phase as advances in molecular analysis on a faster and larger scale have supplemented in vitro and in vivo electrophysiologic and phenotypic characterization.

The current volume sets forth a series of chapter reviews by researchers involved in these advances. It is not meant to be a comprehensive overview of the field of epilepsy research, but rather a composite profile of some of the recent investigations in certain select areas of enquiry.

Yan Yang and *Wayne Frankel* describe a genetic approach to studying seizure disorders in mice using a targeted mutagenesis method to exploit the genetic defects identified in human epilepsy families. This includes both the knock-in introduction of the human mutations into the corresponding mouse gene as well as analysis of seizure phenotype and the effect of mouse strain background on seizure threshold. Genetic mapping and isolation of the affected genes in these seizure-prone models will enable further characterization of molecular pathways involved in seizures.

Christine Gall and *Gary Lynch* review the potential contributions of integrins to epileptogenesis. The concept that ultrastructural alterations may interact with functional processes of synaptic plasticity within neurons is supported by observations that integrin adhesion receptors play crucial roles in stabilizing changes in neuronal plasticity. Seizures and even subseizure neuronal activity can modulate the expression of integrins, matrix ligands and proteases. Seizure-induced integrin modulation may contribute to lasting changes underlying epileptogenesis.

Excess activation of growth factors may contribute to hyperexcitability. *Devin Binder* reviews the biology and pathophysiology of brain-derived neurotrophic factor (BDNF). This ubiquitous neurotrophin is dramatically upregulated following seizures and various studies have shown that BDNF appears to contribute to epileptogenesis. Multiple adult CNS diseases in addition to epilepsy now appear to relate to either a deficiency or excess of BDNF.

Susan Croll, Jeffrey Goodman and *Helen Scharfman* address the role of vascular endothelial growth factor (VEGF) after seizures. VEGF induces angiogenesis, vascular permeability, and inflammation. Interestingly, receptors for VEGF have been localized to neurons and glia as well as to vascular endothelium. Croll and colleagues show that there is a striking increase in VEGF protein in both neurons and glia after status epilepticus thus   VEGF may contribute to blood-brain barrier breakdown and inflammation observed after seizures.

Ionotropic glutamate receptors have long been studied with respect to their contribution to hyperexcitability. Recently, the role of metabotropic (G-protein-coupled) glutamate receptors has also attracted attention. *Robert Wong, Shih-Chieh Chuang* and *Riccardo Bianchi* address the modulation of metabotropic glutamate receptors (mGluRs) in epilepsy. In particular, application of group I mGluR agonists appears to activate voltage-dependent depolarizing currents that lead to epileptogenesis in vitro.

Recent findings shed new light on the long-studied role of the γ-amino-butyric acid (GABA) system in the pathophysiology of epilepsy. *George Richerson* and *Yuanming Wu* review the role of the GABA transporter in seizures. Novel anticonvulsant drugs such as tiagabine appear to block GABA reuptake after synaptic release. Depolarization-induced reversal of the GABA transporter contributing to GABA release is powerfully inhibitory, and is enhanced by the anticonvulsants gabapentin and vigabatrin. Thus, recent data indicate that the GABA transporter plays a critical role not only in tonic inhibition but also in GABA release after seizures.

G nther Sperk , *Sabine Furtinger, Christoph Schwarzer* and *Susanne Pirke* review the neuropharmacology of GABA and its receptors in epilepsy. Epileptogenesis is associated with loss of a subset of GABAergic neurons as well as altered expression of GABA receptor subunits. Altered physiology and pharmacology of both $GABA_A$ and $GABA_B$ receptors may contribute to hyperexcitability in hippocampal and cortical networks, and specific knowledge of receptor subtype pharmacology may suggest novel therapeutic targets.

Kevin Staley discusses the role of the depolarizing GABA response in epilepsy. First described as a depolarizing response to prolonged application of GABA agonists, it can be elicited by focal activation of GABA receptors in dendrites (but not somata) of cultured pyramidal cells. Staley gives a clear explanation of the ionic shifts leading to alteration of the chloride reversal potential and accounting for the depolarizing GABA response. Blockade of the depolarizing response to GABA may be the anticonvulsant mechanism of acetazolamide, and may be an appropriate pharmacologic target.

Roger Traub, Hillary Michelson-Law, Andrea Bibbig, Eberhard Buhl, and *Miles Whittington* summarize the evidence that gap junctions may contribute to epileptogenesis in the hippocampus and cortex. In particular, they describe the discovery of a novel class of axo-axonal gap junctions that electrically interconnect hippocampal principal neurons. The ability of these gap junctions to promote very high-frequency neuronal oscillations may be critical in precipitating seizures, and thus gap junction inhibitors may be effective anticonvulsants.

One field of significant interest to epilepsy is the interaction of nervous and immune systems in the pathophysiology of disease. *Annamaria Vezzani, Daniela Moneta, Cristina Richichi, Carlo Perego* and *Maria Grazia De Simoni* review the potential functional role of pro- and anti-inflammatory cytokines in seizures. Cytokines are polypeptide hormones which interact with both neurons and glia, and are produced after limbic status epilepticus. Vezzani et al note the proconvulsant effects of interleukin-1β and the anticonvulsant effects of its inhibitor interleukin-1 receptor antagonist (IL-1Ra).

Deborah Young and *Matthew During* review studies of the neuroimmunology of epileptic syndromes, in particular recent data on GluR3 autoantibodies in Rasmussen s encephalitis. They have developed a novel approach for epilepsy and stroke treatment using vaccination with NMDAR1, and demonstrate that NMDAR1 vaccination generates autoantibodies to NMDAR1 that under pathologic conditions block injury-induced neuronal cell death. A vaccine approach is novel as well in that it is potentially prophylactic as well as therapeutic.

Malformations of cortical development (MCD) have received a great deal of recent attention. *Philip Schwartzkroin, Steven Roper*, and *H. Jurgen Wenzel* provide a comprehensive overview of MCD syndromes. These syndromes of cortical dysplasia frequently involve chronic seizures. Their categorization and anatomic and histopathologic features are extensively outlined and discussed. Current animal models of cortical dysplasias are described, which allow investigators to examine the developmental mechanisms that give rise to these brain lesions, and the relationship between structural abnormalities and epileptogenesis.

Peter Crino reviews recent progress regarding the genetics of MCD. Genetic analysis has identified genes for MCD including lissencephaly, subcortical band heterotopia, and tuberous sclerosis. The pathogenesis of other MCD such as focal cortical dysplasia, hemimegalencephaly, and polymicrogyria remains unknown, but new genetic techniques will allow characterization of gene expression within MCD.

Helen Scharfman addresses the consequences to epilepsy research of the recent recognition of neurogenesis in the adult brain. She describes the observations of neurogenesis after seizures, especially the newly-born dentate granule cells and their potential role in the hippocampal network and in epileptogenesis.

Roland Bender, Celine Dub and *Tallie Baram* explain the often-cited connection between early febrile seizures and later development of temporal lobe epilepsy (TLE). Their characterization of a novel immature rat model of prolonged febrile seizures has led to insights into the role of complex febrile seizures in hippocampal epileptogenesis.

Tim Benke and *John Swann* describe the tetanus toxin (TT) model of chronic epilepsy. In this model, tetanus toxin is injected into dorsal hippocampus or neocortex, where it is internalized and transported within neurons in which it cleaves synaptobrevin to inhibit release of neurotransmitter. A persistent epileptic state develops from a single unilateral hippocampal injection of TT in infancy.

Jeffrey Goodman reviews a new area that has garnered much attention of late, the use of brain stimulation to treat epilepsy. This area perhaps started with vagus nerve stimulation many years ago, but has recently grown rapidly to include brain targets. These treatments have been remarkably successful and have galvanized a related area of investigation: seizure prediction based on electrographic analysis.

The editors would like to thank all of the authors for their effort and expertise in presenting recent research results, and Ron Landes and Cynthia Conomos for tireless production assistance and advice.

PARTICIPANTS

Tallie Z. Baram
Departments of Anatomy
  and Neurobiology, Pediatrics
  and Neurology
University of California at Irvine
Irvine, California
USA

Roland A. Bender
Departments of Anatomy
  and Neurobiology and Pediatrics
University of California at Irvine
Irvine, California
USA

Timothy A. Benke
Cain Foundation Laboratories
Department of Pediatrics
Baylor College of Medicine
Houston, Texas
USA

Riccardo Bianchi
Department of Physiology/
  Pharmacology
State University of New York Health
  Science Center at Brooklyn
Brooklyn, New York
USA

Andrea E.J. Bibbig
Departments of Physiology,
  Pharmacology and Neurology
SUNY Downstate Medical Center
Brooklyn, New York
USA

Devin K. Binder
Department of Neurological Surgery
University of California
  at San Francisco
Moffitt Hospital
San Francisco, California
USA

Eberhard H. Buhl
School of Biomedical Sciences
University of Leeds
Leeds
England

Shih-Chieh Chuang
Department of Physiology/
  Pharmacology
State University of New York Health
  Science Center at Brooklyn
Brooklyn, New York
USA

Peter B. Crino
Penn Epilepsy Center
  and Department of Neurology
University of Pennsylvania
  Medical Center
Philadelphia, Pennsylvania
USA

Susan D. Croll
Department of Psychology
  and Neuropsychology Doctoral
  Subprogram
Queens College and the Graduate
  Center of the City University
  of New York
Flushing, New York
USA
Department of Neuro
  and Endocrine Biology
Regeneron Pharmaceuticals
Tarrytown, New York
USA

Maria G. De Simoni
Department of Neuroscience
Mario Negri Institute
  for Pharmacology Research
Milano
Italy

Celine M. Dub
Department of Anatomy
  and Neurobiology
University of California at Irvine
Irvine, California
USA

Matthew J. During
CNS Gene Therapy Center
Department of Neurosurgery
Jefferson Medical College
Philadelphia, Pennsylvania
USA

Wayne N. Frankel
The Jackson Laboratory
Bar Harbor, Maine
USA

Sabine Furtinger
Department of Pharmacology
University of Innsbruck
Peter-Mayr-Strasse 1A
Innsbruck
Austria

Christine M. Gall
Department of Anatomy
  and Neurobiology
University of California at Irvine
Irvine, California
USA

Jeffrey H. Goodman
Center for Neural Recovery
  and Rehabilitation Research
Helen Hayes Hospital
New York State Department
  of Health
West Haverstraw, New York
USA

Gary Lynch
Department of Psychiatry
  and Human Behavior
University of California at Irvine
Irvine, California
USA

Hillary Michelson-Law
Departments of Physiology,
  Pharmacology and Neurology
SUNY Downstate Medical Center
Brooklyn, New York
USA

Daniela Moneta
Department of Neuroscience
Mario Negri Institute
  for Pharmacology Research
Milano
Italy

Carlo Perego
Department of Neuroscience
Mario Negri Institute
  for Pharmacology Research
Milano
Italy

Susanne Pirker
Department of Pharmacology
University of Innsbruck
Peter-Mayr-Strasse 1A
Innsbruck
Austria

George B. Richerson
Departments of Neurology and
  Cellular and Molecular Physiology
Yale University and Veterans Affairs
  Medical Center
New Haven, Connecticut
USA

Cristina Richichi
Department of Neuroscience
Mario Negri Institute
  for Pharmacology Research
Milano
Italy

Steven N. Roper
Department of Neurological Surgery
  and McKnight Brain Institute
University of Florida College
  of Medicine
Gainesville, Florida
USA

Helen E. Scharfman
Center for Neural Recovery
  and Rehabilitation Research
Helen Hayes Hospital
New York State Department
  of Health
West Haverstraw, New York
USA
Departments of Pharmacology
  and Neurology
Columbia University
New York, New York
USA

Philip A. Schwartzkroin
Department of Neurological Surgery
University of California at Davis
Davis, California
USA

Christoph Schwarzer
Department of Pharmacology
University of Innsbruck
Peter-Mayr-Strasse 1A
Innsbruck
Austria

G nther Sperk
Department of Pharmacology
University of Innsbruck
Peter-Mayr-Strasse 1A
Innsbruck
Austria

Kevin J. Staley
Departments of Neurology
  and Pediatrics
University of Colorado Health
  Sciences Center
Denver, Colorado
USA

John W. Swann
Department of Pediatrics
Cain Foundation Laboratories
Baylor College of Medicine
Houston, Texas
USA

Roger D. Traub
Departments of Physiology,
   Pharmacology and Neurology
SUNY Downstate Medical Center
Brooklyn, New York
USA

Annamaria Vezzani
Department of Neuroscience
Mario Negri Institute
   for Pharmacology Research
Milano
Italy

H. Jurgen Wenzel
Department of Neurological Surgery
University of California at Davis
Davis, California
USA

Miles A. Whittington
School of Biomedical Sciences
University of Leeds
Leeds
England

Robert K.S. Wong
Departments of Physiology/
   Pharmacology and Neurology
State University of New York Health
   Science Center at Brooklyn
Brooklyn, New York
USA

Yuanming Wu
Departments of Neurology and
   Cellular and Molecular Physiology
Yale University
New Haven, Connecticut
USA

Yan Yang
The Jackson Laboratory
Bar Harbor, Maine
USA

Deborah Young
Functional Genomics and Translational
   Neuroscience Laboratory
Department of Molecular Medicine
   and Pathology
Faculty of Medical and Health
   Sciences
University of Auckland
Auckland
New Zealand

CONTENTS

4. VASCULAR ENDOTHELIAL GROWTH FACTOR (VEGF) IN SEIZURES: A DOUBLE-EDGED SWORD 57

Susan D. Croll, Jeffrey H. Goodman and Helen E. Scharfman

5. PLASTICITY MECHANISMS UNDERLYING mGLUR-INDUCED EPILEPTOGENESIS 69

Robert K.S. Wong, Shih-Chieh Chuang and Riccardo Bianchi

6. ROLE OF THE GABA TRANSPORTER IN EPILEPSY 76

George B. Richerson and Yuanming Wu

12. CORTICAL DYSPLASIA AND EPILEPSY: ANIMAL MODELS 145

Philip A. Schwartzkroin, Steven N. Roper and H. Jurgen Wenzel

13. MALFORMATIONS OF CORTICAL DEVELOPMENT: MOLECULAR PATHOGENESIS AND EXPERIMENTAL STRATEGIES 175

Peter B. Crino

14. FUNCTIONAL IMPLICATIONS OF SEIZURE-INDUCED NEUROGENESIS 192

Helen E. Scharfman

CHAPTER 1

Genetic Approaches to Studying Mouse Models of Human Seizure Disorders

Yan Yang and Wayne N. Frankel

Introduction

Epilepsy, characterized by recurrent spontaneous seizures resulting from abnormal, synchronized discharges of neurons in the brain, is one of the most common neurological problems afflicting humans. Although epilepsy clearly has a large environmental component, genetics is thought to be important in the pathogenesis of at least 50% of cases.[1] While common epilepsy genes have yet to be identified in humans, several genes have now been identified for rarer, monogenic epilepsies through linkage analysis or association studies followed by positional cloning.[2] In parallel, the identification of candidate genes in mouse epilepsy models also facilitates the discovery of human disease genes, as shown for at least one case of idiopathic generalized epilepsy.[3,4] Indeed, many features of seizures and the means by which they are induced are conserved in mammals,[5] indicating common neural substrates or pathways. Mice already contribute significantly to the discovery of all the currently available antiepileptic drugs and remain a critical part of the comprehensive screening process in the search for new anticonvulsant drugs. In general, mice can provide excellent genetic models for epilepsy by permitting systematic dissection of the molecular and pathophysiologic factors that predispose to seizures in large, genetically homogenous populations.[6]

Specifically, the ability to manipulate the mouse germline and the fact that most human disease alleles exhibit dominance makes it conceptually simple to introduce a human disease allele into mice. Experimentally, the most straightforward method is the transgenic approach where the gene of interest is injected into the pronuclear space of single-celled zygotes. However, because the DNA construct integrates randomly into the host cell genome, the expression of a transgene may be influenced by its site of integration and copy number. Therefore, multiple transgenic lines generated by the same DNA construct must be examined to evaluate the contribution of the gene to a particular phenotype. On the other hand, targeted mutagenesis through homologous recombination in mouse embryonic stem (ES) cells provides advantages in that the targeting construct is regulated by the mouse endogenous promoter and gene dosage is not disturbed, making the genotype more similar to that of the corresponding human disease.

In this chapter, we first discuss the potential of one particular gene targeting method – the "knock-in"—to introduce the human mutations into the corresponding mouse gene—an approach that will have increasing value as more common epilepsy alleles are identified in humans. We then discuss phenotypic evaluation procedures and the effect of mouse strain background on seizure threshold, with the goal of arriving at a systematic plan for characterizing mouse models of human genetic epilepsy. Optimizing the genetic and physiological characterization of these models is necessary to improve the chance that these new mouse models will

Recent Advances in Epilepsy Research, edited by Devin K. Binder and Helen E. Scharfman. ©2004 Eurekah.com and Kluwer Academic / Plenum Publishers.

show seizure phenotypes and thus contribute to our understanding of the mechanisms of human epilepsy.

The Knock-In Technique

The knock-in method is a modified version of the standard gene targeting technology in mice which uses homologous recombination to create a null mutation in a gene of interest. A mouse gene can be replaced by the mutated human version found in the corresponding human disorder. This approach provides useful tools to analyze genetic predisposition to disease phenotype as shown in the studies of polyglutamine disorders.[7]

When the respective mouse homologue and the mutations in human epilepsy families have been characterized, one can follow a few steps towards the generation of mice with the endogenous gene replaced by the human disease allele.[8,9] First, the mutation-bearing fragment of the gene as well as a positive selectable marker such as the neomycin resistance (*neo*) gene are cloned into an appropriate targeting vector. A negative selectable marker such as thymidine kinase (*tk*) gene should also be added to flank the gene sequence.

The second step is to introduce the targeting vector into embryonic stem (ES) cells via electroporation. During this step, selection through the negative marker will eliminate those clones where the entire targeting vector was randomly integrated into the ES cells' genome. Positive selection with G418 (an aminoglycoside related to neomycin) will ensure that the integration of the "knock-in" construct was mediated by homologous recombination. Further screening through PCR or Southern blot will be used to identify cells containing correctly integrated construct.

The third step involves microinjecting the "targeted ES cells" into the inner cavity of a blastocyst to produce chimeric embryos. Subsequently, these chimeric embryos are placed back into foster mothers to bring them to term.

The final step is to breed the chimeric mice to test whether the ES cells successfully entered the germ line of the chimeric founders. If some heterozygous animals are produced from the chimeric animal, the genetic part of the experiment is considered a success. Homozygous animals with both copies of the mutated gene can then be obtained, if desired, by intercrossing heterozygotes; if homozygous mutant mice are not lethal, 1/4 of the intercross progeny are expected to be homozygous for the mutation.

A well-designed targeting construct is critical to the successful generation of the "knock-in" mice and a positive selectable marker such as *neo* is routinely used to identify targeted ES-cell clones. However, it has been shown that sometimes the cryptic splice sites in *neo* interfere with normal splicing events and therefore reduce the expression of the targeted gene.[10] A solution to this problem in creating knock-in mice through Cre mediated site-specific recombination has been reported by Kask et al where the *neo* cassette was flanked by two *loxP* sites.[11] The subsequent removal of the *neo* was achieved by mating highly chimeric male to females of the "deleter" strain which express Cre in germ cells.[11]

Choosing a Target

A clear Mendelian mode of inheritance and sufficient numbers of family members carrying key recombination events to narrow chromosomal locations have been fundamental to the gene identification efforts so far in human epilepsy. All genes showed autosomal dominant transmission and most encode missense mutations, the exception being a null mutation in the *SCN1A* gene (Table 1). Thus, introducing these missense mutations into the mouse genome through a "knock-in" approach offers the potential to reproduce the epileptic symptoms in mice. Further, it is easy to obtain mice carrying both copies of the targeted allele, that is, barring the possibility of embryonic lethality. Since it is almost impossible to find patients carrying both copies of the disease allele, mice carrying no normal copies of the disease gene provide further opportunities to study the mutation's involvement in the pathophysiology of epilepsy in vivo.

Table 1. Known mutations in human families with idiopathic epilepsy[a]

Clinical Category	Gene	Gene Product	Reference
JME	*GABRA1*	GABA$_A$ receptor α1 subunit	12
EPT	*LGI1*	leucine-rich glioma-inactivated 1	13, 14
GEFS+3	*GABRG2*	GABA$_A$ receptor γ2 subunit	15, 16
GEFS+	*SCN2A*	sodium channel α2 subunit	17
GEFS+2, SMEI	*SCN1A*	sodium channel α1 subunit	18, 19
GEFS+1	*SCN1B*	sodium channel β1 subunit	20
IGE	*CACNB4*	calcium channel β4 subunit	4
BFNC1	*KCNQ2*	potassium channel subunit 2	21, 22
BFNC2	*KCNQ3*	potassium channel subunit 3	23
ADNFLE1	*CHRNA4*	nicotinic acetylcholine receptor α4 subunit	24
ADNFLE3	*CHRNB2*	nicotinic acetylcholine receptor β2 subunit	25

a) abbreviations: JME= juvenile myoclonic epilepsy, EPT= autosomal dominant lateral temporal epilepsy, GEFS+= generalized epilepsy with febrile seizures plus, SMEI= severe myoclonic epilepsy of infancy, IGE= idiopathic generalized epilepsy, BFNC= benign familial neonatal convulsions, ADNFLE= autosomal dominant nocturnal frontal lobe epilepsy.

It should be noted that the recently identified genetic defects only account for a small proportion of idiopathic epilepsy. For the common syndromes such as juvenile myoclonic epilepsy, childhood absence epilepsy or temporal lobe epilepsy, the underlying mutations still remain at large.[26,27] When more common human epilepsy mutations are identified, the case will be even stronger for generating and studying the corresponding mouse models. Still, creating "knock-in" mouse models based on current knowledge will not only facilitate the individualized anti-epileptic drug screening process but also further the establishment of robust assays to characterize mouse models in the future.

Phenotypic Characterization of Epileptic Mutant Mice

It is expected that the mice carrying human mutations will survive to adulthood since most of the human epilepsy syndromes are not associated with lethality. Once the genotype of a particular mutant line is confirmed, it is essential to put these animals through a set of well-characterized neurological and behavioral tests to examine how well the human epileptic conditions are recapitulated. It should be noted that some of the models may fail to produce equivalent phenotypes as observed in human—as shown in Huntington's disease[7] as well as in some cardiovascular diseases[28]—simply because of species differences between mouse and human. For epileptic seizures in mice, it may be difficult to distinguish whether these failures reside in the lack of occurrence or in our inability to observe seizures in mice. However, even when mutant mice do not show spontaneous seizures, susceptibilities to convulsive stimuli or further electrophysiological and molecular analysis may potentially contribute to the elucidation of the molecular pathways underlying epilepsy in humans. For example, it is difficult to observe naturally occurring benign neonatal convulsions in mouse pups, as would be desired in mouse mutant alleles of the human genes *KCNQ2* and *KCNQ3*, and yet *Kcnq2* knockout mice clearly have a low threshold to induced seizures.[29]

Behavioral Monitoring of Spontaneous Seizures

Behavioral monitoring can enable the detection of overt spontaneous seizures, however it is also important to note that quite often seizures can be readily provoked by routine handling. Although monitoring mice in their home cages can be tedious, video camera or activity monitoring instruments may provide convenient ways to detect aberrant spontaneous behavior in an unprovoked setting. Nevertheless, we emphasize that it takes some experience with mouse mutants to distinguish an unusual behavior that may appear, to the untrained eye, to be an epileptic seizure from a movement disorder or nonparoxysmal motor abnormality. Indeed, many spontaneous seizures do result in overt motor abnormalities. Landmark events during bona-fide seizures include (in increasing severity):

    a) excessive grooming (paw-paddling),
    b) rearing and/or loss of balance,
    c) excessive salivation or defecation,
    d) dorsal or ventral neck flexion,
    e) tonic-clonic jaw or limb extension,
    f) wild-running and jumping,
    g) tonic extension of the hindlimbs, sometimes followed by death.

These events are episodic, i.e., interspersed with normal activity, and mice having a bona-fide seizure are usually nonresponsive to external stimuli during these episodes. A comprehensive examination of the mouse's behavior, including body position, locomotor activity, gait and responses to various stimuli, is also necessary in order to exclude other major neurological abnormalities.

Detecting Seizures in Mice by Electroencephalography (EEG)

As mentioned earlier, it is possible that some of mutant lines may show spontaneous seizures. Although obvious convulsions can be identified through routine handling, the diagnosis of sub-clinical seizures relies on electroencephalography (EEG).[30] Moreover, as in humans, the gold standard for declaring any event a seizure in mice is to correlate the apparent behavioral seizure with the synchronous electrical discharges detected by EEG. Hence, EEG has been used to validate a convulsive seizure, to detect relatively silent petit-mal seizures and also inter-ictal events which are indicators of the state of neuroexcitability. Further, when combined with antiepileptic drug (AED) administration, it provides a tool to evaluate the efficacy of a particular drug. An effective treatment will either reduce the frequency or range of the abnormal discharge or completely abolish the spontaneous epileptiform activity.

However, it is not yet possible to reliably perform EEG noninvasively in mice for the purpose of evaluating seizures as surgery is required for the implantation of electrodes. There are various specialized ways to implant electrodes in living mice, but for the general evaluation and validation of seizure activity one common method is to implant 3-4 electrodes into the cerebral cortex. A differential signal can then be obtained by comparing the signal from one or more of these electrodes to a ground electrode anchored to the skull, or the signal between two electrodes. An alternative surgical procedure, developed by Noebels[31,32] is to slide two or more electrodes between the skull and the brain. The advantages of this are that the surgery is simpler than that of implants and, as in human, the electrodes do not penetrate the brain. However, the disadvantage is that it can be difficult to control or validate the placement of electrodes. Regardless of the method, to one degree or another artifacts stemming from the animal's movement, its breathing, or from vibrations of the electrode leads themselves can provide misleading results. Therefore, while EEG is considered to be a gold standard for establishing seizures in mouse, further technology development can only bring more value to epilepsy research from mouse models.

Seizure Threshold Models

Seizure disorders, whether in human or experimental animals, can be very difficult to study because the onset of an episode is not predictable, even by using EEG. Although it is desirable to have models with spontaneous seizures, if seizures can not be produced and measured "on demand" it can be quite a challenge to study mechanisms and treatments. Thus, induced seizure paradigms are important tools in the epilepsy laboratory. In general, since all mouse strains can be induced into a seizure if the stimulus is strong enough, the induced seizure models essentially examine whether the seizure threshold is lower, higher or the same in a mutant mouse compared to appropriate genetically-matched controls (described in detail below).

Electroconvulsive Threshold (ECT)

ECT is a robust method for testing seizure threshold in mice. ECT is the best-known industry standard model for assessing the efficacy of AEDs, by generating a dose-response curve in the presence of a convulsive current. Although there are a variety of ECT protocols that have been used, in general the endpoints of this test are well defined, the stimulus can be controlled exquisitely and a mouse can often be tested in $\leq$ 2 minutes.[33,34] Typical experimental techniques involve passing an electric current of less than a second's duration from one side of the animal's head to the other through the cornea or the external portion of the ears. Depending on how many times a mouse is tested, ECT response can be evaluated by either a multiple-test paradigm or a single-test paradigm. In a repeated-seizure testing procedure, a mouse is tested daily with a single treatment where the current setting increases in each successive trial until a chosen endpoint, e.g., maximal tonic hindlimb extension seizure, is induced.[35] Alternatively, a mouse is only tested once at a fixed current and the seizure threshold can be determined by testing groups of animals at different current intensities.[33] Recently, a third method with features from both repeated and single-testing procedures has been reported.[36] In this method, the intensity of the current increases linearly during the test until the occurrence of tonic hindlimb extension, allowing determination of the seizure threshold for individual animals in one test.[36]

The most common means of testing seizure susceptibility in mice by ECT is to use high-frequency pulses (e.g., $\geq$ 60 Hz) through transcorneal electrodes. The endpoints of this test are minimal seizures involving clonus of the jaws and forelimbs, and maximal seizures involving tonic hindlimb extension. The brain regions involved in each seizure type vary with stimulus strength and routes of stimulation, but generally, forebrain is more involved in minimal seizures and hindbrain in maximal seizures. Nevertheless, since maximal seizures are generated under the same stimulus route and electrical settings as minimal seizures except for the higher current applied, the difference in current required for these two seizure endpoints may also provide an indicator of the ability of seizures to spread throughout the brain. Although sensitivity to electrical stimuli is not a "clinical" phenotype per se, ECT accommodates experimental modalities such as AED treatment which provide the basis for relating a mouse mutant to the clinically relevant human pharmacology, forming the basis of the screening and evaluation of antiepileptic drugs in the United States and abroad.[34]

Chemoconvulsants

A large number of chemical convulsants, including bicuculline, excitatory amino acids (e.g., kainate, glutamate), gamma-hydroxybutyrate (GHB), beta carbolines, nicotine and pentylenetetrazol (PTZ), have been used to induce seizures in mice. Seizures produced by some of the chemoconvulsants are considered pharmacological models because the electrographic and behavioral features are consistent with those observed in humans. As a result, the PTZ seizure paradigm has been incorporated into the testing protocol of the National Institute of Health-sponsored Anticonvulsant Screening Project which examines the efficacy of potential AEDs in rodents.[37]

PTZ is a noncompetitive $GABA_A$ antagonist which lowers neuronal inhibition. Unlike excitotoxins such as kainic acid, PTZ does not cause cell death and is less likely to induce secondary epilepsies. Also, PTZ does not have to be metabolized to function in vivo and has a rapid turnover rate.[38,39] When administered subcutaneously, PTZ causes mice to display a dose-dependent progression of behavioral patterns, from myoclonic jerks to minimal clonic seizures to tonic extension of forelimbs and hindlimbs.[39] If the variable is a given genetic mutation versus a control, PTZ typically accelerates a seizure's onset or worsens severity. Correlated increases or decreases of the amount of PTZ required to induce a seizure event provide a quantitative measure of its neuroexcitatory potential. In addition to the subcutaneous PTZ test, intravenous PTZ infusion is a more accurate method for assessing seizure threshold.[39] In this test, PTZ is continuously administered to a mouse through its tail vein. The time it takes for a mouse to reach a quantifiable endpoint from the start of the infusion is recorded. Therefore, it offers a sensitive measure to detect changes in individual seizure thresholds. Further, a quantifiable endpoint can be obtained with only 8-10 mice per group. One potential technical difficulty of the intravenous PTZ test is the use of tail vein as the route of administration. It has been noted that finding the tail vein in some pigmented strains (such as C57BL/6) is a challenging task (Dr. H.S.White, personal communication). Nevertheless, chemoconvulsants are popular because they require no special equipment.

Audiogenic Seizures (AGS)

One of the oldest means of inducing seizures in mice is by loud acoustic stimuli, which, prior to the genomics era, was a popular means of examining strain differences to seizure threshold. The initiation and propagation of AGS activity is dependent upon hyperexcitability in the auditory system, particularly the inferior colliculus. Rodent AGS models represent generalized reflex epileptic behaviors in humans.[40] While AGS, like ECT, is a very robust approach because it is a physical stimulus that can be carefully controlled experimentally, its general utility is precluded by the facts that AGS relies on the ability of the mouse to hear: common mouse strains indeed have a range of audiogenic thresholds, including some strains that become deaf at a relatively young age.[41] Moreover, even within a strain, irrespective of the audiogenic threshold there may be particular ages when the mice are more sensitive. For these reasons, AGS is usually an approach that is best left for specialty laboratories.

Genetic Background on Seizure Phenotyping

It has been acknowledged that genetic background may have profound effects on the evaluation of the phenotypes of genetically manipulated mice, and that the exploitation of genetic variation inherent in the common inbred strains of mice provide unique tools, and sometimes obstacles, for neuroscience researchers.[42-44] As described earlier, the first batch of animals obtained from targeted mutagenesis is usually genetically mixed, often a hybrid between a 129/Sv subline and C57BL/6 for the easy identification of the chimeric mice due to the coat-color difference. At this stage, one has to decide on which inbred strain the seizure phenotype will be examined. Inbred mouse strains are generally known to vary in their seizure response to a variety of stimuli; although the earliest examples were shown for audiogenic seizures, where the strain sensitivity to hearing comes into play, just about every carefully studied seizure model since has shown mouse strain diversity to seizure stimuli. For example, studies by Schauwecker and Steward demonstrated that four commonly used inbred mouse strains responded differently to kainic acid (KA)-induced excitotoxic cell death.[45] Peripheral KA administration in rodents results in seizures and subsequently degeneration of neurons in the selective subfields in the hippocampus.[46] Although all four strains exhibited the same class of behavioral seizures, C57BL/6 and BALB/c mice did not show cell death after seizures whereas KA induced pronounced excitotoxic cell death in FVB/N and 129/SvEMS mice. Significantly, the strains involved in gene targeting demonstrated different susceptibility to KA.[45] Further, hybrid mice of 129/SvEMS × C57BL/6 generated through embryonic stems cell injection into blastocyst did not show evidence of cell loss after KA treatment, suggesting that the protection against KA came from the C57BL/6 chromosomes.[45] These findings underscore the importance of genetic background on behavioral phenotype analysis.

More recently, Frankel and colleagues examined the ECT responses in a number of common inbred mouse strains.[33] A broad spectrum of thresholds was observed, indicating that many neuroexcitability alleles exist in these mouse strains. Of the 16 inbred strains surveyed, DBA/2J, CBA/J and FVB/NJ showed a much lower threshold to seizures whereas C57BL/6J and BALB/cByJ were found to have high electroconvulsive threshold, consistent with previous findings in a chemoconvulsant seizure paradigm.[45] Interestingly, FVB ranked second in susceptibility to minimal clonic seizures and psychomotor seizures and first in susceptibility to maximal tonic hindlimb extension seizures. An interesting contrast is provided by CBA/J which was almost as susceptible as FVB to minimal clonic seizures but had a much higher threshold to maximal seizures—providing evidence that seizures spread much more readily in FVB mice.[33] In addition, FVB has been reported to have occasional spontaneous seizures.[47] Histologically, however, FVB mice seem to have similar hippocampal structures and cell counts when compared to other inbred strains (inferred from ref. 45). Taken together, FVB has emerged as an

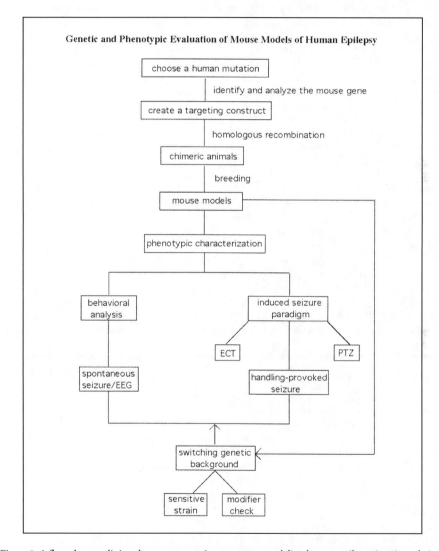

Figure 1. A flow chart outlining the reverse genetics strategy to modeling human epilepsy in mice as being discussed in this chapter. Please note that when the germline transmission of the mutation has been confirmed, one may desire to put the mutation onto different genetic backgrounds through a continuous backcrossing method. Then, seizure phenotyping and breeding can be carried out simultaneously.

intriguing seizure-sensitive strain with normal hippocampal morphology. Therefore, one standard approach that could be taken when evaluating new genetic seizure models that arise on other mouse strains will be to examine the phenotype on FVB background. Given that FVB mice are susceptible to seizure disorders, extensive convulsive phenotype and the related defects may be expected. In preliminary studies, this could be achieved by continuous backcross to FVB until the N_5 backcross generation is reached where, on average > 96% of chromosomal material unlinked to the mutation comes from FVB. If new seizure phenotypes such as spontaneous seizures are seen in mutant N_5 mice but not in littermate wild-type controls, further studies on a purer genetic background could be done after additional backcrosses to FVB.

Another advantage of moving the targeted allele to a different inbred mouse strain is the potential identification of genetic modifiers.[48] Given the wide range of seizure thresholds in the common inbred strains (Table 2), it is likely that the seizure phenotype caused by the human mutation can be suppressed or enhanced. To this end, seizure-sensitive strains such as FVB/NJ and DBA/2J and resistant strains such as C57BL/6J and BALB/cByJ could be considered since they offer genetic diversity and at the same time show different susceptibility to stimuli at least in those two seizure paradigms.[33,45] Continued backcross strategy again will be used to generate mice with the desired genotype. Genes carrying mutations that lead to phenotypes of interest provide one entry point into a biological pathway or pathological process. Any mechanism such as genetic modification by a second-site locus provides useful information on the pathogenesis of the disease phenotype, leading to more potential targets for therapeutic intervention.

Summary

In conclusion, we have discussed a reverse genetics approach to studying seizure disorders in mice (Fig. 1), employing a targeted mutagenesis method to exploit the genetic defects identified in human epilepsy families. After detailed characterization of the nature of the human mutation and the mouse counterpart gene, a targeting vector containing the human disease allele is created. The endogenous mouse gene is replaced by the human disease allele through homologous recombination in ES cells, leading to the generation of chimeric animals. Mice carrying one copy or both copies of the human mutation can be bred to study the phenotypic effect of heterozygous and homozygous mutations. At this stage, one may want to split the newly created mice into two groups. One group will go through seizure phenotyping tests, while the other group will be used to generate disease allele-carrying mice on a different genetic

Table 2. Different sensitivity to major seizure stimuli among five common mouse strains[a]

	ECT[b]	PTZ[c]	KA[d]	KA Excitotoxicity[e]
DBA/2J	Sen.	Sen.	Sen.	N/A
FVB/NJ	Sen.	N/A	N/A	Sen.
129	Sen. (129S1/SvImJ)	N/A	N/A	Sen. (129/SvEMS)
BALB/c	Res. (BALB/cByJ)	Int. (BALB/cJ)	N/A	Res.
C57BL/6J	Res.	Res.	Res.	Res.

a) abbreviations: ECT = electroconvulsive threshold, PTZ = pentylenetetrazol, KA = kainate, Sen. = relatively sensitive, Res. = relatively resistant, Int. = intermediate or paradigm-dependent
b) Ref. 33, minimal or maximal seizures
c) Ref. 49, except for BALB/cJ data inferred from Ref. 50
d) Ref. 51
e) Ref. 45

background. Phenotypic characterization of mice on different inbred strains includes behavioral monitoring and EEG analysis looking for the occurrence of spontaneous seizures, as well as routine cage examination looking for handling-provoked seizure and ECT- and PTZ- induced seizure paradigms looking for sensitivity to these stimuli. A complete evaluation of the seizure phenotype at the whole-animal level establishes the relevance of the mouse model to the human condition. Further investigation including imaging, electrophysiology and AED response in these mouse models will shed light on the mechanistic basis of the convulsive disorder.

Current epilepsy research in mouse genetics offers promise for understanding the molecular mechanisms that underlie epileptogenesis in humans. A large-scale forward genetic effort to create novel mouse mutants with seizure phenotypes by in vivo chemical mutagenesis with ethyl-nitroso urea (ENU) is underway at the Jackson Laboratory (http://www.jax.org/nmf/). Genetic mapping and isolation of the affected genes in these seizure-prone models will provide additional molecular pathways involved in seizures. The mutant mice generated through both forward and reverse genetic approaches will be a valuable resource for the biomedical community to study epilepsy at the molecular level and to characterize the pathological consequences of seizures in the whole organism.

Acknowledgements

We thank Drs. Greg A. Cox and Verity A. Letts for reviewing a preliminary version of this manuscript. We gratefully acknowledge research support from the National Institutes of Health (NS31348 and NS40246 to W.N.F.). Y.Y. is a Jackson Laboratory Fellow supported by the institutional funds of The Jackson Laboratory.

References

1. Anderson VE, Hauser WA, Rich SS. Genetic heterogeneity and epidemiology of the epilepsies. In: Delgado-Escueta AV, Wilson WA, Olsen RW, Porter RJ, eds. Jasper's Basic Mechanisms of the Epilepsies. Philadelphia: Lippincott Williams & Wilkins, 1999.
2. Meisler MH, Kearney J, Ottman R et al. Identification of epilepsy genes in human and mouse. Annu Rev Genet 2001; 35:567-88.
3. Burgess DL, Jones JM, Meisler MH et al. Mutation of the Ca^{2+} channel beta subunit gene *Cchb4* is associated with ataxia and seizures in the lethargic (*lh*) mouse. Cell 1997; 88:385-392.
4. Escayg A, De Waard M, Lee DD et al. Coding and noncoding variation of the human calcium-channel beta4-subunit gene CACNB4 in patients with idiopathic generalized epilepsy and episodic ataxia. Am J Hum Genet 2000; 66: 1531-1539.
5. Krall RL, Penry JK, Kupferberg HJ et al. Antiepileptic drug development: I. History and a program for progress. Epilepsia 1978; 4:393-408.
6. Frankel WN. Detecting genes in new and old mouse models for epilepsy: A prospectus through the magnifying glass. Epilepsy Res 1999; 36:97-110.
7. Gusella JF, MacDonald ME. Molecular genetics: Unmasking polyglutamine triggers in neurodegenerative disease. Nat Rev Neurosci 2000; 2:109-15.
8. Bronson SK, Smithies O. Altering mice by homologous recombination using embryonic stem cells. J Biol Chem 1994; 269:27155-8.
9. Capecchi MR. The new mouse genetics: Altering the genome by gene targeting. Trends Genet 1989; 5:70-6.
10. Lewandoski M. Conditional control of gene expression in the mouse. Nat Rev Genet 2001; 10:743-55.
11. Kask K, Zamanillo D, Rozov A et al. The AMPA receptor subunit GluR-B in its Q/R site-unedited form is not essential for brain development and function. Proc Natl Acad Sci USA 1998; 95:13777-82.
12. Cossette P, Liu L, Brisebois K et al. Mutation of *GABRA1* in an autosomal dominant form of juvenile myoclonic epilepsy. Nature Genet 2002; 31:184-189.
13. Kalachikov S, Evgrafov O, Ross B et al. Mutations in LGI1 cause autosomal-dominant partial epilepsy with auditory features. Nature Genet 2002; 30:335-341.
14. Morante-Redolat JM, Gorostidi-Pagola A, Piquer-Sirerol A et al. Mutations in the *LGI1/Epitempin* gene on 10q24 cause autosomal dominant lateral temporal epilepsy. Hum Mol Genet 2002; 11:1119-1128.

15. Baulac S, Huberfeld G, Gourfinkel-An I et al. First genetic evidence of GABA(A) receptor dysfunction in epilepsy: a mutation in the gamma2-subunit gene. Nat Genet 2001; 28:46-8.
16. Wallace RH, Marini C, Petrou S et al. Mutant GABA(A) receptor gamma2-subunit in childhood absence epilepsy and febrile seizures. Nat Genet 2001; 28:49-52.
17. Sugawara T, Tsurubuchi Y, Agarwala KL et al. A missense mutation of the Na+ channel alpha II subunit gene Na(v)1.2 in a patient with febrile and afebrile seizures causes channel dysfunction. Proc Natl Acad Sci USA. 2001; 98:6384-9.
18. Escayg A, MacDonald BT, Meisler MH et al. Mutations of SCN1A, encoding a neuronal sodium channel, in two families with GEFS+2. Nat Genet 2000; 24:343-5.
19. Claes L, Del-Favero J, Ceulemans B et al. De novo mutations in the sodium-channel gene SCN1A cause severe myoclonic epilepsy of infancy. Am J Hum Genet 2001; 68:1327-32.
20. Wallace RH, Wang DW, Singh R et al. Febrile seizures and generalized epilepsy associated with a mutation in the Na+-channel beta 1 subunit gene SCN1B. Nat Genet 1998; 19:366-70.
21. Singh NA, Charlier C, Stauffer D et al. A novel potassium channel gene, KCNQ2, is mutated in an inherited epilepsy of newborns. Nat Genet 1998; 18:25-9.
22. Biervert C, Schroeder BC, Kubisch C et al. A potassium channel mutation in neonatal human epilepsy. Science 1998; 279:403-6.
23. Charlier C, Singh NA, Ryan Sg et al. A pore mutation in a novel KQT-like potassium channel gene in an idiopathic epilepsy family. Nat Genet 1998; 18:53-5.
24. Steinlein OK, Mulley JC, Propping P et al. A missense mutation in the neuronal nicotinic acetylcholine receptor alpha 4 subunit is associated with autosomal dominant nocturnal frontal lobe epilepsy. Nat Genet 1995; 11:201-3.
25. De Fusco M, Becchetti A, Patrignani A et al. The nicotinic receptor beta 2 subunit is mutant in nocturnal frontal lobe epilepsy. Nat Genet 2000; 26:275-6.
26. Steinlein OK Genes and mutations in idiopathic epilepsy. Am J Med Genet 2001; 106:139-45.
27. Jacobs MP, Fischbach GD, Davis MR et al. Future directions for epilepsy research. Neurology 2001; 57:1536-1542.
28. Dietschy JM, Turley SD. Control of cholesterol turnover in the mouse. J Biol Chem 2002; 277:3801-3804.
29. Watanabe H, Nagata E, Kosakai A et al. Disruption of the epilepsy KCNQ2 gene results in neural hyperexcitability. J Neurochem 2000; 75:28-33.
30. Noebels JL. Single locus mutations in mice expressing generalized spike-wave absence epilepsies. Ital J Neurol Sci 1995; 16:107-11.
31. Noebels JL, Sidman RL. Inherited epilepsy: spike-wave and focal motor seizures in the mutant mouse tottering. Science 1979; 204:1334-1336.
32. Cox GA, Lutz CM, Yang CL et al. Sodium/hydrogen exchanger gene defect in slow-wave epilepsy mutant mice. Cell 1997; 91:139-48.
33. Frankel WN, Taylor L, Beyer B et al. Electroconvulsive thresholds of inbred mouse strains. Genomics 2001; 74:306-12.
34. Peterson SL, Electroshock. In: Peterson SL, Albertson TE, eds. Neuropharmacology Methods in Epilepsy Research. Boca Raton: CRC Press, 1998:1-26.
35. Ferraro TN, Golden GT, Smith GG et al. Mouse strain variation in maximal electroshock seizure threshold Brain Res 2002; 936:82-86.
36. Kitano Y, Usui C, Takasuna K et al. Increasing-current electroshock seizure test: A new method for assessment of anti- and pro-convulsant activities of drugs in mice. J Pharmacol Toxicol Methods 1996; 35:25-9.
37. White HS, Wolf HH, Woodhead JH et al. The National Institutes of Health Anticonvulsant Drug Development Program: Screening for efficacy. Adv Neurol 1998; 76:29-39.
38. Loscher W. New visions in the pharmacology of anticonvulsion. Eur J Pharmacol 1998; 342:1-13.
39. White HS. Chemoconvulsants. In: Peterson SL, Albertson TE, eds. Neuropharmacology Methods in Epilepsy Research. Boca Raton: CRC Press, 1998:27-40.
40. Skradski SL, Clark AM, Jiang H et al. A novel gene causing a mendelian audiogenic mouse epilepsy. Neuron 2001; 31:537-44.
41. Zheng QY, Johnson KR, Erway LC. Assessment of hearing in 80 inbred strains of mice by ABR threshold analyses. Hear Res 1999; 130:94-107.
42. Gerlai R. Gene-targeting studies of mammalian behavior: Is it the mutation or the background genotype? Trends Neurosci 1996; 19:177-81.
43. Bucan M, Abel T. The mouse: Genetics meets behavior. Nat Rev Genet 2002; 2:114-23.
44. Frankel WN. Mouse strain backgrounds: More than black and white. Neuron 1998; 20:183.
45. Schauwecker PE, Steward O. Genetic determinants of susceptibility to excitotoxic cell death: Implications for gene targeting approaches. Proc Natl Acad Sci USA 1997; 94:4103-8.

46. Ben-Ari Y, Cossart R. Kainate, a double agent that generates seizures: Two decades of progress. Trends Neurosci 2000; 23:580-7.
47. Goelz MF, Mahler J, Harry J et al. Neuropathologic findings associated with seizures in FVB mice. Lab Anim Sci 1998; 48:34-37.
48. Cox GA, Mahaffey CL, Frankel WN. Identification of the mouse neuromuscular degeneration gene and mapping of a second site suppressor allele. Neuron 1998; 1327-37.
49. Ferraro TN, Golden GT, Smith GG et al. Mapping loci for pentylenetetrazol-induced seizure susceptibility in mice. J Neurosci 1999; 19:6733-9.
50. Kosobud AE, Cross SJ, Crabbe JC. Neural sensitivity to pentylenetetrazol convulsions in inbred and selectively bred mice. Brain Res 1992; 592:122-8.
51. Ferraro TN, Golden GT, Smith GG et al. Mapping murine loci for seizure response to kainic acid. Mamm Genome 1997; 8:200-8.

CHAPTER 2

Integrins, Synaptic Plasticity and Epileptogenesis

Christine M. Gall and Gary Lynch

Abstract

A number of processes are thought to contribute to the development of epilepsy including enduring increases in excitatory synaptic transmission, changes in GABAergic inhibition, neuronal cell death and the development of aberrant innervation patterns in part arising from reactive axonal growth. Recent findings indicate that adhesion chemistries and, most particularly, activities of integrin class adhesion receptors play roles in each of these processes and thereby are likely to contribute significantly to the cell biology underlying epileptogenesis. As reviewed in this chapter, studies of long-term potentiation have shown that integrins are important for stabilizing activity-induced increases in synaptic strength and excitability. Other work has demonstrated that seizures, and in some instances subseizure neuronal activity, modulate the expression of integrins and their matrix ligands and the activities of proteases which regulate them both. These same adhesion proteins and proteases play critical roles in axonal growth and synaptogenesis including processes induced by seizure in adult brain. Together, these findings indicate that seizures activate integrin signaling and induce a turnover in adhesive contacts and that both processes contribute to lasting changes in circuit and synaptic function underlying epileptogenesis.

Introduction

Hypotheses about the causes of epileptogenesis usually posit that one or more of the following long-lasting changes have occurred at critical sites in brain: increased strength of excitatory connections, a loss or change in strength of inhibitory input, neuronal cell death, and aberrant axonal growth leading to greater excitability. The increased excitation hypothesis found important support with the discovery that high-frequency activity of a type not greatly different from that associated with seizures triggers a long-term potentiation (LTP) of glutamatergic transmission.[1,2] Specifically, in LTP bursts of high frequency or theta burst-like afferent activity lead to an enduring strengthening of transmission that is specific to synapses activated during the inducing phase.[3-5] Later studies demonstrated that brief seizures can themselves induce LTP.[6] Since then, an impressive list of points in common between LTP and epileptogenesis has been compiled.[7] NMDA-type glutamate receptors, for example, appear to be necessary for the induction but not the expression of either phenomenon across a range of experimental circumstances.[8-14] Di-acylglycerol metabolism and the density of voltage sensitive calcium channels are other more recent examples of factors reported to shape the probability of LTP and epileptogenesis. Thus, although not identical,[15,16] there is good evidence that LTP and epileptogenesis share cellular mechanisms. With regard to disturbances in inhibitory connections, there is a large literature showing that changes in the components of GABAergic syn-

Recent Advances in Epilepsy Research, edited by Devin K. Binder and Helen E. Scharfman.
©2004 Eurekah.com and Kluwer Academic / Plenum Publishers.

apses and in inhibitory transmission itself occur in both animal models[17-20] and human epilepsy.[21] This includes recent reports of aberrant binding to GABA receptors in the human condition.[22,23] Most likely related to these changes in inhibition, neuronal cell death has been observed in most experimental paradigms[17,24-28] and in human epilepsy.[25] The loss of specific populations of cells, and particularly of local circuit neurons in hippocampus, may account for at least some of the changes in the balance of excitatory and inhibitory activities and reactive axonal growth with epileptogenesis. Regarding the latter point, several laboratories have described axonal sprouting and aberrant innervation patterns in hippocampus and the dentate gyrus following seizures and in experimental models of epileptogenesis. This was originally described, and has been most fully characterized, for sprouting of the granule cell mossy fiber axons in animal seizure models[6,9,29-31] and in human epilepsy.[32,33] However, more recent studies have found evidence for local sprouting in field CA1 in both animal and human material.[28,34-37] Moreover, the balance of evidence indicates that in both fields sprouting contributes to synchronous neuronal activity and excitability. For example, in field CA1 new axonal growth gives rise to aberrant collaterals to the cell body layer and proximal dendrites, leading to new connections that are likely to underlie increased excitability. This growth is correlated in time with increases in the local response to applied glutamate and the increased probability that stimulation of individual pyramidal cells will elicit spontaneous network bursts.[28,37] Similarly, in the dentate gyrus the post-seizure emergence of aberrant mossy fiber collaterals to the dentate inner molecular layer has been associated with increased glutamate excitability and granule cell burst activity that cannot be accounted for by changes in inhibition.[38,39] Thus, in the better-studied animal models of epileptogenesis, evidence suggests that along with the more acute effects of seizure, there is a reorganization and growth of excitatory connections which potentiates network excitability and contributes to seizure proneness.[28]

As discussed throughout this volume, numerous cellular mechanisms have been proposed as steps leading to the epilepsy-related changes described above. One possibility that has begun to receive increasing attention involves cell adhesion chemistries. In particular, recent studies have shown that integrins, and other classes of adhesion receptors, are required for sustaining different phases of LTP and kindling. Moreover, there is evidence that integrins and their matrix ligands regulate GABAergic transmission[40] in the cortical telencephalon and in some cases mediate seizure-induced cell death.[41] Finally, cell adhesion and extracellular matrix proteins are well known to play critical roles in the motor-mechanisms and guidance of axonal growth and in the stabilization of synaptic contacts: recent studies have shown the same chemistries are active in adult neuroplasticity as well. In all, several phenomena thought to contribute to epileptogenesis are likely to be regulated by cell adhesion. This, combined with growing evidence that seizures activate a battery of adhesion-related enzymes and adhesion protein gene expression, points to the possibility that epileptogenesis involves a local breakdown of normal cell-cell and cell-matrix relationships followed by new synthesis and reactive axonal growth, and the appearance of stable, hyperexcitable arrangements.

The present review covers a selection of recent results relating to the above hypothesis. It begins with an introduction to the structure and cellular functions of integrin receptors, as they have been characterized in non-neural systems, and a description of results demonstrating that integrins are differentially expressed across regions of the adult brain. This will be followed by presentation of evidence that integrins contribute to the consolidation of LTP and kindling. Finally, results demonstrating that adhesive contacts and related chemistries are altered by seizures will be described and the implications of these findings to the development of seizure susceptibility will be discussed.

Integrins: Cell-Matrix and Cell-Cell Adhesion Receptors

There are three major families of transmembrane adhesion receptors throughout the body as well as several more minor groupings.[42] The major families, illustrated in Figure 1, are the integrins, cadherins, and immunoglobulin (Ig) adhesion proteins.[43,44] The cadherins and Ig

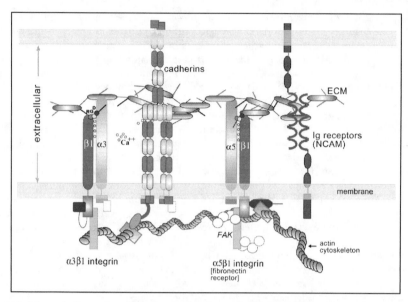

Figure 1. Schematic illustration of the three major classes of adhesion receptors discussed in the text. As shown, the cadherins form dimers which have homophilic interactions with like-cadherins on adjacent cells. The Ig class receptors, for which Neural Cell Adhesion Molecule (NCAM) is the prototypical example, also form homomeric cell-cell interactions. In contrast, the integrins are $\alpha\beta$ heterodimers: their extracellular domain binds extracellular matrix (ECM) proteins as well as proteins on adjacent cells whereas their intracellular domain is linked both (1) to the actin cytoskeleton via a variety of integrin-scaffolding proteins (e.g., vinculin, talin, actinin, spectrin) and (2) to kinase and G-protein signaling cascades. This includes direct associations with focal adhesion kinase (FAK) and Src kinases. Divalent cations (Mg^{+2}, Mn^{+2}, Ca^{+2}) regulate integrin activity for ligand binding through influences on the binding "pocket" between α and β strands.

adhesion proteins are both homophilic receptors that bind the extracellular domain of identical transmembrane proteins on adjacent cells. As shown, the cadherins are homodimers and their binding is calcium-dependent: calcium stabilizes a linear conformation and brings individual strands into the correct alignment for both dimerization and binding to cadherins on adjacent cells. Unlike cadherins and Ig receptors, the integrins are heterophilic receptors that bind extracellular matrix proteins such as osteopontin, collagen, tenascin-C, and reelin as well as transmembrane proteins on adjacent cells (e.g., VCAM-1).[45,46] Thus, integrins mediate both cell-cell and cell-extracellular matrix (ECM) adhesion and are the principal receptors for the ECM throughout the body.

As reviewed in detail elsewhere,[46-49] the integrins are non-covalently bound $\alpha\beta$ heterodimers, each containing one of 18 known α subunits and one of 8 known β subunits to give rise to over 20 dimer combinations. The α subunit has a relatively short cytoplasmic domain and is thought to be of particular importance for determining the ligand specificity of the receptor. The β subunit also participates in ligand recognition but importantly, it has a longer cytoplasmic domain that is associated with both the actin cytoskeleton and with a variety of signaling molecules.[46,50] Although the integrins have no catalytic activity themselves, they are seen as having both "inside out" and "outside-in" signaling functions that depend largely upon the three-dimensional conformation of the receptor.[47] Integrins within the membrane can exist in a latent or resting state, and can be activated by a variety of influences upon their cytoplasmic domain including signals from protein kinase C, calcium/calmodulin-dependent protein kinase II and G proteins.[47,51-53] These activities, as well as divalent cation levels,[45] lead to conformational changes that propagate from the cytoplasmic to the extracellular domain to activate the receptor and increase ligand binding activity. Similarly, ligand binding, and in particular binding of anchored ligands, or in some instances simple integrin receptor clustering, leads to conformational changes that propagate from the outside to the inside of the cell (hence, "outside-in" signaling) leading to changes in the physical arrangements of integrin-associated cytoskeletal proteins and signaling molecules.[46,51] The latter are likely to be of particular importance for functional synaptic plasticity. Integrins bind either directly or indirectly to a number of signaling intermediaries that are increasingly appreciated to play critical roles in synaptic efficacy and long term changes in synaptic function including focal adhesion kinase (FAK) and its homologue proline rich tyrosine kinase 2 (PYK2), Src family kinases and PI(3) kinase to name a few.[50,54-56] Through these and other elements, integrin ligand binding stimulates strong increases in tyrosine phosphorylation,[57] and increases signaling through Ras, Raf, MEK, and small GTPases (Rho, Rac among others).[50,54,58-64] As a consequence, integrins play critical roles in a wide variety of very basic cellular functions including cell cycling, phenotypic differentiation, gene expression, process outgrowth, cell survival and the regulation of intracellular pH and calcium levels.

Finally, regarding specialized neuronal functions to be discussed below, it is important to note that integrins have functional associations with other codistributed receptors and transmembrane proteins.[65] Regarding issues of synaptic plasticity, it is intriguing that functional interactions have been particularly well-demonstrated for integrins and tyrosine kinase growth factor receptors,[54,62-64] potassium channels[66] and voltage-dependent calcium channels.[61,67] As just one example of this, integrin binding has been shown to both stabilize codistributed epidermal growth factor receptor (EGFR) in the cellular membrane and, presumably through effects on receptor densities, to mediate EGF-independent signaling through the EGFR.[62] Findings such as this raise the intriguing possibility that synaptic integrins may interact with synaptic trophic factor receptors in processes mediating activity-induced synaptic plasticity.

Integrin Expression in the Adult CNS

Integrin functions within the CNS have been most extensively studied within the context of development,[68] and during early stages integrins have been shown to be important for fundamental processes of process outgrowth,[69,70] cellular migration,[71-74] and substrate recognition. However, there is now good evidence that integrins, along with the other major classes of adhesion receptors, are highly expressed in the adult brain.

Given the involvement of integrins in very basic adhesive functions, one might expect the integrins to be broadly expressed by virtually all cell types throughout the brain. Contrary to this, localization studies have shown that integrin expression profiles are highly differentiated, being both brain region- and cell type-specific. Moreover, while both neurons and glia clearly express integrins,[75-78] immunocytochemical and in situ hybridization studies have shown that in the adult integrins are most highly expressed by neurons. This point can be illustrated by results of in situ hybridization studies conducted in our laboratories. Distributions and levels of mRNAs for integrin subunits α1-8, αv, and β1-5 were evaluated in adult rat brain.[79,80] Figure 2 shows photomicrographs of autoradiographic labeling obtained with ^{35}S-cRNAs and illustrates some of the basic differences in expression profiles. Labeling patterns were highly varied: some subunits (e.g., α3, α5, αv, β5; Fig. 2B) were broadly expressed across brain with neuronal labeling throughout neocortex and hippocampus while other transcripts (e.g., α4, 1; Fig. 2A) were much more narrowly distributed and detected in only a few discrete cell types. Still others (e.g., β1, α7) were particularly highly expressed by neurons projecting outside of the CNS (e.g., efferent cranial nerve neurons and spinal motor neurons). As can be seen in these examples, integrin subunit expression conformed to neuronal cytoarchitectonics and labeling patterns clearly distinguished cell types within a specific region, and subfields within a particular brain area. The latter features can be seen in the subfield-specific labeling in hippocampus and layer-specific labeling in cerebellum. In hippocampus, integrin α7 expression clearly

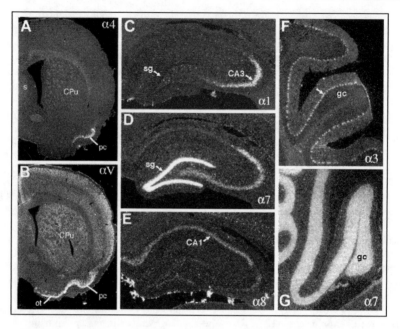

Figure 2. Integrin gene expression is both region- and cell-specific. Photomicrographs show the autorad-iographic localization of ^{35}S-cRNA in situ hybridization labeling of integrin mRNAs in adult rat brain (dark-field microscopy; cellular labeling seen as white). (A,B) Panels A and B show labeling of α4 and αv mRNAs, respectively, and illustrate subunits with very limited (A) and broad (B) distributions. In the plane shown, α4 mRNA is restricted to piriform cortex (pc) whereas αv mRNA is distributed across superficial and deep layers of neocortex, caudate/putamen (CPu), piriform cortex and olfactory tubercle (ot). (C-E) Photomicrographs show subfield-specific integrin expression profiles in hippocampus. As illustrated, α1 mRNA (C) is prominently expressed in CA3 stratum pyramidale alone, α7 mRNA (D) is abundant in stratum granulosum (sg) and CA3 stratum pyramidale, and α8 mRNA (E) is most dense in CA1 stratum pyramidale. (F,G) Photomicrographs of cerebellar cortex (hemispheres) showing α3 mRNA (F) is restricted to the Purkinje cell layer (arrow) while α7 mRNA is highly expressed in the granule cells (gc) as well as the Purkinje cells (Fig. 2G). Note, in panels C-G none of the transcripts are prominently expressed by cells in the molecular layers.

differentiates subfields, being highly expressed in the principal cell layers of the dentate gyrus and field CA3 but not field CA1 (Fig. 2D). In contrast, the α1 and α8 transcripts were most highly expressed in the pyramidal cells of CA1 and CA3, respectively (Fig. 2C,E). The absence of neuronal labeling in the hippocampal molecular layers indicates that expression is greatest in glutamatergic neurons (in the principal cell layers) and low in GABAergic neurons and glia that are scattered across the molecular layers. In cerebellum, α3 mRNA is discretely localized in the granule cells (Fig. 2F) while α7 mRNA is highly expressed in both the granule and Purkinje cells (Fig. 2G). Results of these studies reinforce the conclusion that a given cell type expresses more than one integrin.[80-83] Again, this point is illustrated by consideration of labeling patterns in hippocampus. CA1 pyramidal cells clearly express mRNAs for α subunits 1, 3, 5, 8 and v, and for β subunits 1 and 5. Moreover, PCR amplification of integrin transcripts[84] in adult hippocampus and localization of expression in immature neurons[85] (and Gall, unpublished results) indicates that these cells express β3 and β8 integrins as well. Together these findings indicate that the CA1 pyramidal cells express at least 8 integrins (α1β1, α3β1, α5β1, α8β1, αvβ1, αvβ3, αvβ5 and αvβ8).

The results of in situ hybridization studies (i.e., cell- and region-specific expression profiles, prominent neuronal labeling, multiple integrins expressed by single cells) are largely consistent with immunocytochemical descriptions of integrin expression in brain. Analyses of $\alpha 8$[76,86,87], αv[76], $\alpha 1$[88], $\alpha 5$[75,89,90] and $\beta 1$[91-93] have demonstrated localization in neuronal perikarya, axonal and dendritic processes, and glial cells. Moreover, these immunocytochemical studies indicate that integrins are localized, and in some cases seemingly concentrated, at synapses and limited aspects of dendritic arbors. In central mammalian neurons, $\alpha 1$, $\alpha 8$, αv and $\alpha 3$ have all been localized within synapse-like puncta or synaptic junctions at the light or electron microscopic level, respectively.[76,86,91,94] Our own studies of cultured neurons have demonstrated a concentration of $\alpha 5$-immunoreactivity (ir) at GluR1 and PSD-95 immunoreactive dendritic spines in addition to lower levels of $\alpha 5$-ir in dendritic processes and perikarya (Bernard and Gall, unpublished observations). Similarly, integrin proteins are concentrated at synaptic junctions in *Drosophila*.[95,96] These anatomical observations are consistent with results of neurochemical studies demonstrating that $\alpha 5$, αv, $\alpha 3$ and $\beta 1$ subunits are enriched in synaptic membrane fractions.[97-99]

As mentioned above, there is evidence that some integrins are preferentially distributed across dendritic arbors. In particular, we have observed that in hippocampus and neocortex $\alpha 5$-ir is dense in apical dendrites but below detection in basal dendrites of the great majority of pyramidal cells.[90] This contrasts with labeling of non-pyramidal cells which is clearly distributed across multiple primary dendritic arbors. The apparent apical compartmentalization of $\alpha 5$ is not accounted for by differences in synapse or spine densities. Indeed, given the concentration of $\alpha 5$ at spines and synapses this observation suggests a fundamental neurochemical difference between the apical and basal dendritic spines which could be of consequence to capacities for long term synaptic plasticity. We will return to this point below within the context of $\alpha 5$ involvement in the consolidation of LTP.

Adhesion Proteins Contribute to the Consolidation of Long-Term Potentiation

The argument that epileptogenesis arises from abnormal levels of excitatory input requires: 1) a mechanism whereby remarkable increases in EPSPs can be produced; and 2) a process that could sustain those increases for very long periods. The evidence, derived from studies of LTP reviewed below, suggests that integrins are critical to processes that can satisfy these requirements; i.e., as described, integrins are important for stabilization of enduring activity-induced increases in synaptic efficacy in hippocampal LTP. Having identified this as a plausible candidate cellular mechanism for sustaining hyperexcitability, the question arises as to whether it helps explain specific, poorly understood characteristics of epileptogenesis. By way of addressing this question, the following sections also briefly describe findings indicating that variations in adhesion chemistry may result in developmental and regional differences in the stability of synaptic modifications induced by intense neuronal activity.

LTP Has a Memory-Like Consolidation Period

Functional synaptic plasticity is widely recognized to pass through a 'consolidation' period during which it becomes progressively more resistant to disruption and is converted into a form with extraordinary persistence. This became evident in early studies of the consolidation of memory[100-102] and the principles identified in studies of learning and memory seem to apply to synaptic phenomena and, in particular, to LTP. Consolidation involves multiple processes that seemingly result in ever more enduring forms of encoding. Because of this, there is no single consolidation time although it is clear that with memory the process begins within seconds of learning and produces a reasonably stable memory within 30 minutes. The first evidence that LTP has a memory-like consolidation period came with the discovery that potentiation in anesthetized rats disappears if low-frequency afferent stimulation is applied within a few minutes of induction.[103] Subsequent studies using chronic recording confirmed that LTP does not recover even after delays of 24 hours and that high-frequency stimulation delivered after reversal produces a normal degree of potentiation, as expected if the original effect had in fact been erased.[104] Other work demonstrated that the erasure effect is present throughout the cortical telencephalon.[105,106]

The first in vitro studies of consolidation showed that hypoxia or infusions of adenosine would reverse potentiation if applied within 5 minutes of induction but were without effect 30 minutes later.[107] This time course resembles that described for memory consolidation in any number of paradigms.[100,101] Subsequent slice experiments obtained reversal with naturalistic theta patterns of stimulation[108] and showed that it probably involves adenosine receptors.[109,110] The demonstration that LTP becomes steadily less vulnerable over the 30 min post induction period[111,112] added another and important point of agreement with memory consolidation.

Multiple Adhesion Receptors Contribute to Consolidation of LTP

The logic for investigating the role of adhesion receptors in consolidation is straightforward. There is broad agreement that changes in the shape of synapses and spines accompany changes in synaptic efficacy, although the appropriate description of the change remains controversial. Given that synapses are adhesion junctions, it is unlikely that significant morphological reorganization of the type associated with LTP could occur without extensive modifications to integrins and other adhesion receptors.

Integrins bind discrete amino acid sequences in their matrix targets. Approximately one third of the known integrins bind the arginine-glycine-aspartate, or RGD, amino acid sequence that is shared by many matrix proteins (e.g., fibronectin, vitronectin, tenascin-C, thrombospondin);[45] differences in ligand specificity derive from amino acids flanking the central binding motif.[46,113,114] Small soluble peptides containing the RGD sequence interfere with matrix adhesion for these particular integrins[115] and disrupt a wide variety of integrin-mediated events (e.g., fibroblast attachment, platelet aggregation). To assess the contribution of synaptic RGD-binding integrins to LTP the putative antagonist peptides, or control media, were infused into acute hippocampal slices and effects on baseline physiology and LTP assessed. Infusion of RGD peptide (i.e., GRGDSP) into hippocampal slices had little effect on baseline transmission or the initial expression of LTP but caused potentiation to gradually dissipate over the 60 minutes following induction.[116,117] Additional studies found that RGD peptides were also effective if infused 2 minutes after the induction of LTP but were not effective when applied after delays of 25 minutes or longer.[118] As illustrated in Figure 3 (panels A-C), two snake toxins (a.k.a., disintegrins) that potently block integrin recognition of the RGD ligand-motif also disrupted the consolidation of LTP and, like RGD-peptides, were effective if applied after the inducing stimulation (Fig. 3B).[97] In all, the effects of RGD peptides on the stabilization of LTP support the hypothesis that synaptic stimulation engages integrin activities which then dictate the particular time course over which the increase in synaptic strength becomes stabilized.

As discussed above, hippocampal neurons express an array of integrins and at least some of these ($\alpha3\beta1$, $\alpha5\beta1$, $\alpha v\beta8$) are concentrated in synapses. The above results with antagonists that would be expected to interfere with functions of all RGD-binding integrins do not address the question of which particular receptors are involved in LTP consolidation. Given that integrins and their signaling activities are functionally distinct, identifying the specific dimer(s) subserving consolidation could be an important step towards understanding the manner in which the receptors help stabilize the potentiated state. This problem was addressed using function-blocking antibodies locally infused into hippocampal fields containing synapses that were about to be potentiated. The antibodies were infused during or immediately following theta-burst afferent stimulation and infusion was continued during the post-stimulus interval over which LTP was assessed. Two synaptically localized integrins were in this way implicated in consolidation: $\alpha5\beta1$[97] (Fig. 3D) and $\alpha3\beta1$.[98]

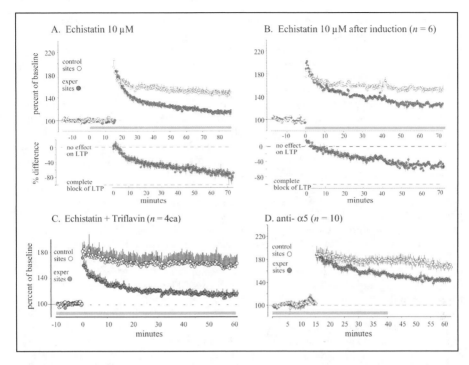

Figure 3. Blocking integrin function disrupts stabilization of LTP. In acute hippocampal slices, a stimulating electrode was placed in field CA1b stratum radiatum between two recording electrodes placed in CA1a and CA1c stratum radiatum. With this arrangement, in untreated slices theta-burst stimulation applied to field CA1b leads to comparable potentiation of synaptic responses at both recording electrodes. In the experimental preparations, the disintegrin echistatin (A,B), echistatin plus a second disintegrin triflavin (C), or a function-blocking antibody to integrin subunit α5 (D) was infused at the experimental recording site during the period indicated by the gray bar; the remaining recording site was used as a control. LTP was induced with trains of 10 theta bursts containing 4 pulses at 100 Hz and a burst separation of 200 ms.[4] The upper aspects of Panels A and B, and Panels C and D, show group average (± SEM) EPSP slopes for single test pulses applied to the same site before and after the LTP-inducing stimulation. In each of these plots, EPSP slope is presented as a percent of baseline values collected before induction of LTP; values from the control and experimental (drug-treated) sites are represented by open and filled circles, respectively. The lower plots in A and B show the percent difference between EPSP slopes for the control and experimental sites. As shown in A, in the presence of the disintegrin echistatin, LTP shows the same initial induction at the control and experimental sites but the experimental site LTP slowly declines and, after 1 hr, potentiation had almost disappeared at the experimental site but was robust and stable at the control site. (B) Echistatin applied after the induction of LTP also led to a loss of potentiation over 1 hr. (C) Treatment with two disintegrins, echistatin and triflavin, led to a similar but more rapid decay in LTP than did treatment with echistatin alone. (D) Treatment with anti-α5 caused a significant but slower decay in LTP; in other experiments the decay was shown to continue over 4 hrs of antibody treatment. For more details, refer to Chun et al, 2001.[97]

The effects produced by the function-blocking antibodies were partial and delayed compared to those achieved with RGD peptides. This opens the interesting possibility that LTP goes through more than one integrin-dependent consolidation stage. An early stage involving integrins other than α5 or α3 would, according to this argument, produce a long-lasting but slowly decremental LTP that can be largely reversed. A slightly later stage of consolidation would be expected to incorporate contributions from the α3β1 and α5β1 integrins and to result in a potentiation that is both stable and extremely resistant to interference. Sequential activation of integrins is not unknown; e.g., cell spreading is reported in some experiments to require activation first of αvβ3 and then of α5β1 receptors.[119] In all, the multiple stages of consolidation described for memory may find integrin-based analogues in long-term potentiation.

Finally, there is evidence that other classes of adhesion receptors in addition to integrins are involved in stabilizing LTP (see 44,120-122 for review). The number of synapses with high concentrations of neural cell adhesion molecule 180 (NCAM 180, the isoform with the longest cytoplasmic domain) increases markedly following LTP induction[123] while potentiation itself is impaired by function-blocking antibodies against NCAM[118] or the related Ig adhesion molecule telencephalin.[125] Moreover, a selective disruption of LTP stabilization is found in mutant mice lacking one of the polysialyltransferases that adds sialic acid to the extracellular domain of NCAM but that retain NCAM itself.[126] This indicates that a specific, well-characterized NCAM state is a necessary ingredient for consolidation. Other studies implicate the third class of synaptic adhesion receptors in LTP as well. Specifically, the number of synapses immunopositive for Neural (N)-cadherin, a cell-cell adhesion protein known to be involved in synaptogenesis during development and that is present in high concentrations in adult synapses, increases after induction of LTP.[127] The authors propose that the increase is part of the synapse building and remodeling that accompanies LTP.[128-130] Moreover, N-cadherin function-blocking antibodies have been shown to disrupt LTP stabilization in a limited number of experiments without disrupting induction or early expression.[127] In agreement with this, Tang et al[131] found that peptides designed to interfere with cadherin binding also interfere with LTP. Finally, function-blocking antibodies to a structurally related synaptic cadherin, arcadlin also impair LTP[132] but as this manipulation also degraded baseline synaptic responses the results do not demonstrate if arcadlin is specifically involved in synaptic plasticity or, alternatively, is important for synaptic transmission in general.

The above evidence makes a convincing argument that activity-driven changes in junctional adhesion can anchor excitatory synapses into a much more potent state. As discussed elsewhere in this review, a separate body of experiments shows that intense synaptic activity and seizures modify the abundance of extracellular matrix and adhesion receptors. Taken together, the results strongly suggest that lasting increases in excitatory drive will be a concomitant of epileptogenesis. Given the evident connection between excitatory transmission and seizure production, this conclusion provides support for the broad hypothesis that such increases in excitatory drive, and cellular processes mediating them, contribute to seizure susceptibility and epileptogenesis.

Adhesion Receptors Are Involved in Multiple Forms of Plasticity Including Kindling

Recently, it has become evident that integrin contributions to plasticity extend well beyond LTP and hippocampus. RGD peptides block activity-dependent synaptic changes in *Drosophila*.[96] Other work with mutant flies established that a variant of the α5 integrin subtype plays an essential role in plasticity.[133,134] Most directly relevant to the present review is evidence that integrin antagonists block the induction of seizure susceptibility as seen in a kindling model. Specifically, RGD peptides were infused into hippocampal slices for one hour prior to attempts to induce kindling with brief periods of high-frequency bursts of afferent stimulation. Baseline synaptic responses and the number of stimuli needed to induce bursting were described as being unaffected by RGD peptides but the later response to kindling stimulation was profoundly altered. That is, the RGD-treated kindled slices had a substantially lower rate of spontaneous bursts but a longer burst period than did slices that had been treated with control peptides. These results establish a direct link between integrins and one form of epileptogenesis.[135]

Variations in Synaptic Adhesion May Account for Regional and Developmental Differences in the Plasticity of Excitatory Synapses

The characteristics and consequences of seizures change during development[136,137] and vary markedly between brain regions. As with any argument about causal factors for epileptogenesis, the enhanced excitation hypothesis needs to address these variations. Accordingly, it may be useful to ask if the adhesion chemistries that appear to stabilize synaptic changes have ontogenetic or regional characteristics that would be predictive of marked differences in the degree to which excitatory inputs can be enhanced.

As described earlier, the brain is regionally differentiated with regard to expression of integrin subtypes and, according to recent results, it does appear that these are predictive of variations in LTP. The extent to which LTP has become consolidated can be assessed by applying low-frequency trains of stimulation at various times after induction and, as described by Karmár et al,[138] LTP in the apical dendrites of field CA1 is little affected by the trains at 30 min post-induction. The basal dendrites of the same CA1 neurons lack the $\alpha5\beta1$ integrin[90] that contributes to the stabilization in their apical counterparts.[97] Recent work has confirmed the prediction that consolidation is also incomplete in the basal dendrites.[138] In the basal dendrites, initial LTP is larger and more easily induced than that in the apical dendrites;[139] despite this, the new results indicate that it is also less stable. In all, there is a possibility that sites at which activity-induced changes in excitatory drive are easily established and readily recorded may not be the sites at which *lasting* adaptations (seizure proneness, LTP) emerge.

The unusual developmental history of the $\alpha5\beta1$ integrin also leads to some interesting predictions about plasticity. As described elsewhere,[90] the integrin does not move into apical dendrites until relatively late in postnatal maturation (i.e., during weeks 2-3 in rat), raising the possibility that integrin-dependent consolidation of LTP does not have the same developmental profile as the other components of synaptic plasticity. This was confirmed in a series of studies testing the reversibility of LTP at different post-natal ages.[138] LTP can be induced in rat hippocampus by postnatal day 9 but, at this age, in contrast to the adult, potentiation is fully erased by low-frequency stimulation even at 60 minutes post-induction. This indicates that a fully competent consolidation process may not emerge until well into the third postnatal week and during that period would be coincident with the appearance of the $\alpha5\beta1$ integrin in these same fields.

In summary, the emergence of adhesion molecules as a consolidation mechanism will likely expand the aberrant excitation hypothesis for epileptogenesis and allow it to make novel predictions.

Seizures Activate Changes in Adhesion Chemistries: Evidence for Turnover in Adhesive Contacts

As described in the introduction, recurrent seizures and kindling are associated with increases in excitability and the formation of aberrant circuitry (e.g., sprouting) and both phenomena have long been suspected to contribute to changes in seizure threshold and epileptogenesis. Knowing the involvement of various adhesion receptors and matrix proteins in facilitating and directing axonal growth during development, a number of investigators have proposed that these proteins may play a role in synaptic rearrangements that occur following seizures and have examined the possibility that seizures modify associations within the adhesive environment. The results of these studies, together with the growing understanding of activity-regulated proteolytic activities in brain, indicate that seizures induce a massive reorganization, or turnover, in adhesive chemistries that clearly facilitates growth and circuit adjustments and, thereby, may play a significant role in epileptogenesis. In particular, results from many laboratories indicate that activity- and seizure-regulated reorganization of adhesive proteins includes both proteolytic cleavage and replacement of integrin-mediated adhesive contacts. However, as this is only part of a much larger story, it is best considered within the broader context of the effects of seizures on adhesive chemistries as they occur, over time, from the onset of the seizure episode. With this in mind, the following paragraphs will describe the effects of seizures on proteolysis of adhesion proteins and on the expression of adhesion receptors and their matrix ligands. We have made an effort to include results that illustrate the magnitude of reorganization induced by seizure but the narrative should not be considered an exhaustive presentation of literature on the subject. Finally, the implications of these perturbations in adhesive chemistries for the functional consequences of seizure activity and epileptogenesis will be discussed.

Activity Regulates Proteolysis of Adhesion Proteins

Recent years have seen a tremendous increase in the understanding of regulated proteolysis in brain and, in particular, there is now very good evidence that seizures, and NMDA receptor-mediated neuronal activity in general, rapidly stimulate the activities of a battery of proteases that modify adhesion proteins and their associations both inside and outside the cell. In the intracellular domain, calcium influx associated with seizures and intense glutamatergic synaptic activation increases local activities of the neutral protease, calpain[140-143] that includes both integrins[144-146] and integrin-associated proteins[143,147-149] among its substrates. As calpain is localized within dendritic spines,[150,151] this proteolysis can be tightly associated with regions of activated synapses and, as a consequence, can be subfield- and possibly even synapse-specific within a given cell.[142] Although calpain-mediated integrin cleavage has not been directly demonstrated to be induced by seizures, Einheber et al[87] found that within the hilus of the dentate gyrus, kainic acid-induced seizures lead to a rapid loss of immunoreactivity for the $\alpha 8$ integrin cytoplasmic domain that is evident first within dendritic processes and only later in the perikaryon of mossy cells. The timing and distribution of change suggests that calpain, or another similarly regulated intracellular protease, accounts for this loss of integrin immunoreactivity.

Neuronal activity has also been shown to stimulate increases in extracellular proteases and proteolysis. Among these, the best studied within the contexts of seizures and synaptic plasticity are the trypsin-like serine proteases including tissue plasminogen activator (tPA) and neuropsin. Gene expression for tPA and neuropsin[152,153] and secretion of tPA[154,155] is stimulated by seizures and/or depolarization. Within the extracellular environment, tPA cleaves the integrin ligand plasminogen to liberate the additional trypsin-like protease, plasmin. Together these and related proteases cleave both matrix proteins and the adhesion receptors themselves. Specifically, seizures or brief NMDA receptor stimulation[156,157] lead to the extracellular cleavage of NCAM at a consensus serine protease site; this proteolysis can be mimicked by the tPA-plasmin system[157,158] and is blocked by a serine protease inhibitor of tPA.[158] Other studies have demonstrated that activities of the tPA-plasmin system also regulate levels of NCAM ligands phosphacan and neurocan,[159] the integrin ligand laminin[160] and, in other systems, the integrin, fibrinogen receptor.[161] These findings suggest that synaptic integrins may also be substrates for activity-regulated serine protease activity in brain but there is, as yet, no information on this point.

In additional to the trypsin-like proteases, seizures increase the expression and activities of metalloproteinases (MMPs) in brain. The MMPs are secreted and both modify the extracellular environment and reveal adhesion-receptor epitopes though proteolysis of ECM proteins.[162] Kainate seizures increase MMP-9 mRNA, protein, and activity levels within hippocampus.[163,164] This is associated with local increases in the expression and extracellular distribution of a natural inhibitor of metalloproteinase (i.e., TIMP-1).[165] Thus, seizures appear to stimulate a controlled MMP-mediated proteolysis of the ECM in adult brain. Given the involvement of MMPs in modifying the extracellular environment to facilitate growth and cellular migration in peripheral tissues and brain,[166] the possibility that these enzymes facilitate axonal growth and, in particular, seizure-induced sprouting and synaptogenesis in hippocampus is particularly attractive.

Seizures Regulate the Expression of Adhesion Receptors

While seizures initiate proteolytic activities that degrade adhesion receptors and matrix proteins, they also initiate changes in the expression of these same target proteins that would be expected to replace those lost by proteolysis but which would also generate additional proteins that may facilitate axonal growth, synaptogenesis and changes in synaptic efficacy that accompany seizures and contribute to epileptogenesis. In particular, in situ hybridization and immunocytochemical studies have shown that seizures, and lesser fluctuations in neuronal activity, differentially regulate the expression of integrins, cadherins, and Ig cell adhesion molecules (CAMs).

Evidence that seizures influence integrin expression comes from in situ hybridization analyses in our laboratory. We have evaluated the effect of recurrent, lesion-induced limbic seizures on mRNA levels for $\beta1$ integrin and its dimer partners, subunits $\alpha1$-8 and αv in adult rat brain. This includes consideration of subunits forming RGD-binding integrins potentially involved in processes of LTP and kindling discussed above (i.e., $\alpha3\beta1$, $\alpha5\beta1$, $\alpha8\beta1$, and $\alpha v\beta1$). In the hilar lesion paradigm used, bilateral recurrent limbic seizures occur intermittently over a period of 8 to 10 hours with the greatest number of seizures clustered early in the episode.[167] As illustrated for select transcripts in Figure 4, we found that each of the integrin mRNAs are influenced by seizures but in distinctly different ways. $\beta1$ mRNA is broadly increased during the seizure episode within both neurons and astroglial cells. Transcript levels increase during the period of seizures and slowly decline thereafter.[79] Among the α subunits, responses are highly transcript-specific.[168] Transcripts for $\alpha1$, $\alpha2$, $\alpha8$ and αv exhibit a monophasic increase either early (e.g., αv; Fig. 4A,D) or late (e.g., $\alpha8$ and $\alpha1$; Fig. 4B,E) in the seizure episode. The $\alpha5$ and $\alpha6$ subunit mRNAs are also induced by seizure but these transcripts are clearly increased in both neurons and glia (Fig. 4C,F). The $\alpha3$ and $\alpha7$ mRNA levels are decreased during the period of recurring seizures but increase above control levels at later intervals. Finally, $\alpha4$ mRNA, which is one of the lesser abundant transcripts in untreated adult rat brain,[80] declines to undetectable levels in entorhinal cortex through the period of seizures.

Together, the changes in integrin gene expression induced by seizures show that neuronal activity regulates integrin expression in very discrete, transcript-specific ways. This includes differential regulation of colocalized transcripts. For example, in neuronal layers of hippocampus, the $\alpha3$ and $\alpha7$ mRNAs are decreased during the same interval in which αv and $\alpha1$ mRNA levels are increased. Integrin subunits compete for dimer pairings. Thus, it is likely that changes in the availability of α partners for the $\beta1$ subunit over time leads to changes in the integrin receptors expressed at the membrane; in this instance we would predict an decrease in $\alpha3\beta1$ and an increase in $\alpha v\beta1$ in the dentate granule cells in the aftermath of seizure activity. Furthermore, given evidence that receptor dimer composition determines the ligand specificity and signaling properties of the receptor, these changes could underlie fundamental differences in the contribution of synaptic integrin receptors to neuronal functions including axonal growth and activity-dependent modifications in synaptic efficacy during, and in the wake of, seizures.

Other investigators have shown that neuronal activity regulates the expression of cadherins and NCAM and have linked these effects with reactive growth and synaptic plasticity. Seizures have been reported to increase expression of N-cadherin[169] and the cadherin family member arcadlin;[132] as described above, both proteins are implicated in stabilizing synaptic potentiation in hippocampus. Moreover, in the case of N-cadherin, pilocarpine seizure-induced increases are rather long-lasting (lasting several weeks) and associated with increases in immunoreactive puncta in the dentate gyrus inner molecular layer[169] suggesting that newly expressed proteins are localized in growing sprouts and nascent synapses of the mossy fibers that are induced by seizure.[29,170,171] Other studies demonstrating induction of increases in synaptic N-cadherin with LTP-inducing cAMP treatment[127] and patterned afferent stimulation[172] suggest that activity induction of this class of adhesion protein may be important to cadherin's contribution to functional synaptic plasticity as well.[44,120] For the Ig CAMs, increased expression has been linked to patterned afferent stimulation,[172] the induction of LTP[123,173] or poten-

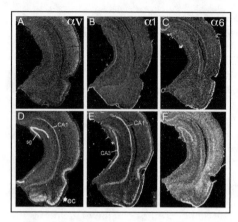

Figure 4. Seizures stimulate integrin gene expression. Photomicrographs show autoradiographic localization of ^{35}S-cRNA in situ hybridization labeling of mRNAs for integrin subunits αv (A,D), α1 (B,E) and α6 (C,D) in coronal sections through caudal hippocampus and cortex in control rats (A-C) and experimental-seizure rats killed 6 hr (D) or 24 hr (E,F) after the onset of recurrent limbic seizures. As shown, seizures induce marked increases in αv (D) and α1 (E) mRNAs that are restricted to neuronal cell layers and are particularly striking in hippocampus (sg, stratum granulosum; CA1 and CA3, stratum pyramidale indicated). In contrast, seizure-induced increases in α6 mRNA (F) are more diffusely distributed and are localized in both neurons and glia.

tiation of AMPA-class glutamate receptor currents with ampakine drugs.[174] Together, these observations indicate that Ig CAM expression can be positively regulated by neuronal activity and, in some instances, with very low threshold.

Seizures Regulate the Expression of Matrix Proteins

Although matrix proteins have proven difficult to study with anatomical techniques in the adult brain, there is evidence that seizures modulate the expression of several including, most particularly, the integrin ligands tenascin-C, vitronectin, and fibronectin. Kainic acid-induced seizures increase the expression of tenascin-C within a few hours of seizure onset; mRNA levels peak at 24 hrs and then decline.[175,176] Within hippocampus, increases in tenascin-C mRNA are prominently localized to the dentate gyrus granule cells, although this protein is expressed by reactive astrocytes as well. Similarly, kainate seizures induce fibronectin mRNA and protein in hippocampus; in situ hybridization and immunocytochemical studies indicate increases are localized to both neurons and astroglia.[177] Following intra-amygdalar kainic acid injection, seizures are associated with neuronal cell death in the hilar and CA3 regions of hippocampus; in rats so treated there is a delayed increase in vitronectin-ir in reactive astroglia in hippocampal fields of damage.[178] These results demonstrate that in adult brain astroglia express vitronectin but the increase observed in this instance may be part of the glial response to local neuronal damage as opposed to the effects of seizure activity.

A more complex response to seizure is exhibited by agrin and the chondroitin sulfate proteoglycan receptor of protein tyrosine phosphatase β (RPTPβ). Agrin is best characterized as a matrix protein at neuromuscular junctions where it is involved in clustering of acetylcholine receptors. However, it is also broadly expressed by neurons in brain[179] and recent studies indicate that for central neurons agrin regulates patterns of dendritic and axonal growth[180] and positively regulates vulnerability to excitotoxic damage with seizure.[181] We found agrin gene expression is regulated in both positive and negative directions by recurrent limbic seizures.[179] In hippocampus agrin mRNA levels are depressed during a recurrent seizure episode and then increase to supranormal levels through 4 days after seizure termination. In contrast, agrin mRNA is reduced in post-seizure intervals within the neocortex. Similarly, RPTPβ-ir, which is normally evident in glia, is reduced in this compartment during status epilepticus but increased in the extracellular environment; this is followed by increased RPTPβ levels in neurons.[182]

Significance of Seizure-Induced Proteolysis and Adhesion Protein Expression to Epileptogenesis

The above paragraphs demonstrate that seizures, and in some instances more modest increases in neuronal activity, lead to proteolysis of adhesion proteins that overlap and are followed by increases in the expression of both the transmembrane adhesion receptors and several of the major matrix proteins. Do these events contribute to the functional sequelae of seizures and to susceptibility for future seizures? Results pertinent to these important questions are sparse but sufficient to answer "yes" in both instances. Perhaps the most complete stories are associated with activity- and seizure-induced trypsin-like proteolytic activity. As described, seizures and LTP stimulation induce increases in tPA expression and secretion, and proteolysis of extracellular adhesion proteins including NCAM and laminin (see above). Several studies indicate this process is important for activity-dependent increases in synaptic strength and, importantly for the present discussion, to kindling. Hoffman et al[158] demonstrated that in hippocampal slices treatment with a serine protease inhibitor (AEBSF), that significantly blocks tPA proteolysis of NCAM, also blocks the formation of stable LTP without disrupting baseline synaptic responses. This accords with the observation that mice lacking tPA have specific deficiencies in the expression of late LTP.[183] In a separate study, Hoffman and colleagues found that systemic treatment with the AEBSF inhibitor of tPA just prior to the induction phase of rapid electrical stimulation-induced kindling, blocked kindling as assessed by measures of stimulated afterdischarge (AD) duration 10 days after the inducing session.[184] In vehicle-treated rats, mean AD duration was 3.5-fold longer 10 days after the single kindling session as compared to baseline responses. In contrast, in rats with protease inhibitor treatment, AD durations were comparable on the first and last days of recording.

Seizures are well known to lead to the death of vulnerable neurons and to axonal sprouting, as most clearly demonstrated in hippocampus. Recent findings link tPA activities to both of these phenomena as well. Kainic acid (KA)-induced seizures are well known to lead to death of hippocampal CA3 neurons. Strickland and colleagues [41,185] have demonstrated that KA-induced necrosis is absent in mice deficient for tPA or the plasmin precursor plasminogen. Moreover, this same group found that KA-seizures normally cause tPA/plasmin dependent reductions in laminin-ir in hippocampus that are coincident with neuronal death and that neutralizing antibody-mediated reductions in laminin bioactivity lead to cell death even with intact tPA/plasmin activities.[160] It is argued that these results demonstrate that integrin-mediated adhesion to ECM laminin is protective and that loss of contact with this matrix substrate, as occurs with seizure-induced proteolysis, causes neuronal cell death. Together with results described in the preceding paragraph, these findings implicate tPA in two adverse phenomena induced by seizure: increases in seizure proneness and neuronal cell death.

It is likely that seizure effects on tPA and the MMPs contribute to axonal sprouting. In tPA-deficient mice there is decreased seizure-induced mossy fiber sprouting and an accumulation of the ECM chondroitin sulfate proteoglycan DSD-1-PG/phosphacan, a matrix protein which can inhibit neurite outgrowth and oppose NCAM function, in the dentate gyrus inner molecular layer.[159] This suggests that the absence of phosphacan proteolysis by tPA leads to an accumulation of a putative growth regulatory protein in the normal target of mossy fiber ingrowth that inhibits or redirects the sprouting response. Although less is known about specific MMP functions following seizures, results from other systems suggest that seizure-induced increases in MMP expression should facilitate matrix remodeling and, in particular, axonal growth through the ECM. For example, with damage to spinal cord[186] or peripheral nerve[187,188] MMP levels are increased in association with growth cones and neurites; MMP enzymatic activities allow the neurites to overcome the growth inhibitory activities of matrix chondroitin sulfate proteoglycans and to grow through the ECM. Similarly, with myelination MMP-9 is expressed on the leading tip of oligodendroglial processes and either blocking MMP-9 gelatinase activity or the absence of MMP-9 in null mutants retards the extension of these processes and myelin formation.[189]

Increases in the expression of the adhesion receptors and matrix proteins are also likely to facilitate axonal growth following seizures. Clearly both classes of molecules are involved in growth processes: the matrix proteins provide the substrate for growth and can facilitate or retard outgrowth depending on the circumstance and the specific adhesion receptors expressed on growing neurites. On the other side of the equation, the adhesion receptors provide mechanisms for surface recognition, the mechanical force needed for process extension, and a means for stabilizing synapses as they mature (68,190,191 for reviews). More specifically for integrins, increases in gene expression have been positively correlated with axonal growth and experimentally induced increases in expression stimulate process outgrowth. With damage to peripheral nerve in adult rats, the ECM proteins laminin[192] and fibronectin[193] are upregulated in the damaged nerve and both proteins facilitate axonal regeneration;[194] this facilitation is mediated by integrins α6β1 and α7β1 that are specific for laminin and are upregulated in motor neurons by nerve damage.[83] With damage to the facial nerve, α7 integrin levels are increased on growth cones of regenerating axons, and transgenic ablation of α7 expression reduces axonal elongation and delays reinnervation.[195] Other studies have shown that overexpression of integrins in cultured cells facilitates axonal growth, most particularly in the presence of chondroitin sulfate proteoglycans thought to limit axonal extension in adult brain.[196] The same adhesion proteins shown to facilitate axonal growth in these lesion studies (i.e., fibronectin; integrin subunits α1, α6, α7, and β1) are induced by seizures and are elevated during periods of seizure-induced axonal growth. Together, these results indicate that coordinated seizure-induced increases in integrin and matrix protein expression contribute both to the remodeling of axonal connections known to occur in these same paradigms and to changes in circuit excitability associated with these aberrant innervation patterns.

Concluding Comments

Studies described here have shown that several classes of adhesion proteins contribute to activity-induced changes in synaptic efficacy and, in particular, to potentiation of glutamatergic transmission. Among these, the integrins are of particular importance for the stabilization of synaptic change in both LTP and kindling, and may determine the time course over which these changes in synaptic efficacy consolidate. Other studies have shown that seizures stimulate integrin and matrix protein expression, and extracellular proteolysis; these processes are in some instances known, and in other instances likely, to facilitate reactive axonal growth and circuit modification following seizures. Together, these results indicate that integrins and their matrix targets are significantly involved in multiple processes that contribute to the development of epilepsy most particularly in adult brain.

Acknowledgements

The authors thank our many laboratory colleagues and collaborators who contributed to research described in this review. This work was supported by grants NS37799 and MH61007.

References

1. Bliss TVP, Lômo T. Long-lasting potentiation of synaptic transmission in the dentate area of the anesthetized rabbit following stimulation of the perforant path. J Physiol 1973; 232:334-356.
2. Lynch G, Bahr BA, Vanderklish PW. Induction and stabilization of LTP. In: Ascher P, Choi DW, Christen Y, eds. Glutamate, cell death and memory. Berlin: Springer Verlag, 1991:45-60.
3. Staubli U, Lynch G. Stable hippocampal long-term potentiation elicited by "theta" pattern stimulation. Brain Res 1987; 435:227-234.
4. Larson J, Lynch G. Theta pattern stimulation and the induction of LTP: the sequence in which synapses are stimulated determines the degree to which they potentiate. Brain Res 1989; 489:49-58.
5. Larson J, Wong D, Lynch G. Patterned stimulation at the theta frequency is optimal for the induction of hippocampal long-term potentiation. Brain Res 1986; 368:347-350.
6. Ben-Ari Y, Represa A. Brief seizure episodes induce long-term potentiation and mossy fiber sprouting in the hippocampus. Trends Neurosci 1990; 13:312-314.
7. Baudry M. Long-term potentiation and kindling: similar biochemical mechanisms. Adv Neurol 1986; 44:401-410.
8. Muller D, Lynch G. Long-term potentiation differentially affects two components of synaptic responses in hippocampus. Proc Natl Acad Sci USA 1988; 85:9346-50.
9. Sutula T, Kock J, Golarai G et al. NMDA receptor dependence of kindling and mossy fiber sprouting: evidence that the NMDA receptor regulates patterning of hippocampal circuits in the adult brain. J Neurosci 1996; 16:7398-7406.

10. Behr J, Heinemann U, Mody I. Kindling induces transient NMDA receptor-mediated facilitation of high-frequency input in the rat dentate gyrus. J Neurosci 2001; 85:2195-2202.
11. Sayin U, Rutecki P, Sutula T. NMDA-dependent currents in granule cells of the dentate gyrus contribute to induction but not permanence of kindling. J Neurophysiol 1999; 81:564-574.
12. Loscher W. Pharmacology of glutamate receptor antagonists in the kindling model of epilepsy. Prog Neurobiol 1999; 54:721-741.
13. Lynch G, Kessler M, Arai A et al. The nature and causes of hippocampal long-term potentiation. Prog Brain Res 1990; 83:333-339.
14. Collingridge GL, Singer W. Excitatory amino acid receptor and synaptic plasticity. Trends Pharm Sci 1990; 11:290-296.
15. Cain DP. Long-term potentiation and kindling: how similar are the mechanisms? Trends Neurosci 1989; 12:6-10.
16. Lynch M, Sayin U, Golarai G et al. NMDA receptor-dependent plasticity of granule cell spiking in the dentate gyrus of normal and epileptic rats. J Physiol 2000; 11:61-68.
17. Bouilleret V, Kiener T, Marescaux C et al. Early loss of interneurons and delayed subunit-specific changes in GABA(A)-receptor expression in mouse model of mesial temporal lobe epilepsy. Hippocampus 2000; 10:305-324.
18. Nusser A, Hajos N, Somogyi P et al. Increased number of synaptic GABA$_A$ receptors underlies potentiation at hippocampal inhibitory synapses. Nature 1998; 395:172-177.
19. Kokaia Z, Kokaia M. Changes in GABA$_B$ receptor immunoreactivity after recurrent seizures in rats. Neurosci Lett 2001; 315:85-88.
20. Morimoto K. Seizure-triggering mechanisms in the kindling model of epilepsy: collapse of GABA-mediated inhibition and activation of NMDA receptors. Neurosci Biobehav Rev 1989; 13:253-260.
21. Olsen RW, Delorey TM, Gordey M et al. GABA receptor function and epilepsy. Adv Neurol 1999; 79:499-510.
22. Billinton A, Baird VH, Duncan TM et al. GABA(B) receptor autoradiography in hippocampal sclerosis associated with human temporal lobe epilepsy. Br J Pharmacol 2001; 132:475-480.
23. Szelies B, Sobesky J, Pawlik G et al. Impaired benzodiazepine receptor binding in peri-lesional cortex of patients with eymptomatic epilepsies studies by [(11)C]-flumazenil PET. Eur J Neurosci 2002; 9:137-142.
24. Sloviter RS. Permanently altered hippocampal structure, excitability, and inhibition after experimental status epilepticus in the rat: the "dormant basket cell" hypothesis and its possible relevance to temporal lobe epilepsy. Hippocampus 1991; 1:41-66.
25. Sutula TP, Pitkanen A. More evidence for seizure-induced neuron loss. Is hippocampal sclerosis both cause and effect of epilepsy? Neurology 2001; 57:169-170.
26. Cavazos JE, Das I, Sutula T. Neuronal loss induced in limbic pathways by kindling: evidence for induction of hippocampal sclerosis by repeated brief seizures. J Neurosci 1994; 14:3106-3121.
27. Bengzon J, Kokaia Z, Elmer E et al. Apoptosis and proliferation of dentate gyrus neurons after single and intermittent limbic seizures. Proc Natl Acad Sci USA 1997; 94:10432-10437.
28. Smith BN, Dudek FE. Short- and long-term changes in CA1 network excitability after kainate treatment in rats. J Neurophysiol 2001; 85:1-9.
29. Sutula TP, Golarai G, Cavazos J. Assessing the functional significance of mossy fiber sprouting. In: Ribak CE, Gall CM, Mody I, eds. The dentate gyrus and its role in seizures. Amsterdam: Elsevier, 1992:251-259.
30. Cronin J, Obenhous A, Houser C et al. Electrophysiology of dentate granule cells after kainate-induced synaptic reorganization of mossy fibers. Brain Res 1992; 573:305-310.
31. Represa A, Ben-Ari Y. Kindling is associated with the formation of novel mossy fibre synapses in the CA3 region. Brain Res 1992; 92:69-78.
32. Babb TL, Kupfer WR, Pretorius JK et al. Synaptic reorganization by mossy fibers in human epileptic fascia dentata. Neuroscience 1991; 42:351-363.
33. Houser CR. Morphological changes in the dentate gyrus in human temporal lobe epilepsy. In: Ribak CE, Gall CM, Mody I, eds. The Dentate Gyrus and its Role in Seizures. Amsterdam: Elsevier, 1992:223-234.
34. Lehmann TN, Gabriel S, Kovacs R et al. Alterations of neuronal connectivity in area CA1 of hippocampal slices from temporal lobe epilepsy patients and from pilocarpine-treated epileptic rats. Epilepsia 2000; 41:S190-190.
35. Esclapez M, Hirsch J, Ben-Ari Y et al. Newly formed excitatory pathways provide a substrate for hyperexcitability in experimental temporal lobe epilepsy. J Comp Neurol 1999; 408:449-460.
36. Perez Y, Morin F, Beaulieu C et al. Axonal sprouting of CA1 pyramidal cells in hyperexcitable hippocampal slices of kainate-treated rats. Eur J Neurosci 1996; 8:736-748.

37. Smith BN, Dudek FE. Network interactions mediated by new excitatory connections between CA1 pyramidal cells in rats with kainate-induced epilepsy. J Neurophysiol 2002; 135:53-65.
38. Okazaki MM, Molnar P, Nadler JV. Recurrent mossy fiber pathway in rat dentate gyrus: synaptic currents evoked in presence and absence of seizure-induced growth. J Neurophysiol 1999; 81:1645-1660.
39. Wuarin J-P, Dudek FE. Excitatory synaptic input to granule cells increases with time after kainate treatment. J Neurophysiol 2001; 1067-1077.
40. Costa E, Davis J, Grayson DR et al. Dendritic spine hypoplasticity and downregulation of reelin and GABAergic tone in schizophrenia vulnerability. Neurobiol Dis 2001; 8:723-742.
41. Tsirka S, Bugge T, Degen JL et al. An extracellular proteolytic cascade promotes neuronal degeneration in the mouse hippocampus. J Neurosci 1997; 17:543-552.
42. Colman D, Filbin MT. Cell adhesion molecules. In: Siegel GJ, Agranoff BW, Albers RW et al., eds.Basic neurochemistry; molecular, cellular and medical aspects, New York: Lippincott Williams and Wilkins, 1998:139-153.
43. Alpin A, Howe A, Alahari S et al. Signal transduction and signal modulation by cell adhesion receptors: the role of integrins, cadherins, immunoglobulin-cell adhesion molecules, and selectins. Pharmacol Rev 1998; 50:197-263.
44. Fields RD, Itoh K. Neural cell adhesion molecules in activity-dependent development and synaptic plasticity. Trends Neurosci 1996; 19:473-480.
45. Plow EF, Haas TA, Zhang L et al. Ligand binding to integrins. J Biol Chem 2000; 275:21785-21788.
46. Hynes R. Integrins, versatility, modulation, and signaling in cell adhesion. Cell 1992; 69:11-25.
47. Mould AP. Getting integrins into shape: recent insights into how integrin activity is regulated by conformational changes. J Cell Sci 1996; 109:2613-2618.
48. Humphries MJ, Delwel GO, Kuikman I et al. Integrin structure. Biochem Soc Trans 2000; 28:311-39.
49. Hynes RO. Cell adhesion: old and new questions. Trends Cell Biol 1999; 9:M33-M37.
50. Giancotti FG, Ruoslahti E. Integrin signaling. Science 1999; 285:1028-1032.
51. Dedhar S, Hannigan G. Integrin cytoplasmic interactions and bidirectional transmembrane signalling. Current Opinion in Cell Biology 1996; 8:657-669.
52. Hughes PE, Pfaff M. Integrin affinity modulation. Trends Cell Biol 1998; 8:359-364.
53. Bouvard D, Molla A, Block MR. Calcium/calmodulin-dependent protein kinase II controls $\alpha 5\beta 1$ integrin mediated inside out signaling. J Cell Sci 1998; 111:657-665.
54. Miranti CK, Brugge JS. Sensing the environment: a historical perspective on integrin signal transduction. Nature Cell Biol 2002; 4:83-90.
55. Huang Y-Q, Lu WY, Ali DW et al. CAKβ/PYK2 kinase is a signaling link for induction of long-term potentiation in CA1 hippocampus. Neuron 2001; 29:485-496.
56. Ali DW, Salter MW. NMDA receptor regulation by Src kinase signalling in excitatory synaptic transmission and plasticity. Curr Opin Neurobiol 2001; 11:336-342.
57. Schlaepfer D, Hunter T. Integrin signalling and tyrosine phosphorylation: just the FAKs? Trends Cell Biol 1998; 8:151-157.
58. Yamada K, Miyamoto S. Integrin transmembrane signaling and cytoskeletal control. Curr Op Cell Biol 1995; 7:681-689.
59. Huhtala P, Humphries MJ, McCarthy JB et al. Cooperative signaling by $\alpha 5\beta 1$ and $\alpha 4\beta 1$ integrins regulated matalloproteinase gene expression in fibroblasts adhering to fibronectin. J Cell Biol 1995; 129:867-879.
60. Howe A, Aplin A, Alahari S et al. Integrin signaling and cell growth control. Curr Opin Cell Biol 1998; 10:220-231.
61. Wu X, Davis GE, Meininger GA et al. Regulation of the L-type calcium channel by $\alpha 5\beta 1$ integrin requires signaling between focal adhesion proteins. J Biol Chem 2001; 276:30285-30292.
62. Moro L, Dolce L, Cabodi S et al. Integrin-induced epidermal growth factor (EGF) receptor activation requires c-Src and p130Cas and leads to phosphorylation of specific EGF receptor tyrosines. J Biol Chem 2002; 277:9405-9414.
63. Yamada KM, Even-Ram S. Integrin regulation of growth factor receptors. Nature Cell Biol 2002; 4:E75-E76.
64. Juliano R. Cooperation between soluable factors and integrin-mediated cell anchorage in the control of cell growth and differentiation. BioEssays 1998; 18:911-917.
65. Porter JC, Hogg N. Integrins take partners: cross-talk between integrins and other membrane receptors. Trends Cell Biol 1998; 8:390-396.
66. Platts SH, Mogford JE, Davis MJ et al. Role of K^+ channels in arteriolar vasodilation mediated by integrin interaction with RGD-containing peptide. Am J Physiol 1998; 275:H1449-H1454.

67. Wildering WC, Herman PM, Bulloch AGM. Rapid neuromodulatory actions of integrin ligands. J Neurosci 2002; 22:2419-2426.
68. Reichardt L, Tomaselli K. Extracellular matrix molecules and their receptors: Functions in neural development. Ann Rev Neurosci 1991; 14:531-570.
69. Yanagida H, Tanaka J, Maruo S. Immunocytochemical localization of a cell adhesion molecule, integrin alpha5beta1, in nerve growth cones. J Orthop Sci 1999; 4:353-360.
70. Grabham PW, Foley M, Umeojiako A et al. Nerve growth factor stimulates coupling of beta1 integrin to distinct transport mechanisms in the filopodia of growth cones. J Cell Science 2000; 113:3003-3012.
71. Dulabon L, Olson EC, Taglienti MG et al. Reelin binds α3β1 integrin and inhibits neuronal migration. Neuron 2000; 27:33-44.
72. Cox EA, Huttenlocher A. Regulation of integrin-mediated adhesion during cell migration. Microsc Res Tech 1998; 43:412-419.
73. Sheetz MP, Felsenfeld DP, Galbraith CG. Cell migration: Regulation of force on extracellular-matrix-integrin complexes. Trends Cell Biol 1998; 8:51-54.
74. Palecek SP, Huttenlocher A, Horwitz AF et al. Physical and biochemical regulation of integrin release during rear detachment of migrating cells. J Cell Sci 1998; 111:929-940.
75. Jones LS. Integrins: possible functions in the adult CNS. Trends Neurosci 1996; 19:68-72.
76. Nishimura SL, Boylen KP, Einheber S et al. Synaptic and glial localization of the integrin αvβ8 in mouse and rat brain. Brain Res 1998; 791:271-282.
77. Tanaka J, Maeda N. Microglial ramification requires nondiffusible factors derived from astrocytes. Exp Neurol 1996; 137:367-375.
78. Martin PT, Kaufman SJ, Kramer RH et al. Synaptic integrins in developing, adult, and mutant muscle: selective association of α1, α7A and α7B integrins with the neuromuscular junction. Dev Biology 1996; 174:125-139.
79. Pinkstaff JK, Lynch G, Gall CM. Localization and seizure regulation of integrin β1 mRNA in adult rat brain. Mol Brain Res 1997; 19:1541-1556.
80. Pinkstaff JK, Detterich J, Lynch G et al. Integrin subunit gene expression is regionally differentiated in adult brain. J Neurosci 1999; 19:1541-1556.
81. Kloss CU, Werner A, Klein MA et al. Integrin family of cell adhesion molecules in the injured brain: regulation and cellular localization in the normal and regenerating mouse facial motor nucleus. J Comp Neurol 1999; 411:162-178.
82. Previtali SC, Feltri ML, Archelos JJ et al. Role of integrins in the peripheral nervous system. Prog Neurobiol 2001; 64:35-49.
83. Hammarberg H, Wallquist W, Piehl F et al. Regulation of laminin-associated integrin subunit mRNAs in rat spinal motoneurons during postnatal development and after axonal injury. J Comp Neurol 2000; 428:294-304.
84. Lee EH, Hsieh YP, Yang CL et al. Induction of integrin-associated protein (IAP) mRNA expression during memory consolidation in rat hippocampus. Eur J Neurosci 2000; 12:1102-1112.
85. Chavis P, Westbrook G. Integrins mediate functional pre- and postsynaptic maturation at a hippocampal synapse. Nature 2001; 411:317-321.
86. Einheber S, Schnapp L, Salzer J et al. The regional and ultrastructural distribution of the α8 integrin subunit in the developing and adult rat brain suggests a role in synaptic plasticity. J Comp Neurol 1996; 370:105-134.
87. Einheber S, Pierce JP, Chow D et al. Dentate hilar mossy cells and somatostatin-containing neurons are immunoreactive for the α8 integrin subunit: characterization in normal and kainic acid-treated rats. Neuroscience 2001; 105:619-638.
88. Murase S-I, Hayashi Y. Integrin α1 localization in murine central and peripheral nervous system. J Comp Neurol 1998; 395:161-176.
89. King VR, McBride A, Priestly JV. Immunohistochemical expression of the α5 integrin subunit in the normal adult rat central nervous system. J Neurocytol 2001; 30:243-252.
90. Bi X, Zhou J, Lynch G et al. Polarized distribution of α5 integrin in the dendrites of hippocampal and cortical neurons. J Comp Neurol 2001; 435:184-193.
91. Schuster T, Krug M, Stalder M et al. Immunoelectron microscopic localization of the neural recognition molecules L1, NCAM, and its isoform NCAM180, the NCAM-associated polysialic acid, beta1 integrin and the extracellular matrix molecule tenascin-R in synapses of the adult rat hippocampus. J Neurobiol 2001; 49:142-158.
92. Grooms S, Terracio L, Jones L. Anatomical localization of β1 integrin-like immunoreactivity in rat brain. Exp Neurol 1993; 122:253-259.
93. Murase S, Hayashi M. Expression pattern of integrin β1 subunit in Purkinje cells of rat and cerebellar mutant mice. J Comp Neurol 1996; 375:225-237.

94. Rodriguez MA, Pesold C, Liu WS et al. Colocalization of integrin receptors and reelin in dendritic spine postsynaptic densities of adult nonhuman primate cortex. Proc Natl Acad Sci USA 2000; 97:3550-3555.
95. Beumer KJ, Bohrbough J, Prokop A et al. A role for PS integrins in morphological growth and synaptic function at the postembryonic neuromuscular junction of Drosophila. Development 1999; 126:5833-5846.
96. Rohrbough J, Grotewiel MS, Davis RL et al. Integrin-mediated regulation of synaptic morphology, transmission and plasticity. J Neurosci 2000; 20:6868-6878.
97. Chun D, Gall CM, Bi X et al. Evidence that integrins contribute to multiple stages in the consolidation of long term potentiation. Neuroscience 2001; 105:815-829.
98. Kramár EA, Bernard JA, Gall CM et al. Alpha3 integrin receptors contribute to the consolidation of long term potentiation. Neuroscience 2002; 110:29-39.
99. Bahr BA, Staubli U, Xiao P et al. Arg-Gly-Asp-Ser-Selective adhesion and the stabilization of long-term potentiation: Pharmacological studies and the characterization of a candidate matrix receptor. J Neurosci 1997; 17:1320-1329.
100. Duncan CP. The retroactive effect of electroshock on learning. J Comp Physiol Psychol 1948; 42:32-44.
101. Popik P, Mamczarz J, Vetulani J. The effect of electroconvulsive shock and nifedipine on spatial learning and memory in rats. Biol Psychiatry 1994; 35:864-869.
102. McGaugh JL. Memory—a century of consolidation. Science 2000; 287:248-251.
103. Barrionuevo G, Schottler S, Lynch G. The effects of repetitive low frequency stimulation on control and "potentiated" synaptic responses in the hippocampus. Life Sciences 1980; 27:2385-2391.
104. Staubli U, Lynch G. Stable depression of potentiated synaptic responses in the hippocampus with 1-5 Hz stimulation. Brain Res 1990; 513:113-118.
105. Martin SJ. Time-dependent reversal of dentate LTP by 5 Hz stimulation. NeuroReport 1998; 9:3775-3781.
106. Burette F, Jay TM, Saroche S. Reversal of LTP in the hippocampal afferent fiber system to the prefrontal cortex in vivo with low-frequency patterns of stimulation that do not produce LTD. J Neurophysiol 1997; 78:1155-1160.
107. Arai A, Larson J, Lynch G. Anoxia reveals a vulnerable period in the development of long-term potentiation. Brain Res 1990; 511:353-357.
108. Larson J, Xiao P, Lynch G. Reversal of LTP by theta frequency stimulation. Brain Res 1993; 600:97-102.
109. Fujii S, Kuroda Y, Ito K et al. Effects of adenosine receptors on the synaptic and EPSP-spike components of long-term potentiation and depotentiation in the guinea-pig hippocampus. J Physiol (Lond) 1999; 521:451-466.
110. Abraham WC. Induction and reversal of long-term potentiation by repeated high frequency stimulation in rat hippocampal slices. Hippocampus 1997; 7:137-145.
111. Huang CC, Liang YC, Hsu KS. A role for extracellular adenosine in time-dependent reversal of long-term potentiation by low-frequency stimulation at hippocampal CA1 synapses. J Neurosci 1999; 19:9728-9738.
112. Staubli U, Chun D. Factors regulating the reversibility of long term potentiation. J Neurosci 1996; 16:853-860.
113. Ruoslahti E. RGD and other recognition sequences for integrins. Ann Rev Cell Dev Biol 1996; 12:697-715.
114. Ruoslahti E, Pierschbacher MD. Arg-Gly-Asp: a versatile cell recognition signal. Cell 1986; 44:517-518.
115. Pierschbacher M, Ruoslahti E. Cell attachment activity of fibronectin can be duplicated by small synthetic fragments of the molecule. Nature 1984; 304:30-33.
116. Staubli U, Vanderklish PW, Lynch G. An inhibitor of integrin receptors blocks LTP. Behav Neural Biol 1990; 53:1-5.
117. Xiao P, Bahr B, Staubli U et al. Evidence that matrix recognition contributes to stabilization but not induction of LTP. NeuroReport 1991; 2:461-464.
118. Staubli U, Chun D, Lynch G. Time-dependent reversal of long-term potentiation by an integrin antagonist. J Neurosci 1998; 18:3460-3469.
119. Yang W, Asakura S, Sakai T et al. Two-step spreading mode of human glioma cells on fibrin monomer: interaction of alpha(v)beta3 with the substratum followed by interaction of alpha5beta1 with endogenous cellular fibronectin secreted in the extracellular matrix. Thromb Res 1999; 93:279-290.
120. Benson DL, Schnapp LM, Shapiro L et al. Making memories stick: cell-adhesion molecules in synaptic plasticity. Trends Cell Biol 2000; 10:473-482.

121. Murase S, Schuman EM. The role of cell adhesion molecules in synaptic plasticity and memory. Curr Opin Cell Biol 1999; 11:549-53.
122. Schachner M. Neural recognition molecules and synaptic plasticity. Curr Opin Cell Biol 1997; 9:627-634.
123. Schuster T, Krug M, Hassan H et al. Increase in proportion of hippocampal spine synapses expressing neural cell adhesion molecule NCAM180 following long-term potentiation. J Neurobiol 1998; 37:359-372.
124. Luthl A, Laurent J, Figurov A et al. Hippocampal long-term potentiation and neural cell adhesion molecules L1 and NCAM. Nature 1994; 372:777-779.
125. Sakurai E, Hashikawa T, Yoshihara Y et al. Involvement of dendritic adhesion molecule telencephalin in hippocampal long-term potentiation. Neuroreport 1998; 9:881-886.
126. Eckhardt M, Bukalo O, Chazal G et al. Mice deficient in the polysialyltransferase ST8SiaIV/PST-1 allow discrimination of the roles of neural cell adhesion molecule protein and polysialic acid in neural development and synaptic plasticity. J Neurosci 2000; 20:5234-5244.
127. Bozdagi O, Shan W, Tanaka H et al. Increasing numbers of synaptic puncta during late-phase LTP: N-cadherin is synthesized, recruited to synaptic sites, and required for potentiation. Neuron 2000; 28:245-259.
128. Lee K, Schottler F, Oliver M et al. Brief bursts of high-frequency stimulation produce two types of structural changes in rat hippocampus. J Neurophysiol 1980; 44:247-258.
129. Chang F-LF, Greenough WT. Transient and enduring morphological correlates of synaptic activity and efficacy change in the rat hippocampal slice. Brain Res 1984; 309:35-46.
130. Toni N, Buchs PA, Nikonenko I et al. LTP promotes formation of multiple spine synapses between a single axon terminal and a dendrite. Nature 1999; 402:421-425.
131. Tang L, Hung CP, Schuman EM. A role for the cadherin family of cell adhesion molecules in hippocampal long term potentiation. Neuron 1998; 20:1165-1175.
132. Yamagata K, Andreasson KI, Sugiura H et al. Arcadlin is a neural activity-regulated cadherin involved in long term potentiation. J Biol Chem 1999; 274:19473-19479.
133. Grotewiel MS, Beck CD, Wu CF et al. Integrin-mediated short-term memory in Drosophila. Nature 1998; 18:7847-7855.
134. Connolly JB, Tully T. Integrins: a role for adhesion molecules in olfactory memory. Curr Biol 1998; 8:R386-R389.
135. Grooms S, Jones L. RGDS tetrapeptide and hippocampal in vitro kindling in rats: Evidence for integrin-mediated physiological stability. Neurosci Lett 1997; 231:139-142.
136. Sankar R, Shin D, Mazarati AM et al. Epileptogenesis after status epilepticus reflects age- and model-dependent plasticity. Ann Neurol 2000; 48:580-589.
137. Lado FA, Sankar R, Lowenstein D et al. Age-dependent consequences of seizures: relationship to seizure frequency, brain damage, and circuitry reorganization. Ment Retard Dev Disabil Res Rev 2000; 6:242-252.
138. Kramár EA, Lynch G. Developmental and regional differences in the consolidation of long term potentiation. Submitted 2002.
139. Arai A, Black J, Lynch G. Origins of the variations in long-term potentiation between synapses in the basal versus apical dendrites of hippocampal neurons. Hippocampus 1994; 4:1-9.
140. Bi X, Chang V, Siman R et al. Regional distribution and time-course of calpain activation following kainate-induced seizure activity in adult rat brain. Brain Res 1996; 726:98-108.
141. del Cerro S, Arai A, Kessler M et al. Stimulation of NMDA receptors activates calpain in cultured hippocampal slices. Neurosci Lett 1994; 167:149-152.
142. Vanderklish PW, Krushel LA, Holst BH et al. Marking synaptic activity in dendritic spines with a calpain substrate exhibiting fluorescence resonance energy transfer. Proc Natl Acad Sci USA 2000; 97:2253-2258.
143. Vanderklish P, Saido TC, Gall C et al. Proteolysis of spectrin by calpain accompanies theta-burst stimulation in cultured hippocampal slices. Mol Brain Res 1995; 32:25-35.
144. Pfaff M, Du X, Ginsberg MH. Calpain cleavage of integrin beta cytoplasmic domains. FEBS Lett 1999; 460:17-22.
145. Huttenlocher A, Palecek SP, Lu Q et al. Regulation of cell migration by the calcium-dependent protease calpain. J Biol Chem 1997; 272:32719-32722.
146. Du X, Saido T, Tsubuki S et al. Calpain cleavage of the cytoplasmic domain of the integrin β3 subunit. J Biol Chem 1995; 270:26146-26151.
147. Schoenwaelder SM, Yuan Y, Cooray P et al. Calpain cleavage of focal adhesion proteins regulates the cytoskeletal attachment of integrin alphaIIbbeta3 (platelet glycoprotein IIb/IIIa) and the cellular retraction of fibrin clots. J Biol Chem 1997; 272:1694-1702.

148. Seubert P, Baudry M, Dudek S et al. Calmodulin stimulates the degradation of brain spectrin by calpain. Synapse 1987; 1:20-24.
149. Siman R, Baudry M, Lynch G. Brain fodrin: substrate for the endogenous calcium-activated protease, calpain I. Proc Natl Acad Sci USA 1984; 81:3572-3576.
150. Perlmutter LS, Siman R, Gall C et al. The ultrastructural localization of calcium-activated protease 'calpain' in rat brain. Synapse 1988; 2:79-88.
151. Perlmutter LS, Gall C, Baudry M et al. Distribution of calcium-activated protease calpain in the rat brain. J Comp Neurol 1990; 296:269-76.
152. Quan A, Gilbert M, Colicos M et al. Tissue-plasminogen activator is induced as an immediate-early gene during seizure, kindling and long-term potentiation. Nature 1993; 361:453-457.
153. Okabe A, Momota Y, Yoshida S et al. Kindling induces neuropsin mRNA in the mouse brain. Brain Res 1996; 728:116-20.
154. Baranes D, Lederfein D, Huang YY et al. Tissue plasminogen activator contributes to the late phase of LTP and to synaptic growth in the hippocampal mossy fiber pathway. Neuron 1998; 21:813-825.
155. Gualandris A, Jones TE, Strickland S et al. Membrane depolarization induces calcuim-dependent secretion of tissue plasminogen activator. J Neurosci 1996; 16:2220-2225.
156. Hoffman KB, Larson J, Bahr BA et al. Activation of NMDA receptors stimulates extracellular proteolysis of cell adhesion molecules in hippocampus. Brain Res 1998; 811:152-155.
157. Endo A, Nagai N, Urano T et al. Proteolysis of neuronal cell adhesion molecule by the tissue plasminogen activator-plasmin system after kainate injection in the mouse hippocampus. Neurosci Res 1999; 33:1-8.
158. Hoffman KB, Martinez J, Lynch G. Proteolysis of cell adhesion molecules by serine proteases: a role in long term potentiation? Brain Res 1998; 811:29-33.
159. Wu YP, Siao CJ, Lu W et al. The tissue plasminogen activator (tPA)/plasmin extracellular proteolytic system regulates seizure-induced hippocampal mossy fiber outgrowth through a proteoglycan substrate. J Cell Biol 2000; 148:1295-1304.
160. Chen A-L, Strickland S. Neuronal death in hippocampus is promoted by plasmin-catalyzed degradation of laminin. Cell 1997; 91:917-925.
161. Rabhi-Sabile S, Pidard D. Exposure of human platelets to plasmin results in the expression of irreversibly active fibrinogen receptors. Thromb Haemost 1995; 73:693-701.
162. Basbaum C, Werb Z. Focalized proteolysis: spatial and temporal regulation of extracellular matrix degradation at the cell surface. Curr Op Cell Biol 1996; 8:731-738.
163. Szklarczyk A, Lapinska J, Rylski M et al. Matrix metalloproteinase-9 undergoes expression and activation during dendritic remodeling in adult hippocampus. J Neurosci 2002; 22:920-930.
164. Zhang JW, Deb S, Gottschall PE. Regional and differential expression of gelatinases in rat brain after systemic kainic acid or bicuculline administration. Eur J Neurosci 1998; 10:3358-3368.
165. Rivera S, Tremblay E, Timsit S et al. Tissue inhibitor of metalloproteinases-1, (TIMP-1) is differentially induced in neurons and astrocytes after seizures; evidence for developmental, immediate early gene, and lesion response. J Neurosci 1997; 17:4223-4235.
166. Binder DK, Berger MS. Proteases and the biology of glioma invasion. J Neuro-Oncol 2002; 56:149-158.
167. Pico RM, Gall CM. Hippocampal epileptogenesis produced by electrolytic iron deposition in the rat dentate gyrus. Epilepsy Res 1994; 19:27-36.
168. Gall CM, Pinkstaff JK. Integrin alpha subunit gene expression is differentially regulated by seizures. Soc Neurosci Abst 2000; 26:123.
169. Shan W, Yoshida M, Wu XR et al. Neural (N-) cadherin, a synaptic adhesion molecule, is induced in hippocampal mossy fiber axonal sprouts by seizure. J Neurosci Res 2002; 69:292-304.
170. Represa A, Le Gall La Salle G, Ben-Ari Y. Hippocampal plasticity in the kindling model of epilepsy in rats. Neurosci Lett 1989; 99:345-350.
171. Sutula T, Cascino G, Cavazos J et al. Mossy fiber synaptic reorganization in the epileptic human temporal lobe. Ann Neurol 1989; 26:321-330.
172. Itoh K, Ozaki M, Stevens B et al. Activity-dependent regulation of N-cadherin in DRG neurons: Differential regulation of N-cadherin, NCAM, and L1 by distinct patterns of action potentials. J Neurobiol 1997; 33:735-748.
173. Fazeli MS, Breen K, Errington ML et al. Increase in extracellular NCAM and amyloid percursor protein following induction of long-term potentiation in the dentate gyrus of anesthetized rats. Neurosci Lett 1994; 169:77-80.
174. Holst BD, Vanderklish PW, Krushel LA et al. Allosteric modulation of AMPA-type glutamate receptors increases activity of the promoter for the neural cell adhesion molecule, N-CAM. Proc Natl Acad Sci USA 1998; 95:2597-2602.

175. Ferhat L, Chevassus-Au-Louis N, Khrestchatisky M et al. Seizures induce tenascin-C mRNA expression in neurons. J Neurocytol 1996; 25:535-546.
176. Nakic M, Mitrovic N, Sperk G et al. Kainic acid activates transient expression of tenascin-C in the adult rat hippocampus. J Neurosci Res 1996; 44:355-362.
177. Hoffman KB, Pinkstaff JK, Gall CM et al. Seizure induced synthesis of fibronectin is rapid and age dependent: implications for long-term potentiation and sprouting. Brain Res 1998; 812:209-215.
178. Niquet J, Jorquera I, Ben-Ari Y et al. Proliferative astrocytes may express fibronectin-like protein in the hippocampus of epileptic rats. Neurosci Lett 1994; 180:13-16.
179. O'Conner L, Lauterborn J, Smith M et al. Expression of agrin mRNA is altered following seizures in adult rat brain. Mol Brain Res 1995; 33:277-287.
180. Mantych KB, Ferreira A. Agrin differentially regulates the rates of axonal and dendritic elongation in cultured hippocampal neurons. J Neurosci 2001; 21:6802-6809.
181. Hilgenberg LG, Ho KD, Lee D et al. Agrin regulates neuronal responses to excitatory neurotransmitters in vitro and in vivo. Mol Cell Neurosci 2002; 19:97-110.
182. Naffah-Mazzacoratti MG, Arganaraz GA, Porcionatto MA et al. Selective alterations of glycosaminoclycans synthesis and proteoglycan expression in rat cortex and hippocampus in pilocarpine-induced epilepsy. Brain Res Bull 1999; 50:229-239.
183. Huang YY, Bach ME, Lipp HP et al. Mice lacking the gene encoding tissue-type plasminogen activator show a selective interference with late-phase long-term potentiation in both Schaffer collateral and mossy fiber pathways. Proc Natl Acad Sci USA 1996; 93:8699-8704.
184. Hoffman KB, Hwee V, Larson J et al. Peripheral administration of a serine protease inhibitor blocks kindling. Brain Res 2000; 861:178-180.
185. Tsirka SE, Gualandris A, Amaral DG et al. Excitotoxin-induced neuronal degeneration and seizure are mediated by tissue plasminogen activator. Nature 1995; 377:340-344.
186. Duchossoy Y, Horvat JC, Stettler O. MMP-related gelatinase activity is strongly induced in scar tissue of injured adult spinal cord and forms pathways for ingrowing neurites. Mol Cell Neurosci 2001; 17:945-956.
187. Ferguson TA, Muir D. MMP-2 and MMP-9 increase the neurite-promoting potential of schwann cell basal laminae and are upregulated in degenerated nerve. Mol Cell Neurosci 2000; 16:157-167.
188. Zuo J, Ferguson TA, Hernandez YJ et al. Neuronal matrix metalloproteinase-2 degrades and inactivates a neurite-inhibiting chondriotin sulfate proteoglycan. J Neurosci 1998; 18:5203-5211.
189. Oh LY, Larsen PH, Krekoski CA et al. Matrix metalloproteinase-9/gelatinase B is required for process outgrowth by oligodendrocytes. J Neurosci 1999; 19:8464-8475.
190. Redies C. Cadherins in the central nervous system. Prog Neurobiol 2000; 61:611-648.
191. Doherty P, Williams G, Williams E-J. CAMs and axonal growth: A critical evaluation of the role of calcium and the MAPK cascade. Mol Cell Neurosci 2000; 16:283-295.
192. Doyu M, Sobue G, Kimata KE et al. Laminin A, B1 and B2 chain gene expression in transected and regenerating nerves: regulation by axonal signals. J Neurochem 1993; 60:543-551.
193. Lefcort F, Venstrom K, McDonald JA et al. Regulation of expression of fibronectin and its receptor, α5β1, during development and regeneration of peripheral nerve. Development 1992; 116:767-782.
194. Wang GY, Hirai KI, Shimada H et al. Behavior of axons, Schwann cells and perineural cells in nerve regeneration within transplanted nerve grants—effects of anti-laminin and anti-fibronectin antisera. Brain Res 1992; 583:216-226.
195. Werner A, Willern M, Jones LL et al. Impaired axonal regeneration in α7 integrin-deficient mice. J Neurosci 2000; 20:1822-1830.
196. Condic ML. Adult neuronal regeneration induced by transgenic integrin expression. J Neurosci 2001; 21:4782-4788.

CHAPTER 3

The Role of BDNF in Epilepsy and Other Diseases of the Mature Nervous System

Devin K. Binder

Abstract

The neurotrophin brain-derived neurotrophic factor (BDNF) is ubiquitous in the central nervous system (CNS) throughout life. In addition to trophic effects on target neurons, BDNF appears to be part of a general mechanism for activity-dependent modification of synapses in the developing and adult nervous system. Thus, diseases of abnormal trophic support (such as neurodegenerative diseases) and diseases of abnormal excitability (such as epilepsy and central pain sensitization) can be related in some cases to abnormal BDNF signaling. For example, various studies have shown that BDNF is upregulated in areas implicated in epileptogenesis, and interference with BDNF signal transduction inhibits the development of the epileptic state. Further study of the cellular and molecular mechanisms by which BDNF influences cell survival and excitability will likely provide novel concepts and targets for the treatment of diverse CNS diseases.

BDNF: Introduction

Brain-derived neurotrophic factor (BDNF) is a member of the "neurotrophin" family of neurotrophic factors. It was originally purified from pig brain due to its survival-promoting action on a subpopulation of dorsal root ganglion neurons.[1] The amino acid sequence of BDNF has a strong homology with nerve growth factor (NGF), the neurotrophin (NT) first described due to its trophic (survival and growth-promoting) effects on sensory and sympathetic neurons. Since the discovery of NGF in the early 1950s by Rita Levi-Montalcini and Viktor Hamburger[2] and the discovery of BDNF by Yves Barde and colleagues in 1982,[1] other members of the NT family such as neurotrophin-3 (NT-3) and neurotrophin-4/5 (NT-4/5) have been described. Each NT appears to have a unique profile of trophic effects on distinct subpopulations of peripheral nervous system and central nervous system neurons.

BDNF Structure

The mature form of human BDNF has been mapped to chromosome 11[3] and shares about 50% amino acid identity with human NGF, NT-3, and NT-4/5. The structure of each NT contains: (1) a signal peptide following the initiation codon; (2) a pro-region containing an N-linked glycosylation site and a proteolytic cleavage site for furin-like pro-protein convertases, followed by the mature sequence; and (3) a distinctive three-dimensional structure containing two pairs of antiparallel β-strands and cysteine residues in a cystine knot motif. Mature NTs are noncovalently-linked homodimers with molecular weight about 28 kDa. Dimerization appears essential for NT receptor activation.

Recent Advances in Epilepsy Research, edited by Devin K. Binder and Helen E. Scharfman.
©2004 Eurekah.com and Kluwer Academic / Plenum Publishers.

BDNF Signaling

Each NT binds one or more high-affinity receptors (the trk receptors) (Kd ~ 10^{-11} M).[4] Trk proteins are transmembrane receptor tyrosine kinases (RTKs) homologous to other RTKs such as the epidermal growth factor (EGF) receptor and insulin receptor family.[5] Signaling by RTKs involves ligand-induced receptor dimerization and dimerization-induced trans-autophosphorylation.[6,7] Receptor autophosphorylation on multiple tyrosine residues creates specific binding sites for intracellular target proteins, which bind to the activated receptor via SH2 domains.[6] For the NT family, these target proteins have been shown to include PLCγ1 (phospholipase C), p85 (the noncatalytic subunit of PI-3 kinase), and Shc (SH2-containing sequence);[4] activation of these target proteins can then lead to a variety of intracellular signalling cascades such as the Ras-MAP (mitogen-activated protein) kinase cascade and phosphorylation of CREB (cyclic AMP response element binding protein).[8-11] Binding specificity is conferred via the juxtamembrane Ig-like domain of the extracellular portion of the receptor in the following pattern:[12] trkA is the high-affinity receptor for NGF (with low-affinity binding by NT-3 in some systems), trkB is the high-affinity receptor for BDNF and NT-4/5 with lower-affinity binding by NT-3, and trkC is the high-affinity receptor for NT-3.[5]

In addition to the high-affinity NT receptors, all of the NTs bind to the low-affinity NT receptor, designated p75^{NTR} (Kd ~ 10^{-9} M).[13] P75^{NTR} has a glycosylated extracellular region involved in ligand binding, a transmembrane region, and a short cytoplasmic sequence lacking intrinsic catalytic activity. It is related to proteins of the tumor necrosis factor (TNFR) superfamily. NT binding to p75^{NTR} is linked to several intracellular signal transduction pathways, including nuclear factor-κB (NF-κB), Jun kinase and sphingomyelin hydrolysis.[14] P75^{NTR} signaling mediates biologic actions distinct from those of the high-affinity trk receptors, notably the initiation of programmed cell death (apoptosis) as well as newly-described roles in the regulation of axonal elongation and synaptic transmission.[15]

Ligand-induced receptor tyrosine phosphorylation is necessary for NT-induced cellular responses.[5] For example, cooperative interaction between tyrosines in trkA mediates the neurite outgrowth effect of NGF.[16] Thus, receptor tyrosine phosphorylation seems a logical measure of the biologic level of NT activity (see below). Tyrosine-490 is phosphorylated following NT application and is known to couple trk receptors to Shc binding and activation of the ras-MAP kinase cascade.[11] Furthermore, recent evidence indicates that activated trk receptors may be endocytosed and retrogradely transported while still tyrosine phosphorylated.[17-23]

Localization, Transport and Release of BDNF

BDNF mRNA as well as the mRNA encoding the high-affinity receptor for BDNF (trkB) has a widespread distribution in the central nervous system, especially in the cerebral cortex, hippocampal formation, and amygdaloid complex.[24-26] Notably, high levels of BDNF and trkB expression are found in brain areas that have been associated with seizure susceptibility, such as hippocampus and entorhinal cortex.[27] Within hippocampus, the granule cells, pyramidal cells, and some hilar GABAergic neurons express mRNA for BDNF and trkB.

In parallel, BDNF protein immunoreactivity is also widespread, and appears to be localized in neuronal cell bodies, axons and dendrites.[24] Like BDNF mRNA, constitutive BDNF protein expression is high in the hippocampus, where the mossy fiber axons of dentate granule cells are intensely immunoreactive for BDNF.[24,28]

Unlike the classical target-derived trophic factor model in which NTs—such as NGF—are retrogradely transported, there is now abundant evidence that BDNF is also anterogradely transported in brain.[24,29-33] Indeed, a recent study using green fluorescent protein (GFP)-tagged BDNF demonstrated direct activity-dependent transneuronal transfer of BDNF to postsynaptic neurons.[34,35] In hippocampus, it appears that BDNF within the hilus and CA3 stratum lucidum is synthesized by the dentate granule cells, anterogradely transported and preferentially stored in mossy fiber terminal boutons.[36]

Biochemical studies suggest that endogenous BDNF may be packaged in a releasable vesicular pool[37] and recent evidence indicates that NTs are released acutely following neuronal depolarization in an intracellular calcium and phospholipase C (PLC)-dependent manner.[38-42]

BDNF Effects in Development

NTs are known to have profound survival, differentiation, and morphoregulatory effects during brain development (leading to formation of appropriately matched functional circuitry).[43-45] The classical view of NT function derived initially from studies of NGF includes effects on growth and survival of neurons, and indeed BDNF has also been shown to be necessary for the survival of some neurons during vertebrate development. Certain peripheral sensory neurons, especially those in vestibular and nodose-petrosal ganglia, depend on the presence of BDNF because BDNF knockout mice (lacking both alleles for BDNF) demonstrate loss of these sensory neurons.[45,46] Unlike NGF, however, sympathetic neurons are not affected, nor are motor neuron pools. BDNF knockout mice fail to thrive, demonstrate lack of proper coordination of movement and balance, and ultimately die by 3 weeks of age. Conversely, provision of BDNF or other NTs to peripheral nerves during development enhances outgrowth,[47] and can support and/or rescue certain sensory neurons.[48,49]

BDNF expression increases in the early postnatal period and then stays high into adulthood, consistent with a role in the mature CNS as well. In vitro and in vivo studies have demonstrated that BDNF has survival- and growth-promoting actions on a variety of CNS neurons, including hippocampal and cortical neurons. Lack or blockade of BDNF leads to death of certain identified forebrain neurons. For example, lack of cortical BDNF leads to death of dorsal thalamic neurons.[50] Similarly, deletion of trkB leads to loss of neocortical neurons.[51]

In addition to its effects on survival, BDNF appears to regulate neuronal morphology and synaptogenesis. BDNF has been shown to enhance axonal branching in cultures of hippocampal neurons[52,53] and also has been shown to have significant differential effects on dendritic branching in cortex[54-56]. Evidence that activity-induced NT expression may modulate axonal sprouting in vivo comes from modulation of retinotectal axon branching by BDNF;[57] inhibition of normal ocular dominance column formation by NT infusion[58,59] or trkB-Fc infusion;[60] and inhibition of pilocarpine-induced cholinergic sprouting in hippocampus by NGF antisera.[61]

BDNF Gene Regulation

A multitude of stimuli have been described that alter BDNF gene expression in both physiologic and pathologic states. Physiologic stimuli are known to increase BDNF mRNA content. For example, light stimulation increases BDNF mRNA in visual cortex,[62] and osmotic stimulation increases BDNF mRNA in the paraventricular hypothalamic nucleus.[63] Other naturalistic behaviors in animals increase BDNF mRNA expression. For example, whisker stimulation increases BDNF mRNA expression in rodent somatosensory barrel cortex;[64] and singing stimulates BDNF expression in the high vocal center (HVC) of adult male canaries.[65] Electrical stimuli that induce long-term potentiation (LTP) in the hippocampus, a cellular model of learning and memory, increase BDNF and NGF expression.[66-68] Even physical exercise has been shown to increase NGF and BDNF expression in hippocampus.[69,70]

This physiologic alteration in BDNF gene expression may be very important in the development of the brain. For example, there is an exciting body of work implicating BDNF in activity-dependent development of the visual cortex.[71] Provision of excess NGF[58] or BDNF[59] or blockade of BDNF signaling[60] leads to abnormal patterning of ocular dominance columns during a critical period of visual cortex development. This suggests a role for BDNF in the patterning of axonal arborizations from the lateral geniculate nucleus (LGN) to the visual cortex during development.

BDNF, Synaptic Plasticity, and Learning

A great deal of evidence now indicates that BDNF and its high-affinity receptor trkB, in addition to modulating neuronal survival and differentiation, are also critically involved in neuronal excitability[72] and modulation of synaptic transmission.[73-75] For example, application of NTs including BDNF has been shown to potentiate synaptic transmission in vitro[76-82] and in vivo.[83] BDNF enhances excitatory (glutamatergic) synaptic transmission[76,79] and reduces inhibitory (GABAergic) synaptic transmission.[84,85] In the hippocampus, a critical level of BDNF/ trkB activation appears to be vital for modulation of synaptic efficacy. Incubation of hippocampal or visual cortical slices with the BDNF scavenger trkB-Fc reduces LTP,[86,87] and hippocampal slices from BDNF knockout animals exhibit impaired LTP induction which is restored by reintroduction of BDNF.[88-90] In addition, antagonists such as K252a block hyperexcitability in hippocampus due to BDNF exposure in vitro.[91]

The site and mechanism of synaptic potentiation by BDNF is not yet clear, but could involve facilitation of transmitter release,[92] phosphorylation of specific NMDA receptor subunits,[93] and/or direct effects on ion channels and conductances.[94-96] Enhanced excitatory transmission may also arise indirectly, because BDNF is known to have effects on the structure and function of inhibitory (GABAergic) neurons.[97] Reduction of trkB has been shown to reduce the ability of tetanic stimulation to induce LTP.[98] A recent study using imaging of dentate granule cells in mouse hippocampal slices identified BDNF-evoked calcium transients in dendritic spines but not at presynaptic sites, suggesting a postsynaptic site for BDNF-induced synaptic potentiation.[99,100]

Learning and memory depend on persistent selective modification of synapses between CNS neurons. Since BDNF appears to be critically involved in activity-dependent synaptic strengthening of the sort observed in the LTP model, there is great interest in its role as a molecular mechanism of learning and memory.

The hippocampus, which is required for many forms of long-term memory in humans and animals, appears to be an important site of BDNF action. Indeed, there is rapid and selective induction of BDNF expression in the hippocampus during contextual learning.[101] In addition, tool-use learning increases BDNF mRNA in monkey parietal cortex.[102] Function-blocking antibodies to BDNF, BDNF knockout,[103] antisense oligonucleotides to BDNF[104] and/or knockout of forebrain trkB signaling in mice[105] impairs spatial learning.

BDNF and Disease

Pathologic states are also associated with alteration in BDNF gene expression. In neurodegenerative diseases, inadequate trophic support may be partially responsible. In conditions such as epilepsy and chronic pain sensitization, excessive activation of excitatory synaptic plasticity may contribute to the disease phenotype.

BDNF and Epilepsy

Epilepsy is a disorder of the brain characterized by periodic and unpredictable occurrence of seizures. Although complex partial epilepsy is the most common type of epilepsy in adults (40% of all cases),[106] seizure control is achieved in only 25% of adults. It is clear that complex partial epilepsy is a major public health problem in that approximately 1 million people in the United States are affected and sufferers experience the periodic and unpredictable occurrence of seizures leading to impairment of consciousness. This handicap severely impairs the performance of many tasks and secondarily the procurement and maintenance of steady employment.

Elucidating the mechanisms of epileptogenesis in cellular and molecular terms may provide novel therapeutic approaches. Seizures have been shown to stimulate the expression of a variety of genes including those encoding transcription factors,[107,108] neuropeptides,[109] GAP-43,[110] proteases[111] and, quite prominently, NTs and trk receptors. The discovery that limbic seizures increase mRNA levels for nerve growth factor[112] led to the idea that seizure-induced expression of neurotrophic factors may contribute to the lasting structural and functional changes under-lying epileptogenesis.[27,113] Indeed, recent in vitro and in vivo findings implicate BDNF in the cascade of electrophysiologic and behavioral changes underlying the epileptic state.[114]

In particular, BDNF, NGF and trkB mRNA levels are increased in kindling and other seizure models whereas NT-3 mRNA content is decreased.[25,109,113,115-125] The magnitude of increase is greatest for BDNF in the hippocampus with BDNF mRNA being markedly upregulated in the dentate gyrus and CA1-CA3 pyramidal cell layers.[27,124]

This mRNA upregulation is accompanied by protein upregulation as well; extracts and in vivo microdialysates from animals after chemical convulsions show marked increases in neu-rotrophic factor activity.[126,127] Increases in BDNF protein content have been described follow-ing hilar lesion-induced limbic seizures, kindling and kainate administration.[128-131]

Seizure-induced increases in BDNF mRNA levels are transient compared to a longer-lasting increase in BDNF protein content. For example, following lesion-induced recurrent limbic seizures, BDNF mRNA levels peak 6 hours after seizure onset and return to control levels by about 12 hours;[113] in contrast, initial increases in BDNF protein content lag behind mRNA changes by 4 hours but remain well elevated over 4 days after the seizure episode.[128]

Following seizures, newly expressed BDNF appears to be anterogradely transported. Using hippocampal microdissection and quantification of BDNF by two-site ELISA, Elmer et al showed that BDNF protein levels after seizures were maximal at 12 hours in the dentate gyrus but at 24 hours in CA3,[129] consistent with anterograde transport of seizure-induced BDNF protein. Other evidence indicates that there is increased BDNF immunoreactivity in dentate granule cells by 4 hours followed by large increases in hilus and CA3 stratum lucidum at 12-24 hours; at the latter time point BDNF immunoreactivity within the granule cell bodies had returned to control levels.[132]

Effects of Inhibition of BDNF/trkB in Seizure Models

Recent studies using the kindling model of epilepsy have functionally implicated BDNF in epileptogenesis. In the kindling model, repeated, focal application of initially subconvulsive electrical stimuli eventually results in intense focal and tonic-clonic seizures.[133-136] Once estab-lished, this enhanced sensitivity to electrical stimulation persists for the life of the animal. The kindling model has been an important tool, since it allows experimental control over seizures and precise quantitation of effects of experimental manipulation on epileptogenesis in vivo.

Funabashi et al[137] and Van der Zee et al[138] found that kindling development was delayed by intraventricular infusion of anti-NGF antisera; however, the lack of specificity of the antisera limited interpretation of these experiments. Kokaia et al[139] reported a significant reduction in the rate of kindling development in BDNF heterozygous (+/-) mutant mice. Both basal and seizure-induced levels of BDNF mRNA were lower in the BDNF +/- compared to wild-type mice. The two-fold reduction in kindling rate in these animals is striking given that there was presumably some reduction but not elimination of trkB receptor signaling. Conversely, transgenic mice overexpressing BDNF display increased seizure severity in response to kainic acid and some display spontaneous seizures.[140] Infusion of BDNF itself into the hippocampus of adult rats leads to spontaneous limbic seizures as well as decreased threshold to chemoconvulsant-induced status epilepticus.[141] Of course, results from both the embryonic BDNF +/- knockouts and the BDNF transgenic mice must be cautiously interpreted in light of potential developmental effects of altered BDNF levels. The availability of conditional knockouts for trkB will enable analysis of the importance of trkB signaling in adult animals de novo.[51]

Recently, we attempted to selectively block trkB receptors during kindling development using trk-specific 'receptor bodies'.[142] These compounds are divalent homodimers that contain the ligand-binding domain of a given trk receptor and thus act as false receptors or 'receptor bodies' that putatively sequester endogenous NT (Figs. 1A, 1B). Intracerebroventricular (ICV) infusion of trkB receptor body (trkB-Fc) inhibited development of kindling in comparison to saline or human IgG controls, trkA-Fc, or trkC-Fc[142] (Figs. 2A, 2B). This effect manifested as

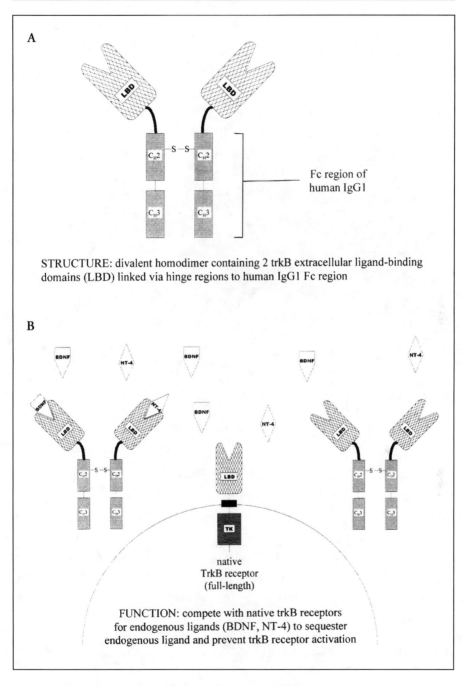

Figure 1. Schematic of structure and function of trkB receptor bodies.

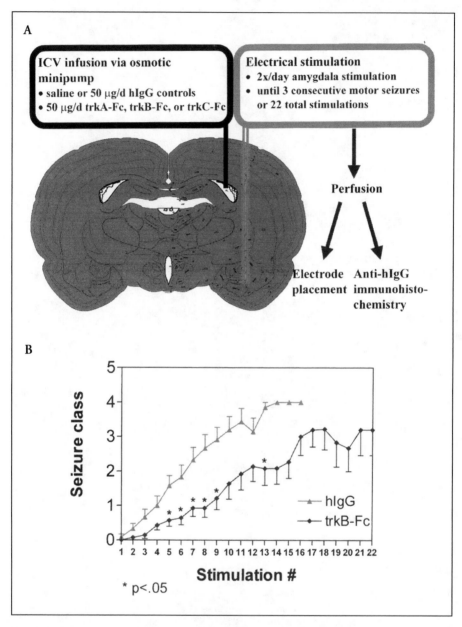

Figure 2A-D. TrkB receptor body inhibits kindling development. For details see Binder et al.[142]
A) Experimental design of ICV protein administration and kindling. B) TrkB-Fc inhibits kindling development compared to human IgG. Reprinted with permission from ref. 142, copyright 1999 Society for Neuroscience.

a reduction in behavioral seizure intensity during kindling development (Fig. 2C). Furthermore, we found that the degree of immunohistochemical penetration of trkB-Fc into hippocampus, but not striatum, septum or other structures correlated with the magnitude of inhibition of kindling development (Fig. 2D).[142]

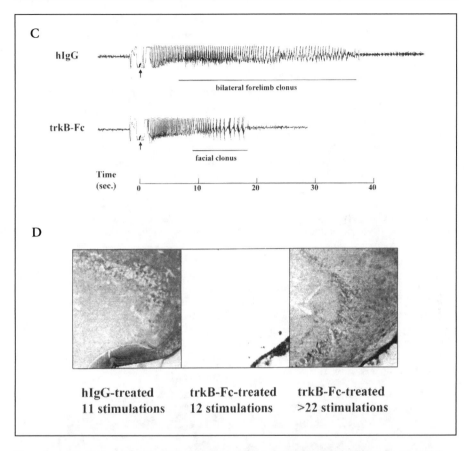

Figure 2, cont'd. C) Representative electroencephalograms from animals at kindling stimulation #15 from human IgG vs. trkB-Fc. Seizure duration and seizure intensity are decreased in trkB-Fc-treated animal. Reprinted with permission from ref. 142, copyright 1999 Society for Neuroscience. D) Hippocampal Fc immunoreactivity correlates with inhibition of kindling by trkB-Fc. Presence of Fc immunoreactivity indicating hippocampal penetration of trkB-Fc correlates with inhibition of kindling development (>22 vs. 12 stimulations to reach kindling criterion).

The finding that ICV trkB-Fc interferes with kindling suggests that BDNF and trkB signaling contributes to the development of kindled seizures. In apparent conflict with these findings, chronic intrahippocampal infusion of BDNF inhibits hippocampal kindling development and reduces electrographic seizure duration.[143] However, prolonged exposure to increased concentrations of BDNF suppresses trkB receptor responsiveness and reduces trkB mRNA and protein levels in vitro;[144,145] likewise, a six-day infusion of BDNF into the adult hippocampus in vivo decreases levels of full-length trkB receptor by 80%.[144] Thus, it is likely that chronic BDNF infusion in these kindling studies led to trkB downregulation and reduced responsiveness; if so, the retarded kindling development observed is consistent with the findings of the trkB-Fc infusion studies and those of BDNF heterozygotes[139] in implicating trkB receptor activation in kindling development. Alternatively, BDNF infusion could have upregulated the inhibitory neuropeptide Y (NPY) in these studies (see below).

Epileptogenesis in transgenic mice overexpressing the truncated form of trkB, a dominant negative receptor for BDNF, has recently been examined.[146] After kainic acid-induced status epilepticus, development of spontaneous seizures was monitored by video-EEG. This study demonstrated that transgenic mice expressing truncated trkB (which would presumably downregulate BDNF signaling through the full-length catalytic trkB receptor) had a lower frequency of spontaneous seizures, and had less severe seizures with later onset and lower mortality.[146]

Activation of trk Receptors after Seizures

The above work suggests that limiting activation of the trkB receptor inhibits epileptogenesis, but this does not address whether or where NT receptor activation occurs during epileptogenesis. Since ligand-induced receptor tyrosine phosphorylation is required for NT-induced cellular responses,[5] receptor tyrosine phosphorylation seems a logical measure of the biologic level of NT activity. Using antibodies that selectively recognize the phosphorylated form of trk receptors (Fig. 3A), we found that in contrast to the low level of phosphotrk immunoreactivity constitutively expressed in the hippocampus of adult rats, phosphotrk immunoreactivity was strikingly increased following partial kindling or kainate-induced seizures.[147] Following seizures, phosphotrk immunoreactivity was selectively increased in dentate hilus and CA3 stratum lucidum of hippocampus (Fig. 3B). This distribution coincides with the 'mossy fiber' pathway arising from the dentate gyrus granule cells (Fig. 3C). This immunoreactivity could be selectively competed with phosphotrk peptide (Fig. 3D).

Interestingly, the anatomic distribution, time course and threshold for seizure-induced phosphotrk immunoreactivity corresponds well to the demonstrated pattern of BDNF upregulation by seizures. That is, both phosphotrk and BDNF immunoreactivities are most prominently increased in hippocampal CA3 stratum lucidum and maximally increased at 24 hours after seizure onset (Fig. 3E).[147] This suggests that the phosphotrk immunoreactivity may be caused by seizure-induced increases in BDNF expression and release. Taken together with the kindling data, these results imply that activation of trkB receptors contributes to the development of kindling, and implicate the hippocampus and in particular the mossy fiber-CA3 synapse as a primary site of trkB action.

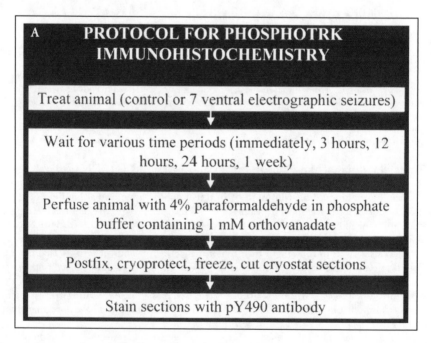

Figure 3. Seizures induce trk phosphorylation in the mossy fiber pathway of adult rat hippocampus. For details see Binder et al.[147] A) Protocol for phosphotrk immunohistochemistry.

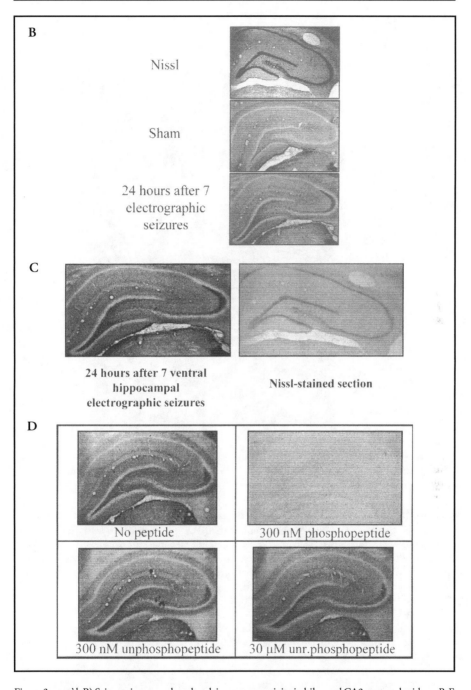

Figure 3, cont'd. B) Seizures increase phosphotrk immunoreactivity in hilus and CA3 stratum lucidum. B-E reprinted with permission from ref. 147, copyright 1999 Society for Neuroscience. C) Comparison of distribution of phosphotrk immunoreactivity with Nissl-stained section demonstrating mossy fiber pathway localization of phosphotrk immunoreactivity. D) Peptide competition of phosphotrk immunoreactivity. Sections are from a rat 24 hours after 7 electrographic seizures.

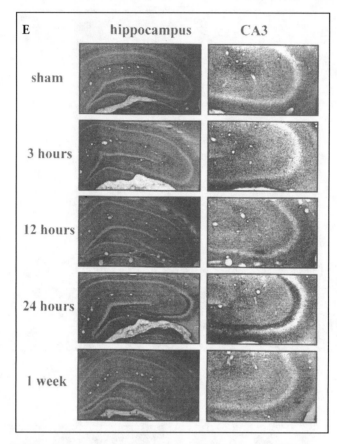

Figure 3, cont'd. E) Time course of phosphotrk immunoreactivity following 7 ventral hippocampal seizures (left: hippocampus, right: CA3). Note phosphotrk immunoreactivity in hilus and CA3 stratum lucidum at the 24-hour time point.

BDNF-Induced Hyperexcitability of the Mossy Fiber-CA3 Synapse

Based on the above data, one might speculate that BDNF upregulation in the adult brain could predispose certain areas to seizures. Indeed, in adult rat hippocampal slices BDNF exposure can produce multiple discharges and spreading depression in area CA3 and the entorhinal cortex upon afferent stimulation.[91] Acute application of exogenous BDNF to hippocampal slices appears to preferentially enhance the efficacy of the excitatory mossy fiber synapse onto CA3 pyramidal cells.[91]

Actions of BDNF have also been examined after pilocarpine-induced status epilepticus and chronic seizures, when sprouting of mossy fiber collaterals occurs. The new collaterals innervate processes in the inner molecular layer, including granule cell dendrites.[148] In hippocampal slices isolated from pilocarpine-treated rats, BDNF enhances responses to stimulation of the mossy fiber collaterals recorded in the inner molecular layer.[149] These effects can be blocked by K252a, a trk inhibitor, and confirm a preferential enhancement of mossy fiber synaptic transmission by BDNF. In addition, BDNF exposure in these epileptic animals led to seizure-like events.[149] Consistent with this are the observations of heightened seizure susceptibility, spontaneous seizures, and hyperexcitability of hippocampal field CA3 in BDNF-overexpressing transgenic mice.[140]

Cellular Model of BDNF-trkB Interaction

The studies summarized above indicate that upregulation of BDNF mRNA, protein and receptor activation occurs during epileptogenesis, that this upregulation is functionally relevant to increased excitability, and that the hippocampus and closely associated limbic structures may be particularly important in the pro-epileptogenic effects of BDNF. A cellular and molecular model of the actions of BDNF in promoting excitability in the hippocampus follows from these studies (Fig. 4). BDNF mRNA upregulation by seizure or perhaps by other stimuli such as ischemia or traumatic brain injury leads to increased BDNF production by the dentate granule cells, heightened anterograde transport and release of BDNF from mossy fiber axons and activation of trkB receptors in hilus and CA3 stratum lucidum. The locus of activation of trkB receptors by released BDNF may be either pre- or postsynaptic.[150,151] TrkB receptor activation could lead to acute depolarization,[96] enhanced glutamatergic synaptic transmission,[79,92] or reduced inhibitory synaptic transmission.[84] Recent data based on the LTP model suggest that BDNF's actions may be primarily postsynaptic.[100] These alterations in synaptic transmission, either alone or in combination with other changes (see below) could be sufficiently long-lived to underlie a permanent hyperexcitability of the hippocampal network (i.e., the epileptic state).

The relevance of results implicating BDNF in modulation of synaptic transmission to epileptogenesis depends critically on whether such modulation occurs in epileptic tissue. Several lines of evidence suggest this is the case. First, BDNF expression is increased in hippocampi of patients with temporal lobe epilepsy (see below). Second, evidence for modulation of ionotropic receptors with epilepsy comes from studies demonstrating altered electrophysiology of dentate granule cells in kindling,[152,153] other animal models [154-156] and human epileptic tissue.[157] Third, increased excitability of CA3 pyramidal cells is observed in kindled animals as detected by increased epileptiform bursting induced by elevated K⁺ or lowered Mg⁺⁺ in isolated hippocampal slices.[158,159] CA3 excitability is also present in other animal models.[160] Fourth,

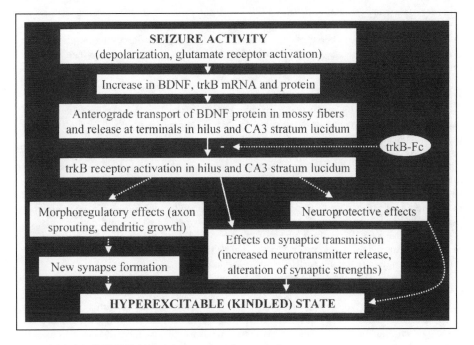

Figure 4. Model of BDNF/trkB involvement in epileptogenesis.

tetanic stimulation of the mossy fiber pathway in hippocampal slices (such as might occur during a seizure) induces synaptic potentiation onto CA3 pyramidal cells while inducing depression onto stratum lucidum interneurons.[161]

While modulation of multiple synaptic stations in the limbic system probably contributes to hyperexcitability following seizures, the pivotal role of the CA3 pyramidal cells in promoting epileptiform activity in the hippocampus; the role of BDNF in hippocampal synaptic transmission; the fact that constitutive and seizure-induced BDNF immunoreactivity within the hippocampus is most intense in the mossy fiber pathway;[24,28] together with the localization of seizure-induced trk receptor activation in CA3 stratum lucidum[147] all suggest that strengthening of the excitatory mossy fiber input onto CA3 pyramidal cells may be a primary mechanism by which BDNF promotes epilepsy.

Other Effects of BDNF

Based on the known effects of BDNF, it is possible that trkB receptor activation could contribute to epileptogenesis not only via synaptic effects on excitability but also by inducing changes in dendritic or axonal sprouting, synaptic morphology and synapse formation on a slower time scale. The most prominent synaptic reorganization known to occur in the epileptic brain is sprouting of the dentate granule cell mossy fibers.[162] Interestingly, mossy fiber sprouting was increased in BDNF +/- compared to +/+ mice despite the inhibition of kindling development in these mutants.[139] In addition, bath-applied trkB-Fc failed to inhibit kainate-induced mossy fiber sprouting in hippocampal explant cultures.[163] BDNF-overexpressing transgenic mice failed to demonstrate increased mossy fiber sprouting.[164] Thus, there is little evidence to date to suggest that BDNF upregulation is responsible for mossy fiber sprouting in the adult brain during epileptogenesis.

BDNF is known to modulate the expression of neurotransmitters and neuropeptides, many of which have potential roles in seizures. Perhaps the best characterized of these is neuropeptide Y (NPY). BDNF but not NGF is known to increase NPY levels.[165] NPY is thought to be inhibitory to seizure generation because NPY knockout animals exhibit increased seizure susceptibility.[166] Interestingly, both kindling and kainate-induced seizures increase NPY immunoreactivity in the mossy fibers[131,167] in a distribution strikingly similar to that of phosphotrk immunoreactivity. This suggests that BDNF-induced trk activation may lead to NPY upregulation in an overlapping anatomic distribution, thereby potentially limiting excitability.[143]

BDNF and Human Epilepsy

Animal models of epilepsy, in particular the kindling model described above, have implicated BDNF in epileptogenesis. What direct evidence is there that BDNF is altered/involved in human epilepsy?

Mathern et al found increased expression of BDNF mRNA in dentate granule cells from patients with temporal lobe epilepsy.[168] Similarly, Murray et al performed in situ hybridization for BDNF mRNA in resected temporal lobe epilepsy specimens and found increased hippocampal BDNF expression compared to autopsy control tissue.[169] Takahashi et al recently showed that protein levels of BDNF but not other NTs were upregulated 2.6-fold in human epilepsy tissue specimens.[170] Interestingly, this study also demonstrated a corresponding upregulation of hippocampal NPY. Recently, Zhu and Roper recorded from hippocampal slices from patients with temporal lobe epilepsy and found that BDNF application enhanced fast excitatory transmission in dentate granule cells.[171]

BDNF may also be involved in tumor-associated epilepsy. Primary brain tumors are often associated with seizures.[172] Interestingly, immunohistochemical expression of BDNF and trkB have been studied in glioneuronal brain tumors. In a study of 40 patients with gangliogliomas and 15 patients with dysembryoplastic neuroepithelial tumors (DNETs), tumors that are associated with chronic medically intractable epilepsy, Aronica et al have recently shown that there is intense immunoreactivity for both BDNF and trkB in these tumors, and furthermore this immunoreactivity is colocalized with NMDAR1 immunoreactivity suggesting a functional interaction potentially contributing to the epilepsy associated with these lesions.[173]

BDNF and Other Diseases of the Adult Nervous System

BDNF and Neurodegenerative Diseases

The idea that degenerative diseases of the nervous system may result from insufficient supply of neurotrophic factors has generated great interest in BDNF as a potential therapeutic agent. Many reports have documented evidence of decreased expression of BDNF in Alzheimer's and Parkinson's disease.[174] Selective reduction of BDNF mRNA in the hippocampus has been reported in Alzheimer's disease specimens and decreased BDNF protein has been demonstrated in the substantia nigra in Parkinson's disease, areas that degenerate in these diseases. BDNF promotes survival of all major neuronal types affected in Alzheimer's and Parkinson's disease, such as hippocampal and neocortical neurons, cholinergic septal and basal forebrain neurons, and nigral dopaminergic neurons. Interestingly, recent work has implicated BDNF in Huntington's disease as well. Huntingtin, the protein mutated in Huntington's disease, upregulates BDNF transcription, and loss of huntingtin-mediated BDNF transcription leads to loss of trophic support to striatal neurons which subsequently degenerate in the hallmark pathology of the disorder.[175] In all of these disorders, provision of BDNF or increasing endogenous BDNF production may conceivably be therapeutic if applied in the appropriate spatiotemporal context.

BDNF and Pain Transmission

BDNF also appears to play an important neuromodulatory role in pain transduction.[176] In particular, BDNF acts as a neuromodulator in small-diameter nociceptive neurons.[177] BDNF is synthesized, anterogradely transported and packaged by these neurons into dense core vesicles at nociceptor (C-fiber) terminals in the dorsal horn, and is markedly upregulated in inflammatory injury to peripheral nerves (along with NGF). Postsynaptic cells in this region express trk receptors,[178] and application of BDNF sensitizes nociceptive afferents and elicits hyperalgesia.[179]

An example of pathologic activity-dependent plasticity somewhat similar to epilepsy is central pain sensitization.[180,181] Central pain sensitization is an activity-dependent increase in excitability of dorsal horn neurons leading to a clinically intractable condition termed 'neuropathic pain' in which normally nonpainful somatosensory stimuli (touch and pressure) become exquisitely painful (allodynia). Like kindling and drug sensitization and dependence, central sensitization is NMDA receptor-dependent and long-lasting.[180]

Furthermore, as in kindling, NTs have been implicated in central sensitization.[182] BDNF is upregulated in dorsal root ganglia and spinal cord following peripheral inflammation.[183] In addition, BDNF reduces $GABA_A$-mediated currents in peripheral afferent fibers, suggesting that it may facilitate nociceptive input into the dorsal horn.[184] Woolf and colleagues have demonstrated that pretreatment with trkB-Fc prevents central sensitization, presumably by competing with endogenous BDNF released at C-fiber terminals onto dorsal horn neurons, thereby preventing activation of trkB receptors on dorsal horn membranes.[185] Inflammation-induced hyperalgesia also appears to be related to NGF/trkA signaling since trkA-Fc or NGF antibodies inhibit the development of hyperalgesia following an inflammatory stimulus.[186-188]

BDNF and Drug Addiction

The neurobiology of drug addiction is rapidly becoming better understood.[189] Emerging evidence indicates that BDNF-related plasticity may also occur in brain structures responsible for drug sensitization and dependence. For example, BDNF has been found to influence the reinforcing and locomotor activating properties of psychostimulants. Repeated injections of amphetamine lead to elevated BDNF mRNA expression in the basolateral amygdala, piriform cortex and paraventricular nucleus of the hypothalamus.[190] This is accompanied by increased

BDNF immunoreactivity in target structures such as the nucleus accumbens, a well-known site related to reinforcing behavior and addiction.[190]

Chronic opiate exposure leads to numerous neurochemical adaptations, in particular in the noradrenergic locus ceruleus (LC). Such adaptations are thought to contribute to physical drug dependence. Now, it appears that opiate administration and withdrawal lead to changes in BDNF expression. Numan et al demonstrate that whereas chronic morphine treatment results in only modest increases in BDNF in the locus ceruleus, withdrawal leads to a marked, rapid and prolonged increase in BDNF and trkB mRNA in the LC.[191] More recently, Akbarian et al[192] have shown that there are dramatic alterations of morphine administration-induced signaling in mice with a conditional deletion of BDNF in postnatal brain. In these mice, there was a three-fold reduction in opiate withdrawal symptoms.[192]

All drugs of abuse increase dopamine in the shell of the nucleus accumbens, and the D3 receptor is thought to be responsible for the reinforcing effects of drugs.[193] A recent study suggested a candidate molecular mechanism for the control of BDNF over behavioral sensitization. Guillin et al demonstrated that BDNF from dopaminergic neurons is responsible for inducing normal expression of the dopamine D3 receptor in nucleus accumbens.[194] Thus, pathologic alterations in BDNF expression may lead to the abnormal D3 expression seen in drug addiction.

BDNF and Affective Behaviors

BDNF signaling may also be involved in affective behaviors.[195,196] BDNF may be dysregulated in depressed individuals.[196] Environmental stresses such as immobilization that induce depression also decrease BDNF mRNA.[197] Conversely, physical exercise is associated with decreased depression and increased BDNF mRNA. Existing treatments for depression are thought to work primarily by increasing endogenous monoaminergic (i.e., serotonergic and noradrenergic) synaptic transmission, and recent studies have shown that effective antidepressants increase BDNF mRNA in the brain. Exogenous delivery of BDNF promotes the function and sprouting of serotonergic neurons in adult rat brains.[195] Thus, new pharmacologic strategies are focused on the potential antidepressant role of BDNF.

A recent study suggests that the BDNF gene may be a susceptibility gene for bipolar disorder, a severe psychiatric disease that affects 1% of the population worldwide and is characterized by recurrent bouts of mania and depression.[198] This study demonstrates linkage disequilibrium between two polymorphisms of the BDNF gene and bipolar disorder in 283 nuclear families.[198]

Summary

BDNF is widespread in the CNS during development and in adulthood, and is regulated in a wide variety of physiologic and pathologic states. Overall, in addition to its trophic effects on target neurons, BDNF appears to constitute a general mechanism for activity-dependent modification of synapses in the developing and adult CNS. Diseases of abnormal trophic support (such as neurodegenerative diseases) and diseases of abnormal excitability (such as epilepsy and central pain sensitization) can be related in some cases to abnormal BDNF signaling.

The evidence implicating BDNF in pathologic activity-dependent plasticity is most clear in the case of epilepsy. BDNF mRNA and protein are markedly upregulated in the hippocampus by seizure activity in animal models, and interference with BDNF/trkB signaling inhibits epileptogenesis. The hippocampus and closely associated limbic structures are thought to be particularly important in the pro-epileptogenic effects of BDNF, and indeed increased BDNF expression in the hippocampus is found in specimens from patients with temporal lobe epilepsy.

It is hoped that understanding of the hyperexcitability associated with BDNF may lead to novel anticonvulsant or antiepileptic therapies. Further study of the cellular and molecular mechanisms by which BDNF influences cell survival and excitability will likely provide novel concepts and targets for the treatment of diverse CNS diseases.

References

1. Barde Y-A, Edgar D, Thoenen H. Purification of a new neurotrophic factor from mammalian brain. EMBO J 1982; 1:549-553.
2. Levi-Montalcini R, Hamburger V. Selective growth-stimulating effects of mouse sarcoma on the sensory and sympathetic nervous system of the chick embryo. J Exp Zool 1951; 116:321-361.
3. Maisonpierre PC, Le Beau MM, Espinosa R et al. Human and rat brain-derived neurotrophic factor and neurotrophin-3: gene structures, distributions, and chromosomal localizations. Genomics 1991; 10:558-68.
4. Patapoutian A, Reichardt LF. Trk receptors: mediators of neurotrophin action. Curr Op Neurobiol 2001; 11:272-280.
5. Barbacid M. The trk family of neurotrophin receptors. J Neurobiol 1994; 25:1386-403.
6. Schlessinger J, Ulrich A. Growth factor signaling by receptor tyrosine kinases. Neuron 1992; 9:381-391.
7. Guiton M, Gunn-Moore FJ, Stitt TN et al. Identification of in vivo brain-derived neurotrophic factor-stimulated autophosphorylation sites on the trkB receptor tyrosine kinase by site-directed mutagenesis. J Biol Chem 1994; 269:30370-30377.
8. Heumann R. Neurotrophin signalling. Curr Op Neurobiol 1994; 4:668-679.
9. Kaplan DR, Stephens RM. Neurotrophin signal transduction by the trk receptor. J Neurobiol 1994; 25:1404-1417.
10. Middlemas DS, Meisenhelder J, Hunter T. Identification of trkB autophosphorylation sites and evidence that phospholipase C-gamma1 is a substrate of the trkB receptor. J Biol Chem 1994; 269:5458-5466.
11. Segal RA, Greenberg ME. Intracellular signaling pathways activated by neurotrophic factors. Annu Rev Neurosci 1996; 19:463-489.
12. Urfer R, Tsoulfas P, O'Connell L et al. An immunoglobulin-like domain determines the specificity of neurotrophin receptors. EMBO J 1995; 14:2795-2805.
13. Chao MV, Hempstead BL. p75 and trk: a two-receptor system. Trends Neurosci 1995; 18:321-326.
14. Roux P, Barker P. Neurotrophin signaling through the p75 neurotrophin receptor. Prog Neurobiol 2002; 67:203.
15. Dechant G, Barde YA. The neurotrophin receptor p75^NTR: novel functions and implications for diseases of the nervous system. Nat Neurosci 2002; 5:1131-1136.
16. Inagaki N, Thoenen H, Lindholm D. TrkA tyrosine residues involved in NGF-induced neurite outgrowth of PC12 cells. Eur J Neurosci 1995; 7:1125-1133.
17. Grimes ML, Zhou J, Beattie EC et al. Endocytosis of activated trkA: evidence that nerve growth factor induces formation of signaling endosomes. J Neurosci 1996; 16:7950-7964.
18. Von Bartheld CS, Williams R, Lefcort F et al. Retrograde transport of neurotrophins from the eye to the brain in chick embryos: roles of the p75NTR and trkB receptors. J Neurosci 1996; 16:2995-3008.
19. Bhattacharyya A, Watson FL, Bradlee TA et al. Trk receptors function as rapid retrograde signal carriers in the adult nervous system. J Neurosci 1997; 17:7007-7016.
20. Riccio A, Pierchala BA, Ciarallo CL et al. An NGF-trkA-mediated retrograde signal to transcription factor CREB in sympathetic neurons. Science 1997; 277:1097-1100.
21. Senger DL, Campenot RB. Rapid retrograde tyrosine phosphorylation of trkA and other proteins in rat sympathetic neurons in compartment cultures. J Cell Biol 1997; 138:411-421.
22. MacInnis BL, Campenot RB. Retrograde support of neuronal survival without retrograde transport of nerve growth factor. Science 2002; 295:1536-9.
23. Barker PA, Hussain NK, McPherson PS. Retrograde signaling by the neurotrophins follows a well-worn trk. Trends Neurosci 2002; 25:379-381.
24. Conner JM, Lauterborn JC, Yan Q et al. Distribution of brain-derived neurotrophic factor (BDNF) protein and mRNA in the normal adult rat CNS—evidence for anterograde axonal transport. J Neurosci 1997; 17:2295-2313.
25. Bengzon J, Kokaia Z, Ernfors P et al. Regulation of neurotrophin and trkA, trkB and trkC tyrosine kinase receptor messenger RNA expression in kindling. Neuroscience 1993; 53:433-46.
26. Fryer RH, Kaplan DR, Feinstein SC et al. Developmental and mature expression of full-length and truncated trkB receptors in the rat forebrain. J Comp Neurol 1996; 374:21-40.
27. Lindvall O, Kokaia Z, Bengzon J et al. Neurotrophins and brain insults. Trends Neurosci 1994; 17:490-496.
28. Yan Q, Rosenfeld RD, Matheson CR et al. Expression of brain-derived neurotrophic factor protein in the adult rat central nervous system. Neuroscience 1997; 78:431-448.
29. Von Bartheld CS, Byers MR, Williams R et al. Anterograde transport of neurotrophins and axodendritic transfer in the developing visual system. Nature 1996; 379:830-833.

30. Altar CA, Cai N, Bliven T et al. Anterograde transport of brain-derived neurotrophic factor and its role in the brain. Nature 1997; 389:856-860.
31. Zhou X-F, Rush RA. Endogenous brain-derived neurotrophic factor is anterogradely transported in primary sensory neurons. Neuroscience 1996; 74:945-951.
32. Fawcett JP, Bamji SX, Causing CG et al. Functional evidence that BDNF is an anterograde neuronal trophic factor in the CNS. J Neurosci 1998; 18:2808-2821.
33. Tonra JR, Curtis R, Wong V et al. Axotomy upregulates the anterograde transport and expression of brain-derived neurotrophic factor by sensory neurons. J Neurosci 1998; 18:4374-4383.
34. Kohara K, Kitamura A, Morishima M et al. Activity-dependent transfer of brain-derived neurotrophic factor to postsynaptic neurons. Science 2001; 291:2419-2423.
35. Nawa H, Takei N. BDNF as an anterophin; a novel neurotrophic relationship between brain neurons. Trends Neurosci 2001; 24:683-684.
36. Smith MA, Zhang LX, Lyons WE et al. Anterograde transport of endogenous brain-derived neurotrophic factor in hippocampal mossy fibers. Neuroreport 1997; 8:1829-1834.
37. Fawcett JP, Aloyz R, McLean JH et al. Detection of brain-derived neurotrophic factor in a vesicular fraction of brain synaptosomes. J Biol Chem 1997; 272:8837-40.
38. Blöchl A, Thoenen H. Characterization of nerve growth factor (NGF) release from hippocampal neurons: evidence for a constitutive and an unconventional sodium-dependent regulated pathway. Eur J Neurosci 1995; 7:1220-8.
39. Blöchl A, Thoenen H. Localization of cellular storage compartments and sites of constitutive and activity-dependent release of nerve growth factor (NGF) in primary cultures of hippocampal neurons. Mol Cell Neurosci 1996; 7:173-90.
40. Goodman LJ, Valverde J, Lim F et al. Regulated release and polarized localization of brain-derived neurotrophic factor in hippocampal neurons. Mol Cell Neurosci 1996; 7:222-238.
41. Griesbeck O, Canossa M, Campana G et al. Are there differences between the secretion characteristics of NGF and BDNF? Implications for the modulatory role of neurotrophins in activity-dependent neuronal plasticity. Microsc Res Tech 1999; 45:262-75.
42. Canossa M, Gartner A, Campana G et al. Regulated secretion of neurotrophins by metabotropic glutamate group I (mGluRI) and Trk receptor activation is mediated via phospholipase C signalling pathways. EMBO J 2001; 20:1640-50.
43. Purves D, Lichtman JW. Principles of neural development. Sunderland, MA: Sinauer, 1985.
44. Barde YA. Trophic factors and neuronal survival. Neuron 1989; 2:1525-1534.
45. Huang EJ, Reichardt LF. Neurotrophins: roles in neuronal development and function. Annu Rev Neurosci 2001; 24:677-736.
46. Jones KR, Farinas I, Backus C et al. Targeted disruption of the BDNF gene perturbs brain and sensory neuron development but not motor neuron development. Cell 1994; 76:989-99.
47. Tucker KL, Meyer M, Barde YA. Neurotrophins are required for nerve growth during development. Nat Neurosci 2001; 4:29-37.
48. Carroll P, Lewin GR, Koltzenburg M et al. A role for BDNF in mechanosensation. Nat Neurosci 1998; 1:42-6.
49. Acheson A, Conover JC, Fandl JP et al. A BDNF autocrine loop in adult sensory neurons prevents cell death. Nature 1995; 374:450-3.
50. Lotto RB, Asavaritikrai P, Vali L et al. Target-derived neurotrophic factors regulate the death of developing forebrain neurons after a change in their trophic requirements. J Neurosci 2001; 21:3904-10.
51. Xu B, Zang K, Ruff NL et al. Cortical degeneration in the absence of neurotrophin signaling: dendritic retraction and neuronal loss after removal of the receptor trkB. Neuron 2000; 26:233-45.
52. Patel MN, McNamara JO. Selective enhancement of axonal branching of cultured dentate gyrus neurons by neurotrophic factors. Neuroscience 1995; 69:763-70.
53. Lowenstein DH, Arsenault L. The effects of growth factors on the survival and differentiation of cultured dentate gyrus neurons. J Neurosci 1996; 16:1759-1769.
54. McAllister AK, Lo DC, Katz LC. Neurotrophins regulate dendritic growth in developing visual cortex. Neuron 1995; 15:791-803.
55. McAllister AK, Katz LC, Lo DC. Neurotrophin regulation of cortical dendritic growth requires activity. Neuron 1996; 17:1057-1064.
56. McAllister AK, Katz LC, Lo DC. Opposing roles for endogenous BDNF and NT-3 in regulating cortical dendritic growth. Neuron 1997; 18:767-78.
57. Cohen-Cory S, Fraser SE. Effects of brain-derived neurotrophic factor on optic axon branching and remodelling in vivo. Nature 1995; 378:192-196.

58. Maffei L, Berardi N, Domenici L et al. Nerve growth factor (NGF) prevents the shift in ocular dominance distribution of visual cortical neurons in monocularly deprived rats. J Neurosci 1992; 12:4651-62.
59. Cabelli RJ, Hohn A, Shatz CJ. Inhibition of ocular dominance column formation by infusion of NT-4/5 or BDNF. Science 1995; 267:1662-6.
60. Cabelli RJ, Shelton DL, Segal RA et al. Blockade of endogenous ligands of trkB inhibits formation of ocular dominance columns. Neuron 1997; 19:63-76.
61. Holtzman DM, Lowenstein DH. Selective inhibition of axon outgrowth by antibodies to NGF in a model of temporal lobe epilepsy. J Neurosci 1995; 15:7062-7070.
62. Castrén E, Zafra F, Thoenen H et al. Light regulates expression of brain-derived neurotrophic factor mRNA in rat visual cortex. Proc Natl Acad Sci USA 1992; 89:9444-8.
63. Castrén E, Thoenen H, Lindholm D. Brain-derived neurotrophic factor messenger RNA is expressed in the septum, hypothalamus and in adrenergic brain stem nuclei of adult rat brain and is increased by osmotic stimulation in the paraventricular nucleus. Neuroscience 1995; 64:71-80.
64. Rocamora N, Welker E, Pascual M et al. Upregulation of BDNF mRNA expression in the barrel cortex of adult mice after sensory stimulation. J Neurosci 1996; 16:4411-4419.
65. Li XC, Jarvis ED, Alvarez-Borda B et al. A relationship between behavior, neurotrophin expression, and new neuron survival. Proc Natl Acad Sci USA 2000; 97:8584-9.
66. Patterson SL, Grover LM, Schwartzkroin PA et al. Neurotrophin expression in rat hippocampal slices: a stimulus paradigm inducing LTP in CA1 evokes increases in BDNF and NT-3 mRNAs. Neuron 1992; 9:1081-8.
67. Castrén E, Pitkanen M, Sirvio J et al. The induction of LTP increases BDNF and NGF mRNA but decreases NT-3 mRNA in the dentate gyrus. Neuroreport 1993; 4:895-8.
68. Bramham CR, Southard T, Sarvey JM et al. Unilateral LTP triggers bilateral increases in hippocampal neurotrophin and trk receptor mRNA expression in behaving rats: evidence for interhemispheric communication. J Comp Neurol 1996; 368:371-382.
69. Neeper SA, Gomez-Pinilla F, Choi J et al. Physical activity increases mRNA for brain-derived neurotrophic factor and nerve growth factor in rat brain. Brain Res 1996; 726:49-56.
70. Oliff HS, Berchtold NC, Isackson P et al. Exercise-induced regulation of brain-derived neurotrophic factor (BDNF) transcripts in the rat hippocampus. Brain Res Mol Brain Res 1998; 61:147-53.
71. Frost DO. BDNF/trkB signaling in the developmental sculpting of visual connections. Prog Brain Res 2001; 134:35-49.
72. Thoenen H. Neurotrophins and neuronal plasticity. Science 1995; 270:593-8.
73. Schuman EM. Neurotrophin regulation of synaptic transmission. Curr Opin Neurobiol 1999; 9:105-9.
74. Thoenen H. Neurotrophins and activity-dependent plasticity. Prog Brain Res 2000; 128:183-91.
75. Poo MM. Neurotrophins as synaptic modulators. Nat Rev Neurosci 2001; 2:24-32.
76. Lohof AM, Ip NY, Poo MM. Potentiation of developing neuromuscular synapses by the neurotrophins NT-3 and BDNF. Nature 1993; 363:350-3.
77. Knipper M, Leung LS, Zhao D et al. Short-term modulation of glutamatergic synapses in adult rat hippocampus by NGF. Neuroreport 1994; 5:2433-6.
78. Lessmann V, Gottmann K, Heumann R. BDNF and NT-4/5 enhance glutamatergic synaptic transmission in cultured hippocampal neurones. Neuroreport 1994; 6:21-5.
79. Kang H, Schuman EM. Long-lasting neurotrophin-induced enhancement of synaptic transmission in the adult hippocampus. Science 1995; 267:1658-62.
80. Stoop R, Poo MM. Synaptic modulation by neurotrophic factors: differential and synergistic effects of brain-derived neurotrophic factor and ciliary neurotrophic factor. J Neurosci 1996; 16:3256-64.
81. Carmignoto G, Pizzorusso T, Tia S et al. Brain-derived neurotrophic factor and nerve growth factor potentiate excitatory synaptic transmission in the rat visual cortex. J Physiol 1997; 498:153-164.
82. Wang X, Poo M. Potentiation of developing synapses by postsynaptic release of neurotrophin-4. Neuron 1997; 19:825-835.
83. Messaoudi E, Bardsen K, Srebro B et al. Acute intrahippocampal infusion of BDNF induces lasting potentiation of synaptic transmission in the rat dentate gyrus. J Neurophysiol 1998; 79:496-499.
84. Tanaka T, Saito H, Matsuki N. Inhibition of GABA$_A$ synaptic responses by brain-derived neurotrophic factor (BDNF) in rat hippocampus. J Neurosci 1997; 17:2959-2966.
85. Frerking M, Malenka RC, Nicoll RA. Brain-derived neurotrophic factor (BDNF) modulates inhibitory, but not excitatory, transmission in the CA1 region of the hippocampus. J Neurophysiol 1998; 80:3383-6.

86. Figurov A, Pozzo-Miller LD, Olafsson P et al. Regulation of synaptic responses to high-frequency stimulation and LTP by neurotrophins in the hippocampus. Nature 1996; 381:706-9.
87. Akaneya Y, Tsumoto T, Kinoshita S et al. Brain-derived neurotrophic factor enhances long-term potentiation in rat visual cortex. J Neurosci 1997; 17:6707-6716.
88. Korte M, Carroll P, Wolf E et al. Hippocampal long-term potentiation is impaired in mice lacking brain-derived neurotrophic factor. Proc Natl Acad Sci USA 1995; 92:8856-60.
89. Korte M, Griesbeck O, Gravel C et al. Virus-mediated gene transfer into hippocampal CA1 region restores long-term potentiation in brain-derived neurotrophic factor mutant mice. Proc Natl Acad Sci USA 1996; 93:12547-52.
90. Patterson SL, Abel T, Deuel TA et al. Recombinant BDNF rescues deficits in basal synaptic transmission and hippocampal LTP in BDNF knockout mice. Neuron 1996; 16:1137-45.
91. Scharfman HE. Hyperexcitability in combined entorhinal/hippocampal slices of adult rat after exposure to brain-derived neurotrophic factor. J Neurophysiol 1997; 78:1082-1095.
92. Takei N, Sasaoka K, Inoue K et al. Brain-derived neurotrophic factor increases the stimulation-evoked release of glutamate and the levels of exocytosis-associated proteins in cultured cortical neurons from embryonic rats. J Neurochem 1997; 68:370-5.
93. Suen P-C, Wu K, Levine ES et al. Brain-derived neurotrophic factor rapidly enhances phosphorylation of the postsynaptic N-methyl-D-aspartate receptor subunit 1. Proc Natl Acad Sci USA 1997; 94:8191-8195.
94. Berninger B, Garcia DE, Inagaki N et al. BDNF and NT-3 induce intracellular Ca^{2+} elevation in hippocampal neurones. Neuroreport 1993; 4:1303-6.
95. Levine ES, Dreyfus CF, Black IB et al. Differential effects of NGF and BDNF on voltage-gated calcium currents in embryonic basal forebrain neurons. J Neurosci 1995; 15:3084-91.
96. Kafitz KW, Rose CR, Thoenen H et al. Neurotrophin-evoked rapid excitation through TrkB receptors. Nature 1999; 401:918-21.
97. Marty S, Berzaghi MP, Berninger B. Neurotrophins and activity-dependent plasticity of cortical interneurons. Trends Neurosci 1997; 20:198-202.
98. Xu B, Gottschalk W, Chow A et al. The role of brain-derived neurotrophic factor receptors in the mature hippocampus: Modulation of long-term potentiation through a presynaptic mechanism involving TrkB. J Neurosci 2000; 20:6888-97.
99. Manabe T. Does BDNF have pre or postsynaptic targets? Science 2002; 295:1651-3.
100. Kovalchuk Y, Hanse E, Kafitz KW et al. Postsynaptic induction of BDNF-mediated long-term potentiation. Science 2002; 295:1729-34.
101. Hall J, Thomas KL, Everitt BJ. Rapid and selective induction of BDNF expression in the hippocampus during contextual learning. Nat Neurosci 2000; 3:533-535.
102. Ishibashi H, Hihara S, Takahashi M et al. Tool-use learning induces BDNF expression in a selective portion of monkey anterior parietal cortex. Brain Res Mol Brain Res 2002; 102:110.
103. Linnarsson S, Bjorklund A, Ernfors P. Learning deficit in BDNF mutant mice. Eur J Neurosci 1997; 9:2581-2587.
104. Ma YL, Wang HL, Wu HC et al. Brain-derived neurotrophic factor antisense oligonucleotide impairs memory retention and inhibits long-term potentiation in rats. Neuroscience 1998; 82:957-967.
105. Minichiello L, Korte M, Wolfer D et al. Essential role for TrkB receptors in hippocampus-mediated learning. Neuron 1999; 24:401-14.
106. Hauser WA, Kurland LT. The epidemiology of epilepsy in Rochester, Minnesota, 1935 through 1967. Epilepsia 1975; 16:1-66.
107. Morgan JI, Curran T. Stimulus-transcription coupling in the nervous system: Involvement of the inducible proto-oncogenes fos and jun. Annu Rev Neurosci 1991; 14:421-451.
108. Kiessling M, Gass P. Immediate early gene expression in experimental epilepsy. Brain Pathol 1993; 3:381-393.
109. Gall C, Lauterborn J, Bundman M et al. Seizures and the regulation of neurotrophic factor and neuropeptide gene expression in brain. Epilepsy Res—Suppl 1991; 4:225-45.
110. Meberg PJ, Gall CM, Routtenberg A. Induction of F1/GAP-43 gene expression in hippocampal granule cells after seizures. Brain Res Mol Brain Res 1993; 17:295-9.
111. Qian Z, Gilbert M, Colicos MA et al. Tissue plasminogen activator is induced as an immediate-early gene during seizure, kindling, and long-term potentiation. Nature 1993; 361:453-457.
112. Gall CM, Isackson PJ. Limbic seizures increase neuronal production of messenger RNA for nerve growth factor. Science 1989; 245:758-61.
113. Gall CM. Seizure-induced changes in neurotrophin expression: Implications for epilepsy. Exp Neurol 1993; 124:150-66.
114. Binder DK, Gall CM, Croll SD et al. BDNF and epilepsy: Too much of a good thing? Trends Neurosci 2001; 24:47-53.

115. Ernfors P, Bengzon J, Kokaia Z et al. Increased levels of messenger RNAs for neurotrophic factors in the brain during kindling epileptogenesis. Neuron 1991; 7:165-76.
116. Isackson PJ, Huntsman MM, Murray KD et al. BDNF mRNA expression is increased in adult rat forebrain after limbic seizures: temporal patterns of induction distinct from NGF. Neuron 1991; 6:937-48.
117. Dugich-Djordjevic MM, Tocco G, Lapchak PA et al. Regionally specific and rapid increases in brain-derived neurotrophic factor messenger RNA in the adult rat brain following seizures induced by systemic administration of kainic acid. Neuroscience 1992; 47:303-15.
118. Dugich-Djordjevic MM, Tocco G, Willoughby DA et al. BDNF mRNA expression in the developing rat brain following kainic acid-induced seizure activity. Neuron 1992; 8:1127-38.
119. Humpel C, Wetmore C, Olson L. Regulation of brain-derived neurotrophic factor messenger RNA and protein at the cellular level in pentylenetetrazol-induced epileptic seizures. Neuroscience 1993; 53:909-18.
120. Merlio JP, Ernfors P, Kokaia Z et al. Increased production of the TrkB protein tyrosine kinase receptor after brain insults. Neuron 1993; 10:151-64.
121. Schmidt-Kastner R, Olson L. Decrease of neurotrophin-3 mRNA in adult rat hippocampus after pilocarpine seizures. Exp Neurol 1995; 136:199-204.
122. Schmidt-Kastner R, Humpel C, Wetmore C et al. Cellular hybridization for BDNF, trkB, and NGF mRNAs and BDNF-immunoreactivity in rat forebrain after pilocarpine-induced status epilepticus. Exp Brain Res 1996; 107:331-47.
123. Mudo G, Jiang XH, Timmusk T et al. Change in neurotrophins and their receptor mRNAs in the rat forebrain after status epilepticus induced by pilocarpine. Epilepsia 1996; 37:198-207.
124. Sato K, Kashihara K, Morimoto K et al. Regional increases in brain-derived neurotrophic factor and nerve growth factor mRNAs during amygdaloid kindling, but not in acidic and basic fibroblast growth factor mRNAs. Epilepsia 1996; 37:6-14.
125. Gall CM, Lauterborn JC, Guthrie KM et al. Seizures and the regulation of neurotrophic factor expression: associations with structural plasticity in epilepsy. In: Seil RJ, ed. Advances in Neurology, vol 72: Neuronal Regeneration, Reorganization, and Repair. Philadelphia: Lippincott-Raven, 1997:9-24.
126. Lowenstein DH, Seren MS, Longo FM. Prolonged increases in neurotrophic activity associated with kainate-induced hippocampal synaptic reorganization. Neuroscience 1993; 56:597-604.
127. Humpel C, Lindqvist E, Soderstrom S et al. Monitoring release of neurotrophic activity in the brains of awake rats. Science 1995; 269:552-4.
128. Nawa H, Carnahan J, Gall C. BDNF protein measured by a novel enzyme immunoassay in normal brain and after seizure: Partial disagreement with mRNA levels. Eur J Neurosci 1995; 7:1527-35.
129. Elmer E, Kokaia Z, Kokaia M et al. Dynamic changes of brain-derived neurotrophic factor protein levels in the rat forebrain after single and recurring kindling-induced seizures. Neuroscience 1998; 83:351-62.
130. Rudge JS, Mather PE, Pasnikowski EM et al. Endogenous BDNF protein is increased in adult rat hippocampus after a kainic acid induced excitotoxic insult but exogenous BDNF is not neuroprotective. Exp Neurol 1998; 149:398-410.
131. Vezzani A, Ravizza T, Moneta D et al. Brain-derived neurotrophic factor immunoreactivity in the limbic system of rats after acute seizures and during spontaneous convulsions: temporal evolution of changes as compared to neuropeptide Y. Neuroscience 1999; 90:1445-1461.
132. Gall CM, Conner JM, Lauterborn JC et al. Cellular localization of BDNF protein after recurrent seizures in rat: evidence for axonal transport of the newly synthesized factor. Epilepsia 1996; 37 Suppl. 5:47.
133. Goddard GV. The development of epileptic seizures through brain stimulation at low intensity. Nature 1967; 214:1020-1021.
134. Goddard GV, McIntyre DC, Leech CK. A permanent change in brain function resulting from daily electrical stimulation. Exp Neurol 1969; 25:295-330.
135. McNamara JO, Bonhaus DW, Shin C. The kindling model of epilepsy. In: Schwartzkroin PA, ed. Epilepsy: Models, mechanisms, and concepts. New York, NY: Cambridge University Press, 1993:27-47.
136. Binder DK, McNamara JO. Kindling: a pathologic activity-driven structural and functional plasticity in mature brain. In: Corcoran ME, Moshe S, eds. Kindling 5. New York: Plenum Press, 1997:245-254.
137. Funabashi T, Sasaki H, Kimura F. Intraventricular injection of antiserum to nerve growth factor delays the development of amygdaloid kindling. Brain Res 1988; 458:132-6.
138. Van der Zee CE, Rashid K, Le K et al. Intraventricular administration of antibodies to nerve growth factor retards kindling and blocks mossy fiber sprouting in adult rats. J Neurosci 1995; 15:5316-23.

139. Kokaia M, Ernfors P, Kokaia Z et al. Suppressed epileptogenesis in BDNF mutant mice. Exp Neurol 1995; 133:215-24.
140. Croll SD, Suri C, Compton DL et al. Brain-derived neurotrophic factor transgenic mice exhibit passive avoidance deficits, increased seizure severity and in vitro hyperexcitability in the hippocampus and entorhinal cortex. Neuroscience 1999; 93:1491-1506.
141. Scharfman HE, Goodman JH, Sollas AL et al. Spontaneous limbic seizures after intrahippocampal infusion of brain-derived neurotrophic factor. Exp Neurol 2002; 174:201-14.
142. Binder DK, Routbort MJ, Ryan TE et al. Selective inhibition of kindling development by intraventricular administration of trkB receptor body. J Neurosci 1999; 19:1424-1436.
143. Larmet Y, Reibel S, Carnahan J et al. Protective effects of brain-derived neurotrophic factor on the development of hippocampal kindling in the rat. Neuroreport 1995; 6:1937-41.
144. Frank L, Ventimiglia R, Anderson K et al. BDNF downregulates neurotrophin responsiveness, trkB protein and trkB mRNA levels in cultured rat hippocampal neurons. Eur J Neurosci 1996; 8:1220-30.
145. Knusel B, Gao H, Okazaki T et al. Ligand-induced down-regulation of trk messenger RNA, protein and tyrosine phosphorylation in rat cortical neurons. Neuroscience 1997; 78:851-862.
146. Lahteinen S, Pitkanen A, Saarelainen T et al. Decreased BDNF signalling in transgenic mice reduces epileptogenesis. Eur J Neurosci 2002; 15:721-34.
147. Binder DK, Routbort MJ, McNamara JO. Immunohistochemical evidence of seizure-induced activation of trk receptors in the mossy fiber pathway of adult rat hippocampus. J Neurosci 1999; 19:4616-4626.
148. Okazaki MM, Evenson DA, Nadler JV. Hippocampal mossy fiber sprouting and synapse formation after status epilepticus in rats: Visualization after retrograde transport of biocytin. J Comp Neurol 1995; 352:515-534.
149. Scharfman HE, Goodman JH, Sollas AL. Actions of brain-derived neurotrophic factor in slices from rats with spontaneous seizures and mossy fiber sprouting in the dentate gyrus. J Neurosci 1999; 19:5619-5631.
150. Wu K, Xu J, Suen P et al. Functional trkB neurotrophin receptors are intrinsic components of the adult brain postsynaptic density. Mol Brain Res 1996; 43:286-290.
151. Drake CT, Milner TA, Patterson SL. Ultrastructural localization of full-length trkB immunoreactivity in rat hippocampus suggests multiple roles in modulating activity-dependent synaptic plasticity. J Neurosci 1999; 19:8009-26.
152. Kohr G, De Koninck Y, Mody I. Properties of NMDA receptor channels in neurons acutely isolated from epileptic (kindled) rats. J Neurosci 1993; 13:3612-27.
153. McNamara JO. Cellular and molecular basis of epilepsy. J Neurosci 1994; 14:3413-3425.
154. Wuarin J-P, Dudek FE. Electrographic seizures and new recurrent excitatory circuits in the dentate gyrus of hippocampal slices from kainate-treated epileptic rats. J Neurosci 1996; 16:4438-4448.
155. Gibbs JW, Shumate MD, Coulter DA. Differential epilepsy-associated alterations in postsynaptic GABA(A) receptor function in dentate granule and CA1 neurons. J Neurophysiol 1997; 77:1924-1938.
156. Okazaki MM, Molnar P, Nadler JV. Recurrent mossy fiber pathway in rat dentate gyrus: Synaptic currrents evoked in presence and absence of seizure-induced growth. J Neurophysiol 1999; 81:1645-1660.
157. Williamson A, Patrylo PR, Spencer DD. Decrease in inhibition in dentate granule cells from patients with medial temporal lobe epilepsy. Ann Neurol 1999; 45:92-99.
158. Behr J, Lyson KJ, Mody I. Enhanced propagation of epileptiform activity through the kindled dentate gyrus. J Neurophysiol 1998; 79:1726-1732.
159. King GL, Dingledine R, Giacchino JL et al. Abnormal neuronal excitability in hippocampal slices from kindled rats. J Neurophysiol 1985; 54:1295-304.
160. Buzsaki G, Ponomareff GL, Bayardo F et al. Neuronal activity in the subcortically denervated hippocampus: A chronic model for epilepsy. Neuroscience 1989; 28:527-538.
161. Maccaferri G, Toth K, McBain CJ. Target-specific expression of presynaptic mossy fiber plasticity. Science 1998; 279:1368-1370.
162. Sutula T, He XX, Cavazos J et al. Synaptic reorganization in the hippocampus induced by abnormal functional activity. Science 1988; 239:1147-50.
163. Routbort MJ, Ryan TE, Yancopoulos GD et al. TrkB-IgG does not inhibit mossy fiber sprouting in an in vitro model. Soc Neurosci Abstr 1997; 23:888.
164. Qiao X, Suri C, Knusel B et al. Absence of hippocampal mossy fiber sprouting in transgenic mice overexpressing brain-derived neurotrophic factor. J Neurosci Res 2001; 64:268-76.
165. Croll SD, Wiegand SJ, Anderson KD et al. Regulation of neuropeptides in adult rat forebrain by the neurotrophins BDNF and NGF. Eur J Neurosci 1994; 6:1343-53.

166. Baraban SC, Hollopeter G, Erickson JC et al. Knock-out mice reveal a critical antiepileptic role for neuropeptide Y. J Neurosci 1997; 17:8927-36.
167. Marksteiner J, Ortler M, Bellmann R et al. Neuropeptide Y biosynthesis is markedly induced in mossy fibers during temporal lobe epilepsy of the rat. Neurosci Lett 1990; 112:143-148.
168. Mathern GW, Babb TL, Micevych PE et al. Granule cell mRNA levels for BDNF, NGF, and NT-3 correlate with neuron losses or supragranular mossy fiber sprouting in the chronically damaged and epileptic human hippocampus. Mol Chem Neuropathol 1997; 30:53-76.
169. Murray KD, Isackson PJ, Eskin TA et al. Altered mRNA expression for brain-derived neurotrophic factor and type II calcium/calmodulin-dependent protein kinase in the hippocampus of patients with intractable temporal lobe epilepsy. J Comp Neurol 2000; 418:411-22.
170. Takahashi M, Hayashi S, Kakita A et al. Patients with temporal lobe epilepsy show an increase in brain-derived neurotrophic factor protein and its correlation with neuropeptide Y. Brain Res 1999; 818:579-82.
171. Zhu WJ, Roper SN. Brain-derived neurotrophic factor enhances fast excitatory synaptic transmission in human epileptic dentate gyrus. Ann Neurol 2001; 50:188-94.
172. Villemure JG, de Tribolet N. Epilepsy in patients with central nervous system tumors. Curr Opin Neurol 1996; 9:424-8.
173. Aronica E, Leenstra S, Jansen GH et al. Expression of brain-derived neurotrophic factor and tyrosine kinase B receptor proteins in glioneuronal tumors from patients with intractable epilepsy: colocalization with N-methyl-D-aspartic acid receptor. Acta Neuropathol (Berl) 2001; 101:383-392.
174. Murer MG, Yan Q, Raisman-Vozari R. Brain-derived neurotrophic factor in the control human brain, and in Alzheimer's disease and Parkinson's disease. Prog Neurobiol 2001; 63:71-124.
175. Zuccato C, Ciammola A, Rigamonti D et al. Loss of huntingtin-mediated BDNF gene transcription in Huntington's disease. Science 2001; 293:493-498.
176. Bennett DL. Neurotrophic factors: Important regulators of nociceptive function. Neuroscientist 2001; 7:13-17.
177. Thompson SW, Bennett DL, Kerr BJ et al. Brain-derived neurotrophic factor is an endogenous modulator of nociceptive responses in the spinal cord. Proc Natl Acad Sci USA 1999; 96:7714-7718.
178. Zhou X-F, Parada LF, Soppet D et al. Distribution of trkB tyrosine kinase immunoreactivity in the rat central nervous system. Brain Res 1993; 622:63-70.
179. Shu XQ, Mendell LM. Neurotrophins and hyperalgesia. Proc Natl Acad Sci USA 1999; 96:7693-7696.
180. Woolf CJ, Thompson SWN. The induction and maintenance of central sensitization is dependent on N-methyl-D-aspartic acid receptor activation: implications for the treatment of post-injury pain hypersensitivity states. Pain 1991; 44:293-299.
181. Woolf CJ, Salter MW. Neuronal plasticity: increasing the gain in pain. Science 2000; 288:1765-9.
182. Millan MJ. The induction of pain: an integrative review. Prog Neurobiol 1999; 57:1-164.
183. Cho HJ, Kim JK, Zhou XF et al. Increased brain-derived neurotrophic factor immunoreactivity in rat dorsal root ganglia and spinal cord following peripheral inflammation. Brain Res 1997; 764:269-72.
184. Oyelese AA, Rizzo MA, Waxman SG et al. Differential effects of NGF and BDNF on axotomy-induced changes in GABA(A)-receptor-mediated conductance and sodium currents in cutaneous afferent neurons. J Neurophysiol 1997; 78:31-42.
185. Mannion RJ, Costigan M, Decosterd I et al. Neurotrophins: peripherally and centrally acting modulators of tactile stimulus-induced inflammatory pain hypersensitivity. Proc Natl Acad Sci USA 1999; 96:9385-90.
186. McMahon SB, Bennett DL, Priestley JV et al. The biological effects of endogenous nerve growth factor on adult sensory neurons revealed by a trkA-IgG fusion molecule. Nat Med 1995; 1:774-80.
187. Dmitrieva N, Shelton D, Rice ASC et al. The role of nerve growth factor in a model of visceral inflammation. Neuroscience 1997; 78:449-459.
188. Ma QP, Woolf CJ. The progressive tactile hyperalgesia induced by peripheral inflammation is nerve growth factor dependent. Neuroreport 1997; 8:807-810.
189. Nestler EJ, Aghajanian GK. Molecular and cellular basis of addiction. Science 1997; 278:58-63.
190. Meredith G, Callen S, Scheuer D. Brain-derived neurotrophic factor expression is increased in the rat amygdala, piriform cortex and hypothalamus following repeated amphetamine administration. Brain Res 2002; 949:218.
191. Numan S, Lane-Ladd SB, Zhang L et al. Differential regulation of neurotrophin and trk receptor mRNAs in catecholaminergic nuclei during chronic opiate treatment and withdrawal. J Neurosci 1998; 18:10700-8.
192. Akbarian S, Rios M, Liu RJ et al. Brain-derived neurotrophic factor is essential for opiate-induced plasticity of noradrenergic neurons. J Neurosci 2002; 22:4153-62.

193. Le Foll B, Schwartz JC, Sokoloff P. Dopamine D3 receptor agents as potential new medications for drug addiction. Eur Psychiatry 2000; 15:140-6.
194. Guillin O, Diaz J, Carroll P et al. BDNF controls dopamine D3 receptor expression and triggers behavioural sensitization. Nature 2001; 411:86-89.
195. Altar CA. Neurotrophins and depression. Trends Pharmacol Sci 1999; 20:59-61.
196. Nestler EJ, Barrot M, DiLeone RJ et al. Neurobiology of depression. Neuron 2002; 34:13-25.
197. Alleva E, Santucci D. Psychosocial vs. "physical" stress situations in rodents and humans: role of neurotrophins. Physiol Behav 2001; 73:313-20.
198. Neves-Pereira M, Mundo E, Muglia P et al. The brain-derived neurotrophic factor gene confers susceptibility to bipolar disorder: evidence from a family-based association study. Am J Hum Genet 2002; 71:651-5.

CHAPTER 4

Vascular Endothelial Growth Factor (VEGF) in Seizures:
A Double-Edged Sword

Susan D. Croll, Jeffrey H. Goodman and Helen E. Scharfman

Abstract

Vascular endothelial growth factor (VEGF) is a vascular growth factor which induces angiogenesis (the development of new blood vessels), vascular permeability, and inflammation. In brain, receptors for VEGF have been localized to vascular endothelium, neurons, and glia. VEGF is upregulated after hypoxic injury to the brain, which can occur during cerebral ischemia or high-altitude edema, and has been implicated in the blood-brain barrier breakdown associated with these conditions. Given its recently-described role as an inflammatory mediator, VEGF could also contribute to the inflammatory responses observed in cerebral ischemia. After seizures, blood-brain barrier breakdown and inflammation is also observed in brain, albeit on a lower scale than that observed after stroke. Recent evidence has suggested a role for inflammation in seizure disorders. We have described striking increases in VEGF protein in both neurons and glia after pilocarpine-induced status epilepticus in the brain. Increases in VEGF could contribute to the blood-brain barrier breakdown and inflammation observed after seizures. However, VEGF has also been shown to be neuroprotective across several experimental paradigms, and hence could potentially protect vulnerable cells from damage associated with seizures. Therefore, the role of VEGF after seizures could be either protective or destructive. Although only further research will determine the exact nature of VEGF's role after seizures, preliminary data indicate that VEGF plays a protective role after seizures.

Introduction

During and after cerebral ischemia in animal models, there is a well-documented breakdown of the blood-brain barrier that peaks 1-3 days after the ischemic insult.[1-8] Edema can be severe after ischemia, and much of the edema is thought to be vasogenic in nature, that is, caused by leakage of plasma fluids into the brain parenchyma. After seizures, blood-brain barrier breakdown has also been described, although the scope and magnitude of the breakdown is substantially milder than that seen after cerebral ischemia.[9-11] Further, ischemic and post-ischemic vasculature upregulates adhesion molecules which cause leukocytes to adhere to the luminal wall of vascular endothelium.[12-18] Chemokine upregulation then leads to the extravasation of leukocytes, which infiltrate the brain parenchyma in an inflammatory reaction. These vascular-mediated post-ischemic responses are likely to contribute to the damage observed after stroke.[15,19-23] Inflammatory cells can also be found in the brain after seizures.[24] There is, in fact, accumulating evidence that inflammatory cytokines are involved in the expression of seizures (see also Vezzani et al, this volume).[25-28] Because of the qualitative, albeit not quantitative, similarities in inflammation and blood-brain barrier breakdown after both stroke and

Recent Advances in Epilepsy Research, edited by Devin K. Binder and Helen E. Scharfman.
©2004 Eurekah.com and Kluwer Academic / Plenum Publishers.

seizures, it is reasonable to investigate the proposed mediators of post-ischemic vascular abnormalities after seizures. Among the protein factors recently implicated in post-ischemic blood-brain barrier breakdown are vascular growth factors.

Vascular Endothelial Growth Factor (VEGF)

In the past decade, there has been a surge of interest in vascular growth factors. During development, these factors have potent effects on endothelial cells, and are thought to regulate proliferation, migration, endothelial tube formation, vascular differentiation, permeability, and regression (for reviews see refs. 29 and 30). Although much still remains to be understood regarding the effects of these factors on adult vasculature, current data suggest that they play similar roles in the changes that occur in both normal and pathological states (for review see ref. 30). Cerebral ischemia is one such pathological state in which vascular changes are striking. During and after cerebral ischemia, alterations in the cerebral vasculature include blood-brain barrier breakdown, endothelial cell apoptosis, upregulation of adhesion molecules, and angiogenesis (the development of new blood vessels from existing blood vessels).[1-7,31-39] Vascular abnormalities, such as blood-brain barrier breakdown, have been observed after seizures also. Because these vascular alterations might contribute to the brain pathology observed after ischemia or seizures, it is important to understand how changes in the levels of various vascular growth factors might contribute to these pathologies.

There are a large number of protein factors that act on vasculature including, but not limited to, fibroblast growth factor, platelet-derived growth factor, transforming growth factor, hepatocyte growth factor (scatter factor), vascular endothelial growth factor (VEGF), and the angiopoietins. Proteins in the VEGF family of factors have potent vascular effects, and will be the topic of this chapter. These factors modulate the structure and function of vasculature in both developing and adult organisms.

VEGF was originally described as vascular permeability factor (VPF) because of its potent permeabilizing effects on endothelium.[40] Since the discovery of VEGF, four additional VEGF-like family members have been described. These additional VEGF-like proteins are placental growth factor (PlGF), VEGFB, VEGFC, and VEGFD (the original VEGF has been termed "VEGFA") (for review see ref. 41; refer to Fig. 1). The receptors currently described for the VEGF family are VEGFR1 (Flt-1), VEGFR2 (Flk-1 or KDR), VEGFR3 (Flt-4), and the neuropilins (see Fig. 1).[41,42,43,44] The primary receptors for VEGFA, VEGFR1 and VEGFR2, are localized predominantly to the vascular endothelium, including cerebral endothelium. However, several recent papers have reported neuronal localization of VEGFR2 in cultured hippocampal or dorsal root ganglion cells.[45,46] In addition, neurons located in peri-infarct regions after focal cerebral ischemia or in VEGF-treated brain express VEGFR2.[47,48] It is possible that VEGFR2, while not normally detectable in resting neurons, is upregulated during neuronal perturbation. One could argue that cultured neurons are in some way "perturbed," having been removed from their normal neural microenvironment. VEGFR2 has also been described on glial cells, particularly after cerebral ischemia.[49] VEGFR1 has been localized almost exclusively to vascular endothelium, but has been described on circulating inflammatory cells and VEGF-treated astroglia.[48,50] In addition to VEGFR1 and VEGFR2, members of the neuropilin receptor family bind to VEGFA. Although neuropilin can be found on vascular endothelium, it is most densely expressed in nonendothelial cells, especially in the nervous system.[51,52]

There are 5 known isoforms of VEGFA in humans which are termed VEGF 121, 145, 165, 189 and 206, corresponding to the number of amino acids found in each isoform (each isoform has 1 amino acid fewer in rodents).[53] While all isoforms of VEGFA bind with high affinity to both VEGFR1 and VEGFR2, binding of VEGFA to the neuropilins is isoform-specific (e.g., the 165 amino acid isoform of VEGFA binds to neuropilin-1, while the 121 amino acid isoform does not).[44] In addition, the various VEGF family members have different receptor specificities (e.g., VEGFA binds to VEGFR1 and 2, but PlGF only binds to VEGFR1; (Fig. 1); for review see ref. 30). There is still much to be learned about the roles of the diverse members of the VEGF family of proteins, as well as of the various VEGFA isoforms. Because more is

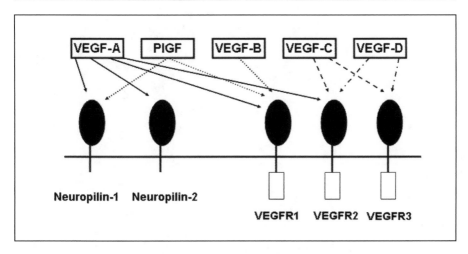

Figure 1. Schematic illustrating the VEGF family of proteins and their receptors.

currently known about VEGFA than its protein relatives, the remainder of this chapter will be devoted to discussions of VEGFA (hereafter referred to simply as "VEGF").

In recent years, most of the in vivo research on VEGF has focused on its role as a potent angiogenic factor, responsible for the development of new vascular sprouts (for reviews see refs. 54 and 55). Gene deletion studies have shown that VEGF is critical for the development of new blood vessels during development. Mutant mice lacking even a single VEGF allele die during embryonic development with a striking lack of secondary vasculature (i.e., deficient angiogenesis).[56,57] Both VEGFR1and VEGFR2 null mutants are early embryonic lethal and show a lack of vasculature (i.e., deficient vasculogenesis).[58,59]

In adult animals, application of recombinant VEGF protein induces the formation of new blood vessels from preexisting blood vessels (angiogenesis) in a variety of tissues, including brain.[48,60-65] However, the blood vessels formed by application of VEGF to adult tissues are grossly abnormal, characterized by profound permeability and a disorganized, dilated, tortuous morphology.[48,63-67]

Because of the leaky nature of the new blood vessels formed by exogenous administration of VEGF, increasing attention has been given to VEGF's originally-described function as a vascular permeabilizing agent. Application of VEGF to adult tissues or cells results in edema and vascular leak. VEGF results in vascular leak in every tissue to which it has been applied including, but not limited to, brain, lung, testis, bladder, skin, duodenum, mesentery, and intestine.[65,66,68-71] VEGF's effects on vascular leak in the brain occur rapidly, within 30 minutes of exposure to VEGF.[68] In ischemic brain, the timing of VEGF mRNA and protein upregulation corresponds closely to the peak of vasogenic edema.[35,47,49,72-79] VEGF expression is increased in both glia and neurons of the ischemic brain, as determined by both immunostaining and in situ hybridization.[47,49,75,76] Presumably, the VEGF is secreted by the neurons and glia, and binds to VEGF receptors on local vasculature to mediate the increases in vascular permeability. Upregulation of VEGF mRNA occurs both in vivo and in vitro under hypoxic conditions.[36,80-83] Hypoxia-induced upregulation of VEGF mRNA is associated with increases in the transcription factor hypoxia inducible factor (HIF)1-α, which is also upregulated after cerebral ischemia.[36,80-82,84] Interestingly, upregulation of VEGF mRNA after cerebral ischemia sometimes occurs in cells not directly affected by the ischemic insult, such as the cingulate cortex and hippocampus, suggesting that triggers other than hypoxia could lead to VEGF upregulation.[47] Possible secondary mechanisms of VEGF upregulation could include damage to neuronal afferents or efferents, pressure effects of edema, or ischemia-related physiological

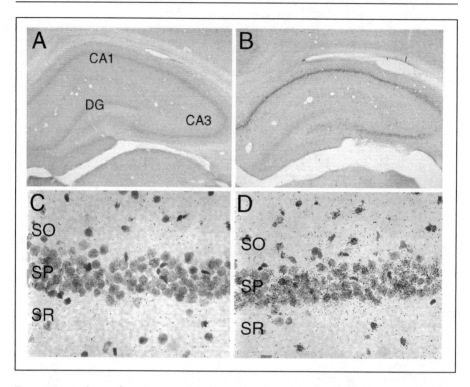

Figure 2. Upregulation of VEGF protein in CA1 of the hippocampus after seizures and VEGF mRNA in CA1 after cerebral ischemia. A) Control brain section stained for VEGF-ir 24h after the animal received a vehicle injection. DG = dentate gyrus. B) Post-status epilepticus (SE) (1 hour of status followed by diazepam injection) brain section stained for VEGF-ir from an animal that was perfused 24h after pilocarpine injection. C) In situ hybridization for VEGF mRNA in CA1 neurons 24h after a sham surgery. SO= stratum oriens, SP= stratum pyramidale, SR = stratum radiatum. D) In situ hybridization for VEGF mRNA in CA1 neurons 24h after MCAO. Note that MCAO does not cause the hippocampus to be ischemic, but rather leads to indirect effects.

phenomena such as cortical spreading depression (for review see ref. 85). Cortical spreading depression has been shown to induce increases in other cytokines.[86] Because some phenomena observed during cerebral ischemia can also be observed in association with seizures, we hypothesized that VEGF might increase in neural cells after seizures. Specifically, post-ischemic cortical spreading depression follows increases in synchronous neuronal firing such as those observed during seizures. Additionally, prolonged seizures can lead to hypoxic states due to breathing compromise during tonus, and hypoxia is the most consistently confirmed trigger of VEGF upregulation.

VEGF Regulation after Seizures

To study the possibility that seizures could lead to increases in VEGF, rats were treated with 380 mg/kg pilocarpine to induce status epilepticus (SE) as previously described.[87] Animals were sacrificed one day, one week, or one to two months after status epilepticus, and their brains were processed for VEGF immunostaining.[88] One day after SE, VEGF protein is dramatically increased both in neurons and glia in the hippocampus and limbic cortex (Fig. 2C and D and Fig. 3). Specifically, what appears to be cytosolic immunostaining for VEGF is observed in the neurons of CA1 and/or CA3 of the hippocampus as well as some of the pyramidal neurons in entorhinal and perirhinal cortex. The increases in neuronal VEGF are similar

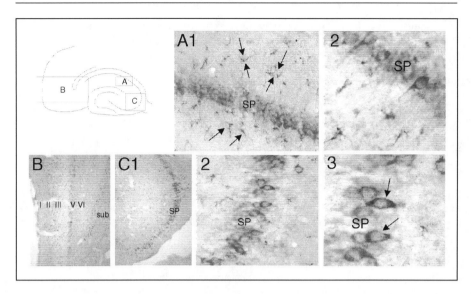

Figure 3. Both neurons and glia upregulate VEGF 24h after SE. All sections are stained for VEGF-ir. A schematic in the upper left shows the locations of parts A-C. A) A low (1) and high (2) magnification view of a post-SE brain section in dorsal CA1. Arrows point to glial VEGF-ir. SP = stratum pyramidale. B) Low magnification view of the entorhinal cortex of the same brain with the major layers of the medial entorhinal cortex marked. Sub = subiculum. C) Low (1), high (2), and higher (3) magnification of ventral CA3 in the same brain. Arrows point to neuronal VEGF-ir.

in intensity to those observed in CA1 pyramidal neurons 24h after middle cerebral artery occlusion (Croll et al, unpublished observations). These increases in neuronal VEGF reflected increased expression of VEGF mRNA in CA1 of the hippocampus, demonstrating that neurons can make VEGF (Fig. 2A and B).[47] Interestingly, the hippocampus does not suffer direct ischemia during middle cerebral artery occlusion, and therefore the increases in VEGF mRNA and protein are likely to be caused by widespread consequences of focal ischemia. One interesting candidate mechanism is spreading depression, a phenomenon characterized by abnormal and synchronous cell firing not unlike that observed with seizures. Glial staining after seizures appears to be at least partially cytosolic, but even more convincing is punctate cell surface staining on glial profiles (Fig. 3). Presumably, this staining pattern represents staining of VEGF protein bound to its receptors. This same punctate cell surface staining can also be observed on vascular endothelial cells in the region of VEGF upregulation.

The trigger for increased neural VEGF after seizure activity is unclear. As previously discussed, hypoxia is the best known trigger of VEGF expression across cell types, and hypoxia can occur during seizures. However, heretofore undescribed triggers of VEGF upregulation are likely, particularly given that the upregulation of neuronal VEGF has been observed after cerebral ischemia in areas not directly affected by the hypoxic insult (Fig. 2C and D).[47] For example, the upregulation of VEGF could result from increased neuronal activity, although there is little prior evidence for activity-dependent increases in VEGF. The paucity of evidence arises not from negative findings, but rather because the question has not been studied yet. Increases in neuronal activity induced by exposure of rats to complex environments increased microvascular density in the brains of these animals.[89] Because VEGF has been shown to be critical for angiogenesis across a wide variety of tissues, one could speculate that active neural cells in stimulated brain could upregulate VEGF in an attempt to induce increased vascular flow, hence metabolically supporting their increased activity. The fact that increased neuronal activity does lead to increased blood flow and an increased need for vascular investment is well-documented.

This premise serves as the basis for much of the current work in human functional brain imaging. Both fMRI and PET scans commonly use increased cerebral blood flow as a surrogate marker of increased neuronal activity.

VEGF As a Neurotrophic Factor

Although all original research, and most current research, on VEGF has focused on its effects on vasculature, there is a growing body of literature studying VEGF as a potential neurotrophic factor (for review, see ref. 90). Many neurons constitutively contain neuropilins, and have been shown under some circumstances, such as after ischemia, during development, and in culture, to express VEGFR2.[46,47,49,91] Glial cells have also been reported to express VEGF receptors. Although the exact roles of these receptors in neural cell function remain to be determined, accumulating evidence exists for such roles. VEGF has been shown to protect neurons across a wide variety of circumstances. For instance, VEGF has been reported to have a neuroprotective effect in the context of cerebral ischemia.[45,92,93] In addition, VEGF supports developing sensory cells in retina and dorsal root ganglia.[46,94,95] Further, VEGF has been shown to speed the recovery of damaged peripheral neurons, although increased vascular density in the injured area could have partially or fully accounted for this effect.[96,97] Elegant evidence for a role of VEGF in protecting adult neurons, either directly or indirectly, came in a recent study showing that deletion of the hypoxia response element from the VEGF promoter resulted in motor neuron disease in adult mice.[98]

Perhaps most relevant to a neuroprotective role in seizure-induced damage is the recent report that VEGF protects cultured hippocampal neurons against glutamate excitotoxicity.[99] Using antisense oligonucleotides, Matsuzaki et al[99] showed that this protective effect of VEGF is mediated through VEGFR2. In addition, blockade of VEGFR2 synthesis blocked induction of the Akt survival pathway in these neurons. Because the Akt survival pathway has been repeatedly demonstrated to be activated after VEGF treatment, it is possible that VEGF directly protects cells from excitotoxic damage by increasing signaling pathways important to survival.[100,101] Because these cultures only contained neurons, a direct protective role of VEGF could be inferred. Further evidence for a direct protective role of VEGF through VEGFR2 and Akt signaling has been presented in cultured hypoxic neurons.[45,93,102] In these hypoxic neurons, caspase-3 levels were increased when VEGF was blocked, providing further evidence for a direct effect of VEGF on pathways mediating cell survival.[93]

One final possibility is that VEGF produces its protective effects, at least in part, through neuropilin-1, which is located on neurons. VEGF and semaphorin 3A compete for binding at neuropilin-1. The semaphorins have traditionally been thought to induce cell death during development. VEGF could block this death pathway by binding to neuropilin-1, hence preventing semaphorin binding.[103] Recent evidence suggests, however, that the relationship between VEGF and semaphorin could be complex.[103] In addition, a recent direct attempt to induce neuroprotection in hypoxic cells with neuropilin-1 activation failed to show protection.[45] Therefore, although the potential role of neuropilin in VEGF-mediated neuroprotection cannot be overlooked, neuroprotection mediated via VEGFR2 activation is currently the most parsimonious explanation.

Potential Effects of VEGF on Seizures and Their Sequelae

VEGF could be a double-edged sword for the epileptic brain, because it results in multiple effects. Its proposed vascular effects might be expected to aggravate seizures and post-seizure brain damage, while its direct effects on neurons could be neuroprotective. VEGF also activates glia, which could potentially impact seizures or their sequelae.[48,65]

As previously discussed, VEGF has dramatic effects on brain vasculature, and results in blood-brain barrier breakdown. At low concentrations, VEGF has minimal angiogenic effects, especially within a short time frame. That is, VEGF's effects on vascular leak and permeability are more striking than its effects on angiogenesis in low amounts or after brief exposures.[65] In

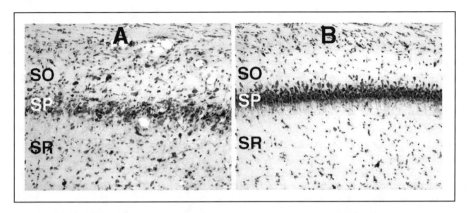

Figure 4. Cresyl violet stained hippocampal sections taken from animals 24 hours after status epilepticus induced by pilocarpine. A) Hippocampus from the median animal in the PBS-infused group showing damage in stratum pyramidale (SP) of area CA1. SO= stratum oriens, SR = stratum radiatum. B) Hippocampus from the median animal in the VEGF-infused group. Note the lack of cell damage in area CA1 of the animal treated with VEGF.

addition to its very potent effects as a vascular permeabilizing agent, VEGF also potently induces inflammation. At the same low doses and early time points which induce blood-brain barrier leak, local application of VEGF causes a striking inflammation characterized by a primarily monocytic infiltrate.[65,66] This effect could be caused by a direct chemoattractant effect on monocytes, which express VEGFR1, but it could also result from the complex pattern of upregulation of multiple inflammatory mediators such as ICAM-1 and Mip-1α, both of which have been observed after VEGF administration to brain.[50,65] Interestingly, VEGF is upregulated by both IL-1 and TNF-α.[104,105,106] These inflammatory cytokines are upregulated after seizures.[25,26,27,28] Evidence accumulated to date suggests that these inflammatory cytokines increase the potential for seizures.[27,28,107] Because both of these cytokines upregulate VEGF, it is possible that VEGF is one common pathway through which these proteins exert their effects. Given the data suggesting detrimental effects of inflammation on seizures and their sequelae, one might hypothesize that VEGF aggravates seizures or seizure-related brain compromise. In cerebral ischemia, VEGF has been shown to worsen post-ischemic edema and VEGF antagonism protects against damage.[108,109] However, there is also evidence that anti-inflammatory agents could lead to more post-seizure cell damage.[110] Therefore, an increased understanding of the role of inflammation in seizures will be necessary before we can fully appreciate the implications of VEGF's pro-inflammatory and vascular (leak) effects on epileptic brain.

In contrast to the potentially detrimental vascular effects of VEGF, we have summarized the developing evidence for a neuroprotective role of VEGF. There is accumulating evidence that VEGF directly protects neurons from damage, including excitotoxic damage.[99] Direct protective effects of VEGF on any cell containing VEGF receptors, including potential induction of VEGFR2 on neurons and glia, is quite possible given VEGF's activation of the Akt intracellular survival pathway.[100,101]

We have conducted preliminary studies to assess the effects of VEGF on cell damage after seizures. By infusing VEGF (30-60 ng/day; continuously for 5 days into adult rat brains via osmotic minipump) before inducing status epilepticus with pilocarpine (350 mg/kg), we have begun to address the role of VEGF upregulation after seizures. These initial experiments have revealed a statistically significant neuroprotective effect of VEGF on cell damage in CA1 and CA3 of the hippocampus, assessed 24 hours following status epilepticus (Fig. 4). Therefore, it seems possible that VEGF, while inducing inflammation, has the ability to protect cells from any damage that could occur as a result of either the inflammation or the excitotoxic insult.

Much additional research needs to be conducted before we can fully understand the implications of VEGF upregulation after seizures. Based on currently available data, VEGF could be a double-edged sword, directly protecting neurons from post-seizure cell death, but simultaneously inducing blood-brain barrier breakdown and inflammation. VEGF could, however, also be the rare pro-inflammatory cytokine that possesses the ability to directly protect neurons from the potentially detrimental effects of the very inflammation that it induces.

Acknowledgments

We thank our many colleagues, especially Drs. Stan Wiegand and John Rudge, for collaborative experiments and stimulating discussions which have helped lead to the premises presented in the current review. In addition, we acknowledge the technical contributions of our students and technicians, Sachin Shah, Ahmed Elkady, and Jamee Nicoletti from Queens College and Anne Sollas and Adam MacLeod from Helen Hayes Hospital, who made possible the generation of the data presented in this chapter.

References

1. Gotoh O, Asano T, Koide T et al. Ischemic brain edema following occlusions of the middle cerebral artery in the rat. I: The time courses of the brain water, sodium and potassium contents and blood-brain barrier permeability to ^{125}I-albumin. Stroke 1985; 16:101-109.
2. Hatashita S, Hoff JT. Role of blood-brain barrier permeability in focal ischemic brain edema. Adv Neurol 1990; 52:327-333.
3. Hatashita S, Hoff JT. Brain edema and cerebrovascular permeability during cerebral ischemia in rats. Stroke 1990; 21:582-588.
4. Nakagawa Y, Fujimoto N, Matsumoto K et al. Morphological changes in acute cerebral ischemia after occlusion and reperfusion in the rat. Adv Neurol 1990; 52:21-27.
5. Kitagawa K, Matsumoto M, Tagaya M et al. Temporal profile of serum albumin extravasation following cerebral ischemia in a newly established reproducible gerbil model for vasogenic brain edema: a combined immunohistochemical and dye tracer analysis. Acta Neuropathol (Berl) 1991; 82:164-171.
6. Menzies SA, Betz AL, Hoff JT. Contributions of ions and albumin to the formation and resolution of ischemic brain edema. J Neurosurg 1993; 78:257-266.
7. Matsumoto K, Lo EH, Pierce AR et al. Role of vasogenic edema and tissue cavitation in ischemic evolution on diffusion-weighted imaging: comparison with multiparameter MR and immunohistochemistry. Am J Neuroradiol 1995; 16:1107-1115.
8. Rosenberg GA. Ischemic brain edema. Prog Cardiovasc Dis 1999; 42:209-216.
9. Ates N, Esen N, Ilbay G. Absence epilepsy and regional blood-brain barrier permeability: the effects of pentylenetetrazole-induced convulsions. Pharmacol Res 1999; 39:305-310.
10. Cornford EM. Epilepsy and the blood brain barrier: endothelial cell responses to seizures. Adv Neurol 1999; 79:845-862.
11. Roch C, Leroy C, Nehlig A et al. Magnetic resonance imaging in the study of the lithium-pilocarpine model of temporal lobe epilepsy in adult rats. Epilepsia 2002; 43:325-335.
12. Feuerstein GZ, Wang X, Barone FC. Inflammatory gene expression in cerebral ischemia and trauma. Potential new therapeutic targets. Ann NY Acad Sci 1997; 825,179-193.
13. del Zoppo GJ, Wagner S, Tagaya M. Trends and future developments in the pharmacological treatment of acute ischaemic stroke. Drugs 1997; 54:9-38.
14. Becker KJ. Inflammation and acute stroke. Curr Opin Neurol 1998; 11:45-49.
15. DeGraba TJ. The role of inflammation after acute stroke: utility of pursuing anti-adhesion molecule therapy. Neurology 1998; 51:S62-68.
16. Jean WC, Spellman SR, Nussbaum ES et al. Reperfusion injury after focal cerebral ischemia: the role of inflammation and the therapeutic horizon. Neurosurgery 1998; 43:1382-1396.
17. del Zoppo GJ, Hallenbeck JM. Advances in the vascular pathophysiology of ischemic stroke. Thromb Res 2000; 98:73-81.
18. Stanimirovic D, Satoh K. Inflammatory mediators of cerebral endothelium: a role in ischemic brain inflammation. Brain Pathol 2000; 10:3-126.
19. Zhang RL, Chopp M, Jiang N et al. Anti-intercellular adhesion molecule-1 antibody reduces ischemic cell damage after transient but not permanent middle cerebral artery occlusion in the Wistar rat. Stroke 1995; 26:1438-1442.

20. Chopp M, Li Y, Jiang N et al. Antibodies against adhesion molecules reduce apoptosis after transient middle cerebral artery occlusion in rat brain. J Cereb Blood Flow Metab 1996; 16:578-584.
21. Chopp M, Zhang ZG. Anti-adhesion molecule and nitric oxide protection strategies in ischemic stroke. Curr Opin Neurol 1996; 9:68-72.
22. Goussev AV, Zhang Z, Anderson DC et al. P-selectin antibody reduces hemorrhage and infarct volume resulting from MCA occlusion in the rat. J Neurol Sci 1998; 161:16-22.
23. Jiang N, Chopp M, Chahwala S. Neutrophil inhibitory factor treatment of focal cerebral ischemia in the rat. Brain Res 1998; 788:25-34.
24. Peltola J, Laaksonen J, Haapala AM et al. Indicators of inflammation after recent tonic-clonic epileptic seizures correlate with plasma interleukin-6 levels. Seizure 2002; 11:44-46.
25. Vezzani A, Conti M, De Luigi A et al. Interleukin-1beta immunoreactivity and microglia are enhanced in the rat hippocampus by focal kainate application: functional evidence for enhancement of electrographic seizures. J Neurosci 1999; 19:5054-5065.
26. De Simoni MG, Perego C, Ravizza T et al. Inflammatory cytokines and related genes are induced in the rat hippocampus by limbic status epilepticus. Eur J Neurosci 2000; 12:2623-2633.
27. Vezzani A, Moneta D, Conti M et al. Powerful anticonvulsant action of IL-1 receptor antagonist on intracerebral injection and astrocytic overexpression in mice. Proc Natl Acad Sci USA 2000; 97:11534-11539.
28. Vezzani A, Moneta D, Richichi C et al. Functional role of inflammatory cytokines and antiinflammatory molecules in seizures and epileptogenesis. Epilepsia 2002; 43(S5):30-35.
29. Tomanek RJ, Schatteman GC, Angiogenesis. New insights and therapeutic potential. Anat Rec 2000; 261:126-135.
30. Yancopoulos GD, Davis S, Gale NW et al. Vascular-specific growth factors and blood vessel formation. Nature 2000; 407:242-248.
31. Bednar MM, Raymond S, McAuliffe T et al. The role of neutrophils and platelets in a rabbit model of thromboembolic stroke. Stroke 1991; 22:44-50.
32. Clark RK, Lee EV, Fish CJ et al. Development of tissue damage, inflammation and resolution following stroke: an immunohistochemical and quantitative planimetric study. Brain Res Bull 1993; 31:565-572.
33. Morioka T, Kalehua AN, Streit WJ. Characterization of microglial reaction after middle cerebral artery occlusion in rat brain. J Comp Neurol 1993; 327:123-132.
34. Okada Y, Copeland BR, Mori E et al. P-selectin and intercellular adhesion molecule-1 expression after focal brain ischemia and reperfusion. Stroke 1994; 25:202-211.
35. Beck H, Acker T, Wiessner C et al. Expression of angiopoietin-1, angiopoietin-2 and tie receptors after middle cerebral artery occlusion in the rat. Amer J Pathol 2000; 157:1473-1483.
36. Marti HJ, Bernaudin M, Bellail A et al. Hypoxia-induced vascular endothelial growth factor expression precedes neovascularization after cerebral ischemia. Am J Pathol 2000; 156:965-976.
37. Schwab JM, Nguyen TD, Postler E et al. Selective accumulation of cyclooxygenase-1-expressing microglial cells/macrophages in lesions of human focal cerebral ischemia. Acta Neuropathol (Berl) 2000; 99:609-614.
38. Wang X, Yue TL, Barone FC et al. Demonstration of increased endothelial-leukocyte adhesion molecule-1 mRNA expression in rat ischemic cortex. Stroke 1995; 26:1665-1668.
39. Zubkov AY, Ogihara K, Bernanke DH et al. Apoptosis of endothelial cells in vessels affected by cerebral vasospasm. Surg Neurol 2000; 53:260-266.
40. Senger DR, Perruzzi CA, Feder J et al. A highly conserved vascular permeability factor secreted by a variety of human and rodent tumor cell lines. Cancer Res 1986; 46:5629-5632.
41. Frelin C, Ladoux A, D'Angelo G. Vascular endothelial growth factors and angiogenesis. Ann Endocrinol (Paris) 2000; 61:70-74.
42. Quinn TP, Peters KG, De Vries C et al. Fetal liver kinase 1 is a receptor for vascular endothelial growth factor and is selectively expressed in vascular endothelium. Proc Natl Acad Sci USA 1993; 90:7533-7537.
43. Soker S, Takashima S, Miao H et al. Neuropilin-1 is expressed by endothelial and tumor cells as an isoform-specific receptor for vascular endothelial growth factor. Cell 1998; 92:735-745.
44. Gluzman-Poltorak Z, Cohen T, Herzog Y et al. Neuropilin-2 and neuropilin-1 are receptors for the 165-amino acid form of vascular endothelial growth factor (VEGF) and of placenta growth factor-2, but only neuropilin-2 functions as a receptor for the 145-amino acid form of VEGF. Biol Chem 2000; 275:18040-5.
45. Jin KL, Mao XO, Greenberg DA. Vascular endothelial growth factor: direct neuroprotective effect in in vivo ischemia. Proc Natl Acad Sci 2000; 97:10242-10247.
46. Sondell M, Sundler F, Kanje M. Vascular endothelial growth factor is a neurotrophic factor which stimulates axonal outgrowth through the flk-1 receptor. Eur J Neurosci 2000; 12:4243-4254.

47. Croll SD, Wiegand SJ. Vascular growth factors and cerebral ischemia. Mol Neurobiol 2001; 23:121-35.
48. Krum JM, Mani N, Rosenstein JM. Angiogenic and astroglial responses to vascular endothelial growth factor administration in adult rat brain. Neuroscience 2002; 110:589-604.
49. Lennmyr F, Ata KA, Funa K et al. Expression of vascular endothelial growth factor (VEGF) and its receptors (Flt-1 and Flk-1) following permanent and transient occlusion of the middle cerebral artery in the rat. J Neuropathol Exp Neurol 1998; 57:874-882.
50. Sawano A, Iwai S, Sakurai Y et al. Flt-1, vascular endothelial growth factor receptor 1, is a novel cell surface marker for the lineage of monocyte-macrophages in humans. Blood 2001; 97:785-791.
51. Kawakami A, Kitsukawa T, Takagi S et al. Developmentally regulated expression of a cell surface protein, neuropilin, in the mouse nervous system. J Neurobiol 1996; 29:1-17.
52. Chédotal A, Del Rio JA, Ruiz M et al. Semaphorins III and IV repel hippocampal axons via two distinct receptors. Develop 1998; 125:4313-4323.
53. Ferrara N, Davis-Smyth T. The biology of vascular endothelial growth factor. Endocr Rev 1997; 18:4-25.
54. Carmeliet P, Collen D. Molecular basis of angiogenesis. Roles of VEGF and VE-cadherin. Ann NY Acad Sci 2000; 902, 249-262.
55. Dvorak HF. VPF/VEGF and the angiogenic response. Semin Perinatol 2000; 24:75-78.
56. Ferrara N, Carver-Moore K, Chen H et al. Heterozygous embryonic lethality induced by targeted inactivation of the VEGF gene. Nature 1996; 380:439-442.
57. Carmeliet P, Ferreira V, Breier G et al. Abnormal blood vessel development and lethality in embryos lacking a single VEGF allele. Nature 1996; 380:435-9.
58. Fong GH, Rossant J, Gertsenstein M et al. Role of the Flt-1 receptor tyrosine kinase in regulating the assembly of vascular endothelium. Nature 1995; 376:66-70.
59. Shalaby F, Rossant J, Yamaguchi TP et al. Failure of blood-island formation and vasculogenesis in Flk-1-deficient mice. Nature 1995; 376:62-66.
60. Bauters C, Asahara T, Zheng LP et al. Physiological assessment of augmented vascularity induced by VEGF in ischemic rabbit hindlimb. Am J Physiol 1994; 267:H1263-1271.
61. Takeshita S, Zheng LP, Brogi E et al. Therapeutic angiogenesis. A single intraarterial bolus of vascular endothelial growth factor augments revascularization in a rabbit ischemic hind limb model. J Clin Invest 1994; 93:662-670.
62. Pearlman JD, Hibberd MG, Chuang ML et al. Magnetic resonance mapping demonstrates benefits of VEGF-induced myocardial angiogenesis. Nat Med 1995; 1:1085-1089.
63. Rosenstein JM, Mani N, Silverman WF et al. Patterns of brain angiogenesis after vascular endothelial growth factor administration in vivo and in vivo. Proc Natl Acad Sci USA 1998; 95:7086-7091.
64. Springer ML, Chen AS, Kraft PE et al. VEGF gene delivery to muscle: potential role for vasculogenesis in adults. Mol Cell 1998; 2:549-558.
65. Kasselman LJ, Ransohoff RM, Cai N et al. Vascular endothelial growth factor (VEGF)-mediated inflammation precedes angiogenesis in adult rat brain. Soc Nsci Abstr 2002; in press.
66. Proescholdt MA, Heiss JD, Walbridge S et al. Vascular endothelial growth factor (VEGF) modulates vascular permeability and inflammation in rat brain. J Neuropathol Exp Neurol 1999; 58:613-627.
67. Carmeliet P. VEGF gene therapy: stimulating angiogenesis or angioma-genesis? Nat Med 2000; 6:1102-1103.
68. Dobrogowska A, Lossinsky AS, Tarnawski M et al. Increased blood-brain barrier permeability and endothelial abnormalities induced by vascular endothelial growth factor. J Neurocytol 1998; 27:63-173.
69. Zhao L, Zhang MM, Ng K. Effects of vascular permeability factor on the permeability of cultured endothelial cells from brain capillaries. J Cardiovasc Pharmacol 1998; 32:1-4.
70. Kaner RJ, Ladetto JV, Singh R et al. Lung overexpression of the vascular endothelial growth factor gene induces pulmonary edema. Am J Respir Cell Mol Biol 2000; 22:657-664.
71. Thurston G, Rudge JS, Ioffe E et al. Angiopoietin-1 protects the adult vasculature against plasma leakage. Nat Med 2000; 6:460-463.
72. Kovacs Z, Ikezaki K, Samoto K et al. VEGF and flt: expression time kinetics in rat brain infarct. Stroke 1996; 27:1865-1873.
73. Hayashi T, Abe K, Suzuki H et al. Rapid induction of vascular endothelial growth factor gene expression after transient middle cerebral artery occlusion in rats. Stroke 1997; 28:2039-2044.
74. Cobbs CS, Chen J, Greenberg DA et al. Vascular endothelial growth factor expression in transient focal cerebral ischemia in the rat. Neurosci Lett 1998; 249:79-82.
75. Issa R, Krupinski J, Bujny T et al. Vascular endothelial growth factor and its receptor, KDR, in human brain tissue after ischemic stroke. Lab Invest 1999; 79:417-425.

76. Lee M-Y, Ju W-K, Cha J-H et al. Expression of vascular endothelial growth factor mRNA following transient forebrain ischemia in rats. Neurosci Lett 1999; 265:107-110.
77. Pichiule P, Chavez JC, Xu K et al. Vascular endothelial growth factor upregulation in transient global ischemia induced by cardiac arrest and resuscitation in rat brain. Brain Res Mol Brain Res 1999; 74:83-90.
78. Plate KH, Beck H, Danner S et al. Cell type specific upregulation of vascular endothelial growth factor in an MCA-occlusion model of cerebral infarct. J Neuropathol Exp Neurol 1999; 58:654-666.
79. Lin TN, Wang CK, Cheung WM et al. Induction of angiopoietin and tie receptor mRNA expression after cerebral ischemia-reperfusion. J Cereb Blood Flow Metab 2000; 20:387-395.
80. Chiarugi V, Magnelli L, Chiarugi A et al. Hypoxia induced pivotal tumor angiogenesis control factors including p53, vascular endothelial growth factor and the NFKB-dependent inducible nitric oxide synthase and cyclo-oxygenase-2. J Cancer Res Clin Oncol 1999; 125:525-528.
81. El Awad B, Kreft B, Wolber EM et al. Hypoxia and interleukin-1 stimulate vascular endothelial growth factor production in human proximal tubular cells. Kidney Intl 2000; 58:43-50.
82. Tsuzuki Y, Fukumura D, Oosthuyse B et al. Vascular endothelial growth factor (VEGF) modulation by targeting hypoxia-inducible factor-10-hypoxia response element-VEGF cascade differentially regulates vascular response and growth rate in tumors. Cancer Res 2000; 60:6248-6252.
83. Yuan HT, Yang SP, Woolf AS. Hypoxia up-regulates angiopoietin-2, a Tie-2 ligand, in mouse mesangial cells. Kidney Int 2000; 58:1912-1919.
84. Bergeron M, Yu AY, Solway KE et al. Induction of hypoxia-inducible factor-1 (HIF-1) and its target genes following focal ischaemia in rat brain. Eur J Neurosci 1999; 11:4159-4170.
85. Hossmann KA. The hypoxic brain. Insights from ischemia research. Adv Exp Med Biol 1999; 474:155-169.
86. Jander S, Schroeter M, Peters O et al. Cortical spreading depression induces proinflammatory cytokine gene expression in the rat brain. J Cereb Blood Flow Metab 2001; 21:218-225.
87. Scharfman HE, Goodman JH, Sollas AL et al. Spontaneous limbic seizures after intrahippocampal infusion of brain-derived neurotrophic factor. Exp Neurol 2002; 174:201-214.
88. Croll SD, Goodman JH, Sollas A et al. Vascular endothelial growth factor (VEGF) is upregulated after pilocarpine-induced seizures in rats. Soc Nsci Abs 2000; in press.
89. Sirevaag AM, Greenough WT. Differential rearing effects on rat visual cortex synapses. III. Neuronal and glial nuclei, boutons, dendrites and capillaries. Brain Res 1987; 424:320-332.
90. Carmeliet P, Storkebaum E. Vascular and neuronal effects of VEGF in the nervous system: implications for neurological disorders. Semin Cell Dev Biol 2002; 13:39-53.
91. Ogunshola OO, Antic A, Donoghue MJ et al. Paracrine and autocrine functions of neuronal vascular endothelial growth factor (VEGF) in the central nervous system. J Biol Chem 2002; 277:11410-11415.
92. Hayashi T, Abe K, Itoyama Y. Reduction of ischemic damage by application of vascular endothelial growth factor in rat brain after transient ischemia. J Cereb Blood Flow Metab 1998; 18:887-895.
93. Jin K, Mao XO, Batteur SP et al. Caspase-3 and the regulation of hypoxic neuronal death by vascular endothelial growth factor. Neuroscience 2001; 108:351-358.
94. Yourey PA, Gohari S, Su JL et al. Vascular endothelial cell growth factors promote the in vivo development of rat photoreceptor cells. J Neurosci 2000; 20:6781-6788.
95. Robinson GS, Ju M, Shih SC et al. Nonvascular role for VEGF: VEGFR-1, 2 activity is critical for neural retinal development. FASEB J 2001; 15:1215-1217.
96. Hobson MI, Green CJ, Terenghi G. VEGF enhances intraneural angiogenesis and improves nerve regeneration after axotomy. J Anat 2000; 197:591-605.
97. Schratzberger P, Schratzberger G, Silver M et al. Favorable effect of VEGF gene transfer on ischemic peripheral neuropathy. Nat Med 2000; 6:405-413.
98. Oosthuyse B, Moons L, Storkebaum E et al. Deletion of the hypoxia-response element in the vascular endothelial growth factor promoter causes motor neuron degeneration. Nat Genet 2001; 28:131-138.
99. Matsuzaki H, Tamatani M, Yamaguchi A et al. Vascular endothelial growth factor rescues hippocampal neurons from glutamate-induced toxicity: signal transduction cascades. FASEB J 2001; 15:1218-1220.
100. Mazure NM, Chen EY, Laderoute KR et al. Induction of vascular endothelial growth factor by hypoxia is modulated by a phosphatidylinositol 3-kinase/Akt signaling pathway in Ha-ras-transformed cells through a hypoxia-inducible factor-1 transcriptional element. Blood 1997; 90:3322-3331.
101. Gerber HP, McMurtrey A, Kowalski J et al. Vascular endothelial growth factor regulates endothelial cell survival through the phosphatidylinositol 3'-kinase/Akt signal transduction pathway. Requirement for Flk-1/KDR activation. J Biol Chem 1998; 273:30336-30343.

102. Wick A, Wick W, Waltenberger J et al. Neuroprotection by hypoxic preconditioning requires sequential activation of vascular endothelial growth factor receptor and Akt. J Neurosci 2002; 22:6401-6407.
103. Bagnard D, Vaillant C, Khuth ST et al. Semaphorin 3A-vascular endothelial growth factor-165 balance mediates migration and apoptosis of neural progenitor cells by the recruitment of shared receptor. J Neurosci 2001; 21:3332-3341.
104. Li J, Perrella MA, Tsai JC et al. Induction of vascular endothelial growth factor gene expression by interleukin-1 beta in rat aortic smooth muscle cells. J Biol Chem 1995; 270:308-312.
105. Ryuto M, Ono M, Izumi H et al. Induction of vascular endothelial growth factor by tumor necrosis factor alpha in human glioma cells. Possible roles of SP-1. J Biol Chem 1996; 271:28220-28228.
106. Jung YD, Liu W, Reinmuth N et al. Vascular endothelial growth factor is upregulated by interleukin-1 beta in human vascular smooth muscle cells via the P38 mitogen-activated protein kinase pathway. Angiogenesis 2001; 4:155-162.
107. Shandra AA, Godlevsky LS, Vastyanov RS et al. The role of TNF-alpha in amygdala kindled rats. Neurosci Res 2002; 42:147-153.
108. Zhang ZG, Zhang L, Jiang Q et al. VEGF enhances angiogenesis and promotes blood-brain barrier leakage in the ischemic brain. Clin Invest 2000; 106:829-838.
109. van Bruggen N, Thibodeaux H, Palmer JT et al. VEGF antagonism reduces edema formation and tissue damage after ischemia/reperfusion injury in the mouse brain. J Clin Invest 1999; 104:1613-1620.
110. Baik EJ, Kim EJ, Lee SH et al. Cyclooxygenase-2 selective inhibitors aggravate kainic acid induced seizure and neuronal cell death in the hippocampus. Brain Res 1999; 843:118-129.

CHAPTER 5

Plasticity Mechanisms Underlying mGluR-Induced Epileptogenesis

Robert K.S. Wong, Shih-Chieh Chuang and Riccardo Bianchi

Abstract

Transient application of group I metabotropic glutamate receptor (mGluR) agonists to hippocampal slices produces ictal-like discharges that persist for hours after the removal of the agonist. This effect of group I mGluR stimulation—converting a 'normal' hippocampal slice into an 'epileptic-like' one—may represent a form of epileptogenesis. Because this epileptogenic process can be induced in vitro and it occurs within hours, it has been possible to examine the cellular and transduction processes underlying the generation and long-term maintenance of ictal-like bursts. $I_{mGluR(V)}$, a voltage-dependent depolarizing current activated by group I mGluR agonists, appears to play an important role in the expression of the ictal-like bursts. Long-term activation of $I_{mGluR(V)}$ following mGluR stimulation is a possible plastic change that enables the long-term maintenance of ictal discharges. Induction of $I_{mGluR(V)}$ may represent a cellular event underlying the mGluR-induced epileptogenesis.

Introduction

Glutamate is the major excitatory neurotransmitter in the central nervous system. The excitatory effects are mediated by three types of ionotropic glutamate receptors (the NMDA, kainate and AMPA receptors) classified according to their sensitivity to agonists. The effects of glutamate are also mediated by G-protein-coupled receptors. These receptors are termed metabotropic because they affect neuronal activities via G-protein-coupled intracellular pathways. Through metabotropic glutamate receptors (mGluRs), glutamate exerts short-term as well as long-term influences on the excitability of central neurons.

Eight genes for mGluRs have been cloned.[11,13] The receptors are classified into three groups. Group I receptors (mGluR1 and mGluR5) are coupled to $G_{q/11}$-proteins. Activation of this receptor group elicits excitation. Group II receptors (mGluR2 and mGluR3) and group III receptors (mGluR4, mGluR6, mGluR7 and mGluR8) are coupled to $G_{i/o}$-protein. Stimulation of these receptor groups elicits inhibition, often through presynaptic mechanisms. Recent studies show that there are exceptions to this generalization. For example, activation of group II mGluRs may also elicit excitation via a postsynaptic action.[20]

Activation of mGluRs produces robust effects on epileptiform activities in the cortex. Consistent with the overall effects of mGluR groups on neuronal activities, group I mGluR agonists have been shown to be proconvulsant[4,5] and group II and III mGluR agonists are anticonvulsant.[2,12] A particularly interesting aspect of mGluR effects is that their actions are often long lasting. For example, the application of a group III mGluR agonist suppresses epileptiform discharges in hippocampal slices. This anticonvulsant effect, once induced, persists even when the agonist is washed away.[22] The convulsant action of group I mGluRs is also long-lasting.

Recent Advances in Epilepsy Research, edited by Devin K. Binder and Helen E. Scharfman.
©2004 Eurekah.com and Kluwer Academic / Plenum Publishers.

Ictal-like discharges elicited by group I mGluRs in hippocampal slices persist for hours after the washout of the agonist.[17] This chapter summarizes progress-to-date on the epileptogenic effects of group I mGluR activation.

Cellular Mechanisms for Group I mGluR-Induced Excitation in Hippocampal Neurons

Application of group I mGluR agonists (e.g., (S)-3,4-dihydrophenylglycine; DHPG) produces depolarization of hippocampal neurons via two broad mechanisms—a suppression of K^+ conductances and an activation of cationic conductances.[1] Based on the sensitivity to transmembrane voltage, cationic conductances activated by group I mGluRs can be further distinguished into voltage-independent and voltage-dependent types. The voltage-independent cationic current is elicited in CA1 pyramidal cells.[10] The current is gated by increases in intracellular Ca^{2+} (I_{CAN}) and its amplitude responds linearly to changes in membrane potential. The second type of cationic current differs from I_{CAN} mainly in that the amplitude of the current exhibits a nonlinear relationship with the membrane voltage. The current-voltage relationship shows a pronounced negative slope conductance region at some potential range more depolarized than the resting potential. At present two species of the voltage-dependent cationic current have been described. Activation of one of the currents appears to be G-protein-independent,[14] whereas activation of the other one is dependent on phospholipase C activation and is therefore presumably G-protein-dependent.[7] Other than a difference in G-protein-dependency, the two current species have remarkably similar properties. The G-protein-independent current has a reversal potential of about -9 mV and exhibits negative slope conductance in the membrane voltage range of -80 to -40 mV. The G-protein-dependent current reverses at about -12 mV and has a negative slope conductance in the membrane potential range of -70 to -30 mV. Both currents are recorded in CA3 pyramidal cells and have peak amplitudes of about -200 pA. Furthermore, preliminary data suggest that both currents are activated by mGluR1 but not mGluR5 stimulation.[9] At present, the possibility remains that a similar current is described by the two studies since the main distinguishing feature—G-protein-dependency of the current—has not yet been rigorously tested. Detailed analysis on one of the currents ($I_{mGluR(V)}$) suggests that the current markedly modifies single cell behavior and may thereby contribute to the generation of epileptiform discharges.

Role of $I_{mGluR(V)}$ in the Neuronal Activities of Single Pyramidal Cells

$I_{mGluR(V)}$ is elicited in CA3 (but not CA1) pyramidal cells by DHPG[8,9] (50 μM). $I_{mGluR(V)}$ is shut off completely at hyperpolarized membrane potentials and is activated by depolarizations positive to -70 mV. The current does not inactivate and is probably carried by a mixture of cations. The specific charge carriers for the current have not yet been determined, however a reversal potential of -12 mV indicates Na^+, K^+, and Ca^{2+} all as likely candidates.

Three interesting roles for $I_{mGluR(V)}$ have been demonstrated in the control of single CA3 pyramidal cell activities that are fundamental to understanding its contribution to network function. These functions are summarized in the following.

Determination of Resting Potential

Because $I_{mGluR(V)}$ does not inactivate and has an activation threshold (-70 mV) hyperpolarized to the normal resting potential (-65 mV) of CA3 pyramidal cells, the current, once induced, is tonically active at the resting state of CA3 pyramidal cells. The tonic activation of $I_{mGluR(V)}$ depolarizes the resting potential of pyramidal cells to about -50 mV. This depolarized resting state can be reset to a more hyperpolarized level when $I_{mGluR(V)}$ is deactivated by an imposed hyperpolarizing pulse or by an inhibitory postsynaptic potential. Thus, $I_{mGluR(V)}$ contributes to generate a bi-stable resting potential condition.[8]

Regulation of Time Courses of Inhibitory and Excitatory Postsynaptic Potentials (IPSPs and EPSPs)

$I_{mGluR(V)}$ is sensitive to small changes in membrane potentials at the resting level. Thus, an impinging IPSP often turns off a proportion of $I_{mGluR(V)}$, causing a lengthening of the inhibitory event. Conversely, an EPSP activates $I_{mGluR(V)}$, causing a more pronounced depolarization.

Control of Rhythmic Oscillations

Perhaps the most interesting aspect of $I_{mGluR(V)}$ function is its contribution to the generation of pacemaker potentials for rhythmic bursting in single hippocampal pyramidal cells. In preparations where individual CA3 pyramidal cells are synaptically isolated by ionotropic glutamate receptor (iGluR) blockers and GABA receptor blockers, group I mGluR agonists often elicit rhythmic bursting activities in individual cells that last for 10-20 s and are separated by intervals of about 7-15 s (e.g., Fig. 1Aa and ref. 7).

The bursts are presumably generated in the following fashion. As mentioned above, $I_{mGluR(V)}$ is active at the resting potential range of CA3 pyramidal cells. Thus, the current depolarizes pyramidal cells to about -50 mV at which point it is counterbalanced by a K^+-mediated outward current activated by the depolarization. The cell fires action potentials at this depolarized state. Ca^{2+} entry associated with the action potentials causes the cumulative activation of K^+ currents (the portion that is not suppressed by group I mGluR activation[6]). Burst firing terminates when the total outward current exceeds the inward current $I_{mGluR(V)}$. During the interval, K^+ current amplitude gradually decreases and eventually becomes smaller than $I_{mGluR(V)}$, allowing the latter to generate a gradual depolarization (pacemaker potential) leading to another bursting phase.

Thus, $I_{mGluR(V)}$ plays two roles in the patterning of single cell bursting in CA3 pyramidal cells: it generates the pacemaker potential that leads to the burst and it provides the depolarization that sustains the burst.

The pattern of single cell bursts is qualitatively similar to that of the prolonged ictal-like synchronized bursts also elicited by group I mGluR agonists when applied to hippocampal slices with intact synaptic connectivity (Fig. 1Aa and Ba). The properties of the network bursts and the plausible relationship between the single cell and network activities are discussed in the following.

Role of $I_{mGluR(V)}$ in the Generation of Network Activity

Group I mGluR agonists also elicit periodic bursting activities in a synaptically intact ('control') hippocampal slice[17,23] (Fig. 1Ba). The bursts have comparable durations (up to 15 s) and intervals (up to 20 s) as the bursts elicited by group I mGluR agonists in single pyramidal cells of synaptically isolated preparations. The major difference between the two types of bursting is that the activities recorded in the control slices are synchronized – burst events always occur simultaneously in all the pyramidal cells of the slice.

Recently, a role of $I_{mGluR(V)}$ in the generation of the synchronized bursting in normal slices has been proposed[7] and is briefly described in the following. The interaction of $I_{mGluR(V)}$ with intrinsic currents of pyramidal cells imparts on single pyramidal cells resonance frequency to burst at a preferred frequency and duration (see Fig. 1Aa). Similar resonance frequency is elicited in a population of CA3 pyramidal cells since the neurons have comparable intrinsic properties including input resistance, time constant, firing characteristics,[27] and conductances,[19,24] as well as similar amplitudes of inducible maximum[9] $I_{mGluR(V)}$. In a synaptically intact 'normal' preparation, single cell activities can be synchronized by recurrent excitatory synapses to give rise to the synchronized network bursts.[18,25] Similarity in the resonance frequency of the pyramidal cells induced by group I mGluR agonists should facilitate the synchronization process and play a part in determining the frequency and duration of the network bursts.

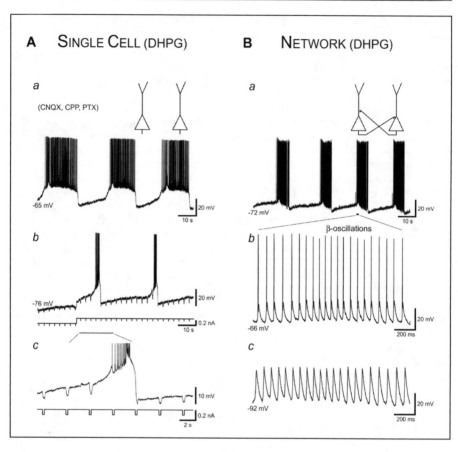

Figure 1. The group I mGluR agonist DHPG induces rhythmic intrinsic bursts and ictal-like network discharges in CA3 pyramidal cells of hippocampal slices. A, intracellular recordings from a hippocampal CA3 pyramidal cell 1 h after addition of (S)-3,4-dihydrophenylglycine (DHPG; 50 μM) to a solution containing blockers of ionotropic glutamate receptors (CNQX and CPP; 20 μM each) and of GABA$_A$ receptors (picrotoxin; 50 μM). Aa, the scheme (inset) illustrates that under these conditions glutamatergic synaptic transmission was suppressed. At -65 mV, rhythmic intrinsic bursts with firing preceded by gradual depolarizations were recorded. Ab, records (top trace) from the same cell as in Aa. Hyperpolarizing DC current (bottom trace: current) held the cell at -76 mV (initial segment of the recording). The cell was then depolarized to about -70 mV. This depolarization caused a gradual depolarization (pacemaker potential) to develop that led to a burst. The gradual depolarization is due to the activation of I$_{mGluR(V)}$. Hyperpolarizing current pulses (-0.1 nA; 300 ms) were applied every 3 s. At an appropriate time during the burst, a short pulse could abruptly terminate the burst. Ac, expansion of the voltage segment indicated by a bar in Ab to illustrate the gradual depolarization (pacemaker potential) preceding the firing and the abrupt hyperpolarization caused by a short pulse during the burst. The increasing amplitude of the voltage responses to the short hyperpolarizing pulses during the pacemaker depolarization was in part due to deactivation of the developing I$_{mGluR(V)}$. B, intracellular recordings from another CA3 pyramidal cell 70 min after addition of DHPG (50 μM) to normal perfusing solution. Ba, in these conditions synaptic transmission was not blocked (inset) and DHPG induced rhythmic network discharges, each of 6-7 s in duration. Bb, the segment of the burst indicated by the bar in Ba is shown on a faster time base. At resting membrane potential, action potentials occurred at a regular frequency of 14-15 Hz (*β-oscillations*). Bc, hyperpolarization of the same cell to –92 mV revealed that EPSPs occurring at β-frequency underlie the rhythmic firing during a network discharge.

Thus, single cell bursts driven by $I_{mGluR(V)}$ appear to be fundamental events underlying the generation of network bursts. Two additional observations support this notion. First, CA3 pyramidal cells from phospholipase Cβ1 knockout mice do not generate $I_{mGluR(V)}$ when exposed to group I mGluR agonists. Group I agonists also do not elicit single cell and network bursts in hippocampal slices prepared from the mutant.[7] Second, CA3 pyramidal cells from mGluR1 knockout mice are also incapable of generating $I_{mGluR(V)}$. Group I mGluR agonist application to slices prepared from the mGluR1 -/- mice also failed to elicit single cell and network bursts.[9] In both types of aforementioned mutants, group I agonists retained their effects of suppressing the 'leak' K^+ conductance and the Ca^{2+}-activated K^+ current that is associated with the post-spike afterhyperpolarization. As noted above, these remaining effects did not give rise to burst firing at the single cell and network levels.

Network bursts elicited by group I mGluR activation have been compared to the ictal bursts recorded in cortical tissues in epilepsy. Both events involve the hypersynchronized discharge of a large population of neurons for periods exceeding 1 s.[26] Furthermore, the events are similar in their synchronized oscillatory patterns. Within a period of group I mGluR-induced network bursts, oscillating synaptic depolarizations occurred at 12 – 27 Hz (Fig. 1B, *β-oscillations*). Similar oscillatory depolarizations also accompany ictal discharges. mGluR-induced network responses thus serve as a model for epileptic discharges. From here on, group I mGluR-dependent network bursts will be termed ictal-like bursts or ictal-like discharges.

Plasticity Mechanisms Underlying Group I mGluR-Induced Network Burst Discharges

The mGluR-induced ictal-like bursts have a particularly interesting feature relevant to epileptogenesis. The ictal-like bursts elicited by mGluR stimulation are long-lasting. The events, once induced by group I mGluR agonist, persist for hours after the agonist has been washed out.[17] This effect of group I mGluR stimulation—transforming a 'normal' hippocampal slice into an 'epileptic-like' one—may represent a form of epileptogenesis. The phenomenon, as in most plasticity responses, proceeds in two temporal stages—induction and maintenance. During induction, biochemical and/or molecular activities modify synaptic and/or ionic channel (cellular) processes. Ictal-like bursts are then expressed by the activation of the modified cellular processes. Induction is followed by maintenance, during which the cellular modifications are stabilized for the long-term expression of the ictal-like bursts. Data regarding the induction, maintenance, and expression of the ictal-like bursts are presented.

Induction

The induction of group I mGluR-dependent ictal-like discharges appears to be protein-synthesis-dependent. Pretreatment of hippocampal slices with a protein synthesis inhibitor (anisomycin or cycloheximide) prevented the appearance of ictal-like discharges.[16] When group I mGluR agonists were applied to hippocampal slices pretreated with iGluR blockers, ictal-like discharges were not expressed. And yet when the group I mGluR agonist was washed away together with iGluR blockers, ictal-like discharges ensued.[15] The data suggest that the induction of ictal-like discharges can occur in silence—i.e., it does not require activation of NMDA or AMPA receptors nor does it depend on the expression of ictal-like discharges. Activation of group I mGluRs seems to be the only requirement for the induction of ictal-like discharges.

Maintenance

Ictal-like discharges, once induced, are no longer affected by protein synthesis inhibitors.[16] Apparently, new protein(s) synthesized in response to group I mGluR stimulation can function for hours to maintain ictal-like bursts. Ictal-like bursts persisting after the washout of the applied group I mGluR agonist are blocked by group I mGluR antagonists.[17] Since endogenous glutamate is the only agonist available during the maintenance phase, ictal-like discharges are most likely maintained by synaptically released glutamate.

Expression

The cellular responses activated synaptically to maintain the ictal-like discharges have not yet been identified. $I_{mGluR(V)}$ is a likely candidate. As mentioned above, $I_{mGluR(V)}$ is elicited by group I mGluR agonists and plays an important role in the generation of ictal-like discharges. The current may also be required for ictal-like discharges generation during the maintenance phase in the absence of applied agonist. $I_{mGluR(V)}$ may then be activated by synaptically released glutamate. At present, the contribution of $I_{mGluR(V)}$ in the long-term maintenance of ictal-like discharges and its role in epileptogenesis remains unclear.

Conclusion

A challenge to the mGluR model of epileptogenesis is whether the induction mechanism studied in vitro is one that is engaged in epilepsy. The data show that ongoing ictal-like discharges can be maintained by synaptic processes endogenous to the hippocampal network. However, the model depends on the application of group I mGluR agonists for the induction process. It is unclear whether conditions exist where synaptic activation of group I mGluRs in pyramidal cells can induce epileptogenesis. One possible example is the kindling model where glutamatergic synapses are excessively activated to induce seizures and epileptogenesis. A role of the NMDA receptors in the induction of epileptic activities in kindling has been emphasized.[3,21] To the extent that high frequency stimulations of glutamatergic input can activate postsynaptic group I mGluRs,[1] mGluRs may also contribute to the induction and maintenance of epileptogenesis in the kindling model.

References

1. Anwyl R. Metabotropic glutamate receptors: electrophysiological properties and role in plasticity. Brain Res Rev 1999; 29:83-120.
2. Attwell PJ, Singh KN, Jane DE et al. Anticonvulsant and glutamate release-inhibiting properties of the highly potent metabotropic glutamate receptor agonist (2S,2R,3R)-2-(2,3-dicarboxycyclopropyl)glycine (DCG-IV). Brain Res 1998; 805:138-143.
3. Behr J, Heinemann U, Mody I. Kindling induces transient NMDA receptor-mediated facilitation of high-frequency input in the rat dentate gyrus. J Neurophysiol 2001; 85:2195-2202.
4. Camon L, Vives P, De Vera N et al. Seizures and neuronal damages induced in the rat by activation of group I metabotropic glutamate receptors with their selective agonist 3,5-dihydroxyphenyglycine. J Neuroscie Res 1998; 51:339-348.
5. Chapman AG, Nanan K, Meldrum BS. Anticonvulsant activity of the metabotropic glutamate group I antagonists selective for mGluR5 receptor: 2-methyl-6-(phenylethynyle)-pyridine (MPEP), and (E)-6-methyl-2-styryl-pyridine (SIB 1983). Neuropharmacology 2000; 39:1567-1574.
6. Charpak S, Gähwiler BH, Do KQ et al. Potassium conductances in hippocampal neurons blocked by excitatory amino acid transmitters. Nature 1990; 347:765-767.
7. Chuang SC, Bianchi R, Kim D et al. Group I metabotropic glutamate receptors elicit epileptiform discharges in the hippocampus through PLCβ1 signaling. J Neurosci 2001; 21:6387-6394.
8. Chuang SC, Bianchi R, Wong RKS. Group I mGluR activation turns on a voltage-dependent inward current in hippocampal pyramidal cells. J Neurophysiol 2000; 83:2844-2853.
9. Chuang SC, Zhao W, Young SJ et al. Activation of group I mGluRs elicits different responses in murine CA1 and CA3 pyramidal cells. J Physiol 2002; 541:113-121.
10. Congar P, Leinekugel X, Ben-ari Y et al. A long-lasting calcium-activated nonselective cationic current is generated by synaptic stimulation or exogenous activation of group I metabotropic glutamate receptors in CA1 pyramidal neurons. J Neurosci 1997; 17:5366-5379.
11. Conn PJ, Pin J-P. Pharmacology and functions of metabotropic glutamate receptors. Ann Rev Pharmacol Toxicol 1997; 37:205-237.
12. Gasparini F, Bruno V, Battaglia G et al. (R,S)-4-phosphonophenylglycine, a potent and selective group III metabotropic glutamate receptor agonist, is anticonvulsive and neuroprotective in vivo. J Pharmacol Exp Ther 1999; 289:1678-1687.
13. Hayashi Y, Sekiyama N, Nakanishi S et al. Analysis of agonist and antagonist activities of phenylglycine derivatives for different cloned metabotropic glutamate receptor subtypes. J Neurosci 1994; 14:3370-3377.
14. Heuss C, Scanziani M, Gahwiler BH et al. G-protein independent signaling mediated by metabotropic glutamate receptors. Nature Neurosci 1999; 2:1070-1077.

15. Merlin LR. Group I mGluR-mediated silent induction of long-lasting epileptiform discharges. J Neurophysiol 1999; 82:1078-1081.
16. Merlin LR, Bergold PJ, Wong RKS. Requirement of protein synthesis for group I mGluR-mediated induction of epileptiform discharges. J Neurophysiol 1998; 80:989-993.
17. Merlin LR, Wong RKS. Role of Group I metabotropic glutamate receptors in the patterning of epileptiform activities in vitro. J Neurophysiol 1997; 78:539-544.
18. Miles R, Wong RKS. Single neurones can initiate synchronized population discharge in the hippocampus. Nature 1983; 306:371-373.
19. Numann RE, Wadman WJ, Wong RKS. Outward currents of single hippocampal cells obtained from the adult guinea-pig. J Physiol 1987; 393:331-353.
20. Otani S, Daniel H, Takita M et al. Long-term depression induced by postsynaptic group II metabotropic glutamate receptors linked to phospholipase C and intracellular calcium rises in rat prefrontal cortex. J Neurosci 2002; 22:3434-3444.
21. Sayin U, Rutecki PA, Sutula T. NMDA-dependent currents in granule cells of the dentate gyrus contribute to induction but not permanence of kindling. J Neurophysiol 1999; 81:564-574.
22. Shrestha A, Staley KJ. Induction of LTD by activation of group III metabotropic glutamate receptor in CA3. Society for Neuroscience Abstract 2001; 27:388.8. (Abstract).
23. Taylor GW, Merlin LR, Wong RKS. Synchronized oscillations in hippocampal CA3 neurons induced by metabotropic glutamate receptor activation. J Neurosci 1995; 15:8039-8052.
24. Thompson SM, Wong RKS. Development of calcium current subtypes in isolated rat hippocampal pyramidal cells. J Physiol 1991; 439:671-689.
25. Traub RD, Wong RKS. Cellular mechanism of neuronal synchronization in epilepsy. Science 1982; 216, 745-747.
26. Wong RKS, Bianchi R, Taylor GW et al. Role of metabotropic glutamate receptors in epilepsy. Adv Neurol 1999; 79:685-689.
27. Wong RKS, Prince DA. Afterpotential generation in hippocampal pyramidal cells. J Neurophysiol 1981; 45:86-97.

CHAPTER 6

Role of the GABA Transporter in Epilepsy

George B. Richerson and Yuanming Wu

Abstract

The GABA transporter plays a well-established role in reuptake of GABA after synaptic release. The anticonvulsant effect of tiagabine appears to result largely from blocking this reuptake. However, there is another side to the GABA transporter, contributing to GABA release by reversing in response to depolarization. We have recently shown that this form of GABA release is induced by even small increases in extracellular [K^+], and has a powerful inhibitory effect on surrounding neurons. This transporter-mediated GABA release is enhanced by the anticonvulsants gabapentin and vigabatrin. The latter drug also potently increases ambient [GABA], inducing tonic inhibition of neurons. Here we review the evidence in support of a physiological role for GABA transporter reversal, and the evidence that it is increased by high-frequency firing. We postulate that the GABA transporter is a major determinant of the level of tonic inhibition, and an important source of GABA release during seizures. These recent findings indicate that the GABA transporter plays a much more dynamic role in control of brain excitability than has previously been recognized. Further defining this role may lead to a better understanding of the mechanisms of epilepsy and new avenues for treatment.

Introduction

The GABA system plays a pivotal role in the pathophysiology of epilepsy, and is the target of many different anticonvulsant drugs. At GABAergic synapses, termination of the postsynaptic response results from diffusion of GABA away from the synaptic cleft and reuptake of GABA by GABA transporters in the plasma membrane of neurons and glia. In recent years, our understanding of these GABA transporters has been drastically altered. Rather than simply being viewed as GABA vacuum cleaners responsible for GABA reuptake after an inhibitory postsynaptic potential (IPSP), it is now known that the GABA transporter reverses surprisingly easily. The GABA transporter is thus a complex component of the inhibitory system with dual roles in both uptake and release. Some anticonvulsants target the GABA transporter, but blocking transporter function could involve either anticonvulsant or proconvulsant effects. Therefore, the actions of these agents are not yet fully defined. Understanding the normal and pathophysiological roles of the GABA transporter will aid in defining the mechanisms of these drugs, and may help to develop new agents that target specific aspects of transporter function.

Diversity of GABA Transporters

There are four cloned isoforms of the GABA transporter in rats and humans, termed GAT-1, GAT-2, GAT-3 and BGT-1 (a.k.a. GAT-4).[5] It should be noted that the nomenclature for these transporters is different in mice, and that there was not agreement on this nomenclature in the rat in early papers. For example, rat GAT-3 was once also called GAT-B.

Recent Advances in Epilepsy Research, edited by Devin K. Binder and Helen E. Scharfman.
©2004 Eurekah.com and Kluwer Academic / Plenum Publishers.

It was originally believed that there was a "neuronal" and a "glial" form of the GABA transporter, but data obtained since the GABA transporters were cloned does not support this conclusion. In the rat and the human brain, GAT-1 is localized primarily on presynaptic terminals of GABAergic neurons, but is also present on axons as well as astrocytic processes.[11,15,28,37,40] In contrast, GAT-3 is located primarily on glia,[29] but is also present on neurons.[5] GAT-3 does not appear to be restricted to perisynaptic regions. GAT-2 is located primarily in the meninges, whereas BGT-1 is primarily in glia.[15] However, all four isoforms have been reported to be in both neurons and glia in various species.[5,15] GAT-1 and GAT-3 are restricted to the nervous system, whereas GAT-2 and BGT-1 are also found in nonneural tissue.[5] BGT-1 also transports betaine, which appears to be its primary function in the kidney. The full significance of this diversity, and the specific roles of each isoform, have not been fully defined.

The study of GABA transporters has been aided by the characterization of many antagonists. Tiagabine, SKF-89976a and NO-711 are all GABA transport inhibitors that are highly selective for GAT-1.[5] There are not such highly specific antagonists of GAT-3, although β-alanine is relatively selective for GAT-2 and GAT-3 compared to GAT-1. There are also a variety of nonspecific antagonists, some of which are substrates for the transporter. Nipecotic acid is an example, and not only blocks the transporter but is also taken up by it. When applied rapidly, it first induces heteroexchange GABA release (see below), followed by blockade of transport.

Each of the GABA transporter isoforms are members of the family of Na^+- and Cl^--coupled transporters. GAT-1 cotranslocates two sodium ions and one chloride ion with each uncharged GABA molecule,[15,20] and is therefore electrogenic, carrying one positive charge into the cell with each thermodynamic reaction cycle. GAT-3 is thought to be similar.[10] In contrast, the stoichiometry of BGT-1 is different, cotranslocating three sodium ions and either one or two chloride ions with each GABA.[26]

Role of the GABA Transporter in GABA Reuptake

The GABA transporter clears GABA from the extracellular space after it is released at synaptic terminals. Evidence for this includes the finding that transporters are concentrated at synaptic terminals of GABAergic neurons[39] (see above). In addition, GABA transporter antagonists prolong the duration of GABAergic IPSPs (Fig. 1).[12,18,41,51] However, this effect of GABA transporter blockade is often relatively small, especially on GABAergic IPSPs from low-intensity presynaptic activity.[12,18,30,34] In contrast, blocking GABA transport induces a much larger increase in ambient [GABA], increasing the tonic form of GABAergic inhibition that occurs via high-affinity $GABA_A$ receptors.[30,49]

Reversal of the GABA Transporter

It is often assumed that the GABA transporter is so efficient that it clears all GABA from the extracellular space after it is released during synaptic transmission—often being viewed simply as a highly efficient GABA vacuum cleaner. However, $[GABA]_o$ at steady-state cannot drop below a finite nonzero level that is determined by the stoichiometry of the GABA transporter. Baseline $[GABA]_o$ has been reported to be $0.1 - 0.8$ μM,[22,52] whereas the [GABA] in the cytosol of the somata of GABAergic cerebellar Purkinje neurons has been reported to be 6.6 mM.[32] The driving force for transporting GABA into the cytoplasm up this steep concentration gradient comes from the membrane potential and the concentration gradients for Na^+ and Cl^-. If $[GABA]_o$ drops below the level that is supported by the driving force, then the GABA transporter will reverse in order to raise $[GABA]_o$. All ion-coupled transporters can reverse like this, and will transport their substrate equally well either forwards or backwards, the direction simply being dependent upon the thermodynamic conditions.

The GABA transporter will reverse in response to depolarization or when there is an increase in intracellular Na^+ or Cl^-.[43,45] For example, an increase in K^+ to 56 mM causes depolarization of cultured striatal neurons, which leads to GABA release that can be measured using HPLC (Fig. 2).[38] About 25% of this K^+-induced GABA release is unaffected by removal of

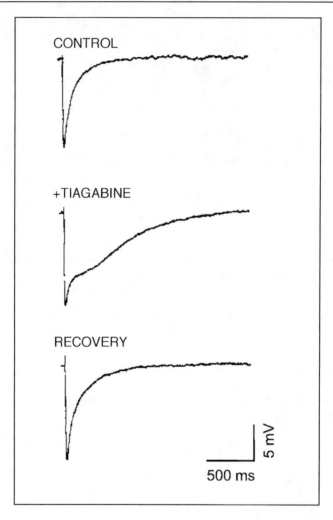

Figure 1. Blockade of the GABA transporter prolongs IPSPs. Shown are recordings from a neuron in a hippocampal slice culture. Top trace—Electrical stimulation evoked a monosynaptic GABAergic IPSP after blocking glutamatergic neurotransmission. Middle trace—Tiagabine in the bath solution induces prolongation of the IPSP without changing the peak amplitude. Bottom trace—The response to tiagabine was reversible. From Thompson and Gahwiler, 1992 (ref. 51).

extracellular calcium or pretreatment with tetanus toxin. This form of GABA release can also be induced by veratridine, ouabain, and glutamate, each of which raise intracellular [Na$^+$]. Similarly, calcium-independent release of preloaded, radiolabeled GABA can be induced from horizontal cells of the toad retina by 41.5 mM K$^+$,[45] from cultured cerebral cortical neurons by 55 mM K$^+$ or 100 μM glutamate,[2] and from rat brain synaptosomes by 60-150 mM K$^+$.[53] In each of these cases, the nonvesicular GABA release is due to reversal of the GABA transporter, since it is blocked by GABA transporter antagonists.

An alternative way to assay GABA transporter reversal is to measure current flow through the transporter. In most native cell types, the density of GABA transporters is too low to allow transporter currents to be measured directly. However, the GABA transporter is present at high

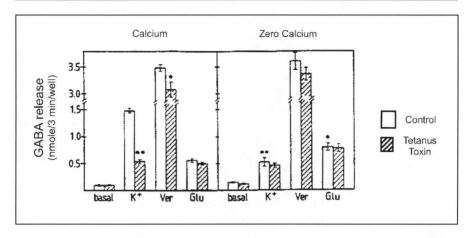

Figure 2. Transporter-mediated GABA release can be induced after blocking vesicular fusion. GABA release was measured from cultures of the mouse striatum using HPLC. Left panel—There is basal release of GABA that is not prevented by pretreatment with tetanus toxin (5 μg/ml) for 24 hours. Depolarization with 56 mM K^+ induces an increase in GABA release that is only partially reduced by tetanus toxin. Treatment with veratridine (10 μM) induces a large amount of GABA release that is insensitive to tetanus toxin. GABA release induced by glutamate (100 μM) is unaffected by tetanus toxin. Right panel—In the absence of extracellular calcium, each of the stimuli induce an increase in GABA release that is roughly equivalent to that induced after treatment with tetanus toxin in the presence of calcium. In each case, further treatment with tetanus toxin has no additional effect.[38]

density in catfish horizontal cells. Using whole-cell voltage clamp recordings, an efflux current is present when these neurons are filled with GABA, and this outward current increases with depolarization.[8] The outward current is associated with release of GABA, which can be detected by measuring the current induced by GABA$_A$ receptor activation. GABA transporter currents have also been measured directly from GAT-1 artificially expressed in HEK293 cells[7] and in *Xenopus* oocytes.[25] These experiments also confirmed that the GABA transporter can work in reverse in response to depolarization, or when the concentration gradients for Na^+, Cl^- or GABA favor reversal.

Functional Significance of GABA Transporter Reversal

The work described above demonstrates that the GABA transporter will reverse, and when it does there can be a substantial amount of carrier-mediated GABA flux. However, the stimuli used in most cases were so extreme that GABA transporter reversal has often been viewed as a phenomenon that only occurs under pathological conditions. In addition, these results do not indicate whether the GABA that is released produces any physiological effect on surrounding neurons.

We have recently shown that carrier-mediated GABA release can be induced by small, physiologically relevant increases in extracellular $[K^+]$.[16,57] When neurons fire action potentials, efflux of K^+ causes an increase in extracellular $[K^+]$, which can reach up to 12-15 mM[21,47] (Fig. 3). This in turn causes depolarization of surrounding neurons and glia. To determine whether GABA transporter reversal occurs in response to this K^+-induced depolarization, we used patch clamp recordings from hippocampal neurons in culture. In response to an increase in extracellular [GABA], GABA$_A$ receptors are activated in recorded neurons, allowing us to use these recordings as a biological assay of GABA release (Fig. 4A). In the absence of extracellular calcium, an increase in $[K^+]_o$ from 3 mM to 12 mM induces a large current in recorded neurons

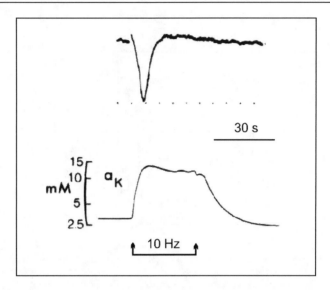

Figure 3. Neuronal firing induces a large increase in extracellular [K⁺] in vivo. Shown are a field potential (top trace) and a recording of extracellular [K⁺] (a_K) in the CA3 region of the hippocampus of an anesthetized rat in vivo. The fimbria was stimulated at 10 Hz during the time marked with arrows. The neuronal activity led to an increase in extracellular [K⁺] from 3 mM to more than 12 mM.[21]

that is blocked by bicuculline (Fig. 4B). This current reverses at the Nernst potential for chloride,[16] and therefore is due to activation of GABA$_A$ receptors as a result of nonvesicular GABA release. The increase in [GABA]$_o$ was prevented by blockade of the GABA transporter with either SKF-89976a or NO-711,[16] indicating that the nonvesicular GABA release was due to reversal of the GABA transporter. More recently, we have shown that reversal of the GABA transporter can be induced by an increase in [K⁺]$_o$ to as little as 6 mM.[57] These results indicate that the GABA transporter reverses in response to a physiologically relevant stimulus. They also show that the resulting nonvesicular GABA release causes potent inhibition of neighboring neurons.

Other experiments also point to a functional role for GABA transporter reversal. For example, reversal of the GABA transporter can be induced by electrical stimulation. Horizontal cells of the fish retina release GABA onto bipolar cells in response to depolarization in the absence of extracellular calcium (Fig. 5A).[46] This GABA release is blocked by nipecotic acid (Fig. 5B & 5C). These neurons are deficient in synaptic vesicles, so that it appears that they utilize reversal of the GABA transporter as a primary means of interneuronal communication. Although it would be easy to dismiss this observation as being irrelevant to mammalian brain function, there is no evidence to prove that this is the case. In fact, high-frequency electrical stimulation induces release of tritiated GABA from striatal slices in the absence of extracellular calcium.[3]

The results from these diverse experiments point to two different mechanisms of GABA release, whose properties are contrasted in (Table 1). Calcium-dependent vesicular release may be most important for rapid point-to-point communication between neurons under normal conditions. As shown above, calcium-independent GABA release can occur via GABA transporter reversal under physiological conditions. However, transporter reversal may become most important under conditions of metabolic stress and/or during high-frequency firing. It may also lead to more diffuse effects on surrounding cells. Thus, transporter-mediated GABA release may play its most critical role as a fail-safe method of feedback inhibition, preventing runaway excitation.

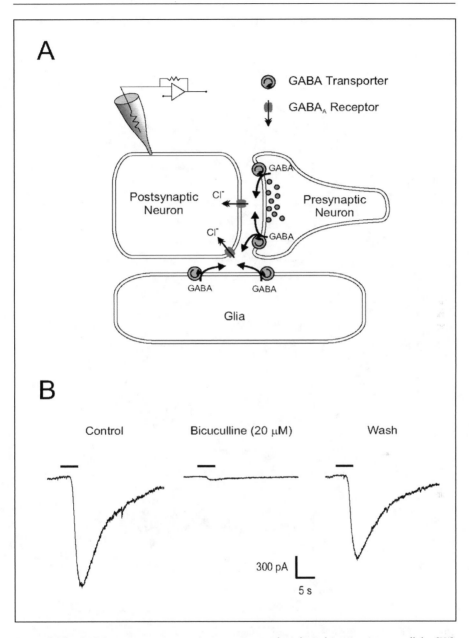

Figure 4. The GABA transporter reverses in response to a physiological increase in extracellular $[K^+]$. A—Experimental setup for Figures 4, 6 & 7. Patch clamp recordings were made from hippocampal neurons. When the GABA transporter on neighboring neurons and glia reverses, the resulting nonvesicular GABA release activates GABA$_A$ receptors on the recorded neuron. This approach is a highly sensitive bioassay of transporter-mediated GABA release. B—Left: In the absence of extracellular calcium, an increase in extracellular $[K^+]$ from 3 mM to 12 mM led to a large increase in extracellular [GABA] that induced an inward current. Middle: This current was blocked by bicuculline. Right: Recovery. The increase in extracellular [GABA] could be prevented with SKF-89976a or NO-711 (data not shown).[16]

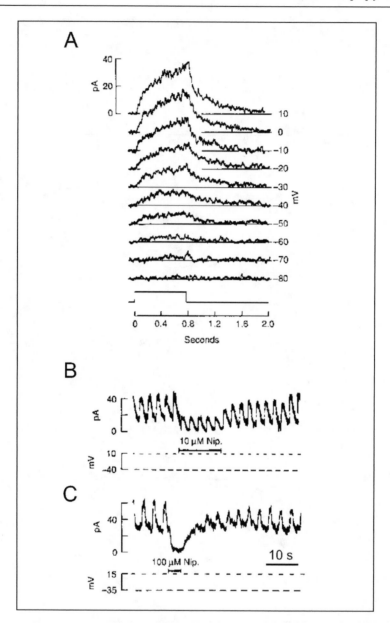

Figure 5. GABA transporter reversal is used for interneuronal communication between catfish retinal neurons. Electrophysiological recordings were made from pairs of neurons isolated from the catfish retina. The postsynaptic response in bipolar cells was measured in response to presynaptic depolarization of horizontal cells. A—Current traces measured in a bipolar cell in response to step depolarization of a horizontal cell to different voltages (shown to the right of each trace). Presynaptic depolarization (bottom trace) led to a graded postsynaptic response whose amplitude was dependent on the presynaptic voltage. The postsynaptic response was inhibited by picrotoxin. B—Repeated voltage steps from -40 mV to 10 mV in a presynaptic horizontal cell (bottom trace) led to postsynaptic current responses in a bipolar cell (top trace). Bath application of 10 μM nipecotic acid decreased the postsynaptic response. C—Same experiment as in B. 100 μM nipecotic acid completely blocked the postsynaptic response.[46]

Table 1. Differences between vesicular and transporter-mediated GABA release

	Vesicular Release	Transporter-Mediated Release
Cell depolarization	Increased	Increased
Low $[Ca^{2+}]_o$	Blocked	No effect
Tetanus or botulinum toxin	Blocked	No effect
Increased $[Na^+]_i$	No effect	Increased
Cellular energy depletion	Decreased	Increased
Source	Neurons	Neurons & glia
Location of GABA pool	Synaptic vesicles	Free cytosolic GABA
Competitive transporter blocker (e.g. nipecotic acid)	Prolong IPSPs	Heteroexchange release, then blockade
Noncompetive transporter blockers (e.g. SKF-89976a, NO-711, tiagabine)	Prolong IPSPs	Blockade

The issues addressed here regarding different mechanisms of GABA release should not be confused with the separate issue of how GABA affects neurons after it is released. Activation of GABA$_A$ receptors can be either inhibitory or excitatory,[9,27,48] but the type of response is dependent upon properties of the postsynaptic target, not the mechanism of GABA release. Thus, under some conditions transporter-mediated GABA release, just like vesicular GABA release, could lead to excitation of surrounding neurons. When this occurs, both forms of GABA release will have a proconvulsant effect.

The GABA Transporter As a Target of Anticonvulsants

The dual role of the GABA transporter in both GABA reuptake and in nonvesicular GABA release suggests that drugs that target the GABA transporter could influence neuronal excitability in different ways. Tiagabine is the anticonvulsant that is most widely recognized as acting on the GABA transporter. This agent blocks the GABA transporter, which as shown above (Fig. 1) prolongs IPSPs by preventing GABA reuptake, thus enhancing inhibition. It is not known whether the increase in inhibition induced by blockade of GABA reuptake by tiagabine is offset to some degree by blockade of nonvesicular GABA release. It is possible that the anticonvulsant effect of the former is partially offset by a proconvulsant effect of the latter. It is also possible that tiagabine or other GABA transporter antagonists block GABA reuptake better than they block GABA release, which would mean that the anticonvulsant effect would dominate.

Vigabatrin also acts via the GABA transporter. Vigabatrin is an antagonist of GABA transaminase, the enzyme that degrades GABA. It is a "suicide inhibitor," causing irreversible inactivation of the enzyme that requires resynthesis before recovery can occur.[19] Thus, acute exposure causes an effect that is delayed in onset and recovery. As would be predicted, vigabatrin causes a slow increase in total brain [GABA].[35] It has been assumed that the increase in brain GABA leads to increased neuronal inhibition, but the specific mechanism of inhibition was not known. The assumption had been that vigabatrin increases the size of IPSPs, but one study showed only a small increase in amplitude of miniature inhibitory postsynaptic conductances (mIPSCs),[14] and another showed a small decrease.[33] We performed similar experiments and found no effect of vigabatrin on the amplitude of mIPSCs.[58] In addition, we used patch-clamp recordings from pairs of neurons and found that vigabatrin induces a small decrease in GABAergic IPSCs induced by presynaptic stimulation.[58]

In contrast to the lack of an increase in IPSP amplitude, we have found that vigabatrin potently increases tonic inhibition by inducing spontaneous and continuous nonvesicular GABA release.[57] Treatment of rat hippocampal cultures with 100 μM vigabatrin induces a large increase in [GABA]$_o$ (Fig. 6A). The increase in [GABA]$_o$ is blocked by the GABA transporter antagonist SKF-89976a[57] and NO-711,[58] indicating that tonic inhibition is due to release of GABA via reversal of the GABA transporter. This effect requires many days to reach a peak (Fig. 6B). In response to 100 μM vigabatrin for 4 days, there is an increase in tonic inhibition of neurons averaging 9 nS, corresponding to a [GABA]$_o$ of more than 0.5 μM. Vigabatrin is extremely potent, inducing a response at a concentration as low as 50 nM (Fig. 6C). The large increase in ambient [GABA], and the resultant tonic inhibition, would have a powerful influence on neuronal excitability and seizure threshold. An increase in tonic GABAergic inhibition has since been confirmed by two other groups.[14,33]

The anticonvulsant gabapentin also increases GABA transporter reversal. We found that gabapentin increases heteroexchange GABA release.[17] As described above, heteroexchange GABA release occurs in response to rapid application of the GABA analogue nipecotic acid, and results when the GABA transporter simultaneously takes up nipecotic acid and releases GABA. When hippocampal neurons are treated with gabapentin, there is a slow potentiation of hetereoexchange GABA release (Fig. 7). The time course of this effect parallels the time course of the anticonvulsant effect in whole animals and lags the increase in drug levels in the brain,[54] suggesting that the drug has an indirect mechanism of action.

The mechanism by which gabapentin increases heteroexchange GABA release is unknown. However, gabapentin and vigabatrin both induce an increase in brain GABA levels in humans, as measured using magnetic resonance spectroscopy.[35,36] In addition, vigabatrin shares the effect of enhancing heteroexchange GABA release.[57] Therefore, it is possible that these two anticonvulsants share the same common final pathway of increasing cytosolic GABA levels and increasing the driving force for GABA transporter reversal. In contrast to vigabatrin, gabapentin has multiple other effects, so that it remains unclear which one(s) is responsible for the anticonvulsant action or the effect on neuropathic pain.[50] Further work characterizing the magnitude, the sensitivity and significance of each of these effects may help to answer this question. However, the enhancement of nonvesicular GABA release by both gabapentin and vigabatrin, as well as the enhancement of tonic inhibition by vigabatrin, would be predicted to lead to potent neuronal inhibition. Thus, these effects are excellent candidates for the mechanism of both of these drugs, and represent a novel mechanism of anticonvulsant action.

It has recently been shown that gabapentin and the related compound pregabalin both cause redistribution of GAT-1 from intracellular locations to the plasma membrane.[55] This is significant because an increase in the number of GABA transporters in the plasma membrane could explain, by itself, why heteroexchange release is increased by gabapentin. In addition, an increase in tonic extracellular [GABA] also decreases internalization of GABA transporters,[4] so that the increase in GAT-1 in the plasma membrane could either be induced by gabapentin directly, or could be secondary to an increase in ambient [GABA] induced by some other primary mechanism. Either way, the end result is an increase in surface expression of GABA transporters, which would lead to a larger amount of nonvesicular GABA release during high-frequency neuronal firing.

Contribution of the GABA Transporter to Tonic GABAergic Inhibition

It is generally believed that the most important form of neuronal inhibition is fast GABAergic synaptic transmission due to presynaptic action potentials. However, it is widely recognized that there is normally a finite [GABA] in the extracellular space.[23,54] Under some conditions this ambient GABA reaches levels sufficient to tonically activate GABA$_A$ receptors.[6] The tonic inhibitory current that is induced can be surprisingly large, exceeding the total charge transfer induced by fast GABAergic synaptic transmission.[6,30] The GABA responsible for tonic inhibi-

Figure 6. Vigabatrin induces a large increase in ambient [GABA] due to reversal of the GABA transporter. A—After treatment of cultured hippocampal neurons with 100 μM vigabatrin for 4 days, there was a large increase in inward holding current when neurons were voltage clamped at -60 mV. This current could be blocked by bicuculline, indicating that it resulted from an increase in extracellular [GABA]. The majority of the increase in ambient [GABA] was due to calcium-independent GABA release, since there was little difference in the response to bicuculline with (gray trace) or without (black trace) extracellular calcium. The increase in ambient [GABA] could be prevented by SKF-89976a (data not shown). B—The increase in ambient [GABA] required 4-5 days to reach a peak. Shown is the amplitude of the bicuculline-sensitive leak current after treatment of hippocampal cultures for variable lengths of time. The slow increase in current is consistent with a slow inactivation of GABA transaminase followed by a slow buildup of GABA in the cytosol of neurons and/or glia. C—Ambient [GABA] was highly sensitive to vigabatrin concentration. Hippocampal cultures were treated with different concentrations of vigabatrin for 3-4 days. As little as 50 nM vigabatrin was sufficient to induce a response. The GABA$_A$ receptor-mediated current increased progressively with increasing concentrations of vigabatrin. From Wu et al, 2001 (ref. 57).

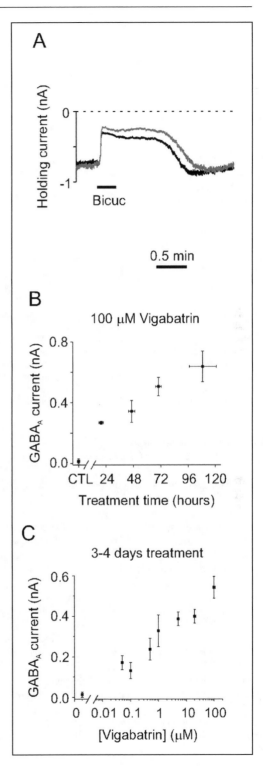

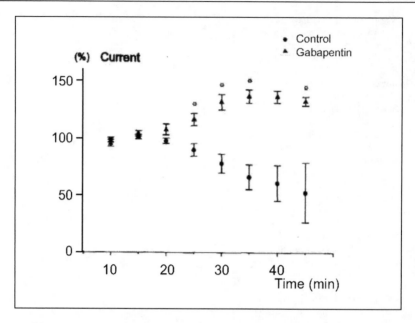

Figure 7. Gabapentin enhances heteroexchange GABA release mediated by reversal of the GABA transporter. Shown are the responses of neurons in hippocampal slices to rapid application of nipecotic acid, which causes transporter-mediated GABA release due to heteroexchange. In control neurons, there was a slow decrease in the response, presumably due to rundown of $GABA_A$ receptors during whole-cell recordings. In neurons exposed to gabapentin (100 μM) starting at the onset of whole-cell breakthrough (time = 0), there was a slow increase in the response to nipecotic acid. After 45 minutes, heteroexchange GABA release in gabapentin-treated neurons was 3.2 times that seen in control neurons.[17]

tion comes in part from spillover of synaptic GABA that has escaped the synaptic cleft before it can be taken back up by GABA transporters.[6,42] In addition, there is also a contribution from the continuous barrage of spontaneous miniature IPSPs that occurs in the absence of action potential-dependent GABA release.[31] Our finding that vigabatrin leads to continuous reversal of the GABA transporter indicates that nonvesicular GABA release also contributes to ambient GABA.

We view tonic GABAergic inhibition as a natural consequence of the stoichiometry of the GABA transporter. In the absence of vesicular release, the GABA transporter will be the major determinant of ambient GABA. It is often assumed that the GABA transporter acts like a vacuum cleaner to eliminate all GABA from the extracellular space. However, this is not actually the case. Instead, the GABA transporter establishes an equilibrium GABA gradient that is determined by the membrane potential, and the concentration gradients for Na^+ and Cl^-. Thus, there will always be a finite, nonzero $[GABA]_o$, and if $[GABA]_o$ drops below this level the transporter will operate in reverse to raise $[GABA]_o$ to the equilibrium level. If the prevailing conditions dictate an equilibrium level of $[GABA]_o$ that is sufficiently high, there will be tonic activation of GABA receptors. Based on the stoichiometry of the transporter, the GABA gradient at equilibrium under normal conditions can be calculated to be approximately 25,000-fold.[58] Assuming that $[GABA]_i$ is 6 mM[32] that would lead to an ambient [GABA] of 0.24 μM at steady state, which is high enough to activate high-affinity $GABA_A$ receptors.[24,44] Ambient [GABA] could never drop below this equilibrium value, because if it did the transporter would reverse until $[GABA]_o$ increased up to the equilibrium value. Thus, it is not necessary to postulate a continuous source of GABA efflux to explain tonic activation of GABA receptors, because when the GABA transporter is at equilibrium $[GABA]_o$ would be a sufficiently high to activate high-affinity $GABA_A$ receptors, even in the absence of vesicular release. Under normal conditions in vivo in which there is continuous vesicular release, the system would not be at steady-state. Therefore, there would be net influx of GABA via the GABA transporter. Under these conditions, ambient [GABA] would be determined by the equilibrium $[GABA]_o$ for the GABA transporter, combined with the rate of vesicular efflux and the rate of influx via the transporter.

These theoretical considerations for the GABA transporter contrast with predictions for the glutamate transporter,[59] which cotransports three Na^+ ions and one proton with each glutamate molecule, while counter-transporting one K^+ ion. This stoichiometry, with greater dependence on the Na^+ gradient, the additional energy from the K^+ gradient, and the transport of two positive charges, leads to a transmembrane glutamate gradient exceeding 1,000,000-fold under physiological conditions, in contrast to the 25,000-fold gradient that is the maximum theoretical limit for the GABA transporter. Therefore, the theoretical lower limit for extracellular [glutamate] is 4.6 nM.[59] Thus, there is a fundamental difference between these two systems, with the equilibrium for the glutamate transporter at an extremely low level that is way below the affinity of its receptors, whereas the equilibrium $[GABA]_o$ is high enough that it would tonically activate GABA receptors. Tonic glutamatergic excitation would be detrimental in the hippocampus—leading to excitotoxicity under normal conditions. In contrast, there may be an advantage to maintaining a high tonic [GABA]—providing a mechanism to control the level of neuronal activity. These predictions are also consistent with the hypothesis that reversal of the glutamate transporter is a phenomenon that occurs during pathological conditions,[1,23] whereas reversal of the GABA transporter occurs under physiological conditions.[16,57]

Role of GABA Transporters in Epilepsy

Since GABA transporters play such a prominent role in maintaining $[GABA]_o$, it follows that changes in GABA transporter function might be linked to the pathophysiology of seizures. Although this aspect of the GABAergic system has received relatively less investigation than others, there is evidence for defects in transporter function in human epilepsy. For example, in patients with temporal lobe seizures, there is a decrease in calcium-independent GABA release in the epileptogenic hippocampus compared to the contralateral hippocampus.[13] In amygdala-kindled rats, there was a similar decrease in nonvesicular GABA release that was associated with a decrease in binding of nipecotic acid.[13] The conclusion that there is impairment of GABA transport in sclerotic hippocampal tissue is supported by the finding that there are prolonged responses to application of GABA in brain slices from epileptogenic human hippocampus compared to control (tumor-related) hippocampus, and that GABA responses in the epileptogenic hippocampus are minimally affected by NO-711.[56]

Based on the physiology of the GABA transporter and the effects of anticonvulsants on both nonvesicular GABA release and ambient [GABA], we have developed a model of the role of the GABA transporter in seizures (Fig. 8). Under normal conditions, the GABA transporter takes up GABA after it is released from the presynaptic terminal. The ambient [GABA] is determined by the numbers and isoforms of GABA transporters present in glial and neuronal membranes, and the driving force for reuptake. Ambient [GABA] will rise when either the rate of vesicular GABA release exceeds the ability of the existing transporters to keep up with the demand, or when there is an increase in the equilibrium $[GABA]_o$ for the GABA transporter. The latter is sufficient, by itself, to explain an increase in ambient [GABA] even in the absence of vesicular GABA release, and occurs in response to vigabatrin due to an increase in intracellular [GABA]. The redistribution of GABA transporters to the plasma membrane in response to increased $[GABA]_o$[4] may be a mechanism to keep local ambient [GABA] near the equilibrium $[GABA]_o$ in the face of a high rate of vesicular efflux. During high-frequency neuronal firing, there is an increase in the outward driving force for the GABA transporter due to the combination of neuronal firing, a rise in extracellular $[K^+]$, and a rise in intracellular $[Na^+]$.

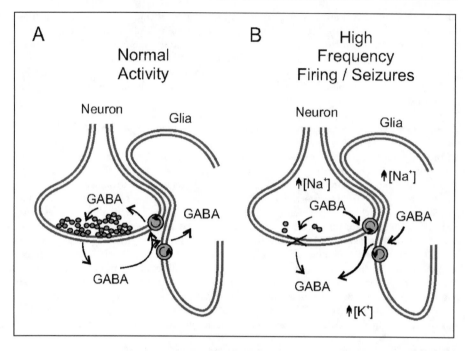

Figure 8. Model of the dual roles of the GABA transporter. A—Under normal conditions, the GABA transporter operates in the forward direction, taking GABA back up after vesicular release. The driving force for uptake also determines the lower limit of ambient [GABA], which in some cases is sufficiently high to activate extrasynaptic high-affinity $GABA_A$ receptors. B—During high frequency firing, the increase in neuronal firing, extracellular $[K^+]$ and intracellular $[Na^+]$ all lead to an increase in the equilibrium $[GABA]_o$ for the GABA transporter. This leads to reversal of the GABA transporter, maintaining GABAergic inhibition at a time that vesicular GABA release decreases. Nonvesicular GABA release may come from neurons, glia, or both.[57]

Under these conditions, transporter-mediated GABA release would occur. This nonvesicular GABA release may serve a critical role in neuroprotection during seizures, because it would maintain neuronal inhibition at a time that energy stores are depleted and vesicular inhibition might fail.

This model for GABAergic inhibition could explain the paradox of why some anticonvulsants increase the number of GABA transporters,[55] and there is a decrease in the number of transporters in epileptic tissue.[13,56] These observations are difficult to reconcile with the traditional view of the GABA transporter, which would predict that an increase in GABA transporters would lead to less inhibition and be proconvulsant. In contrast, our model could explain these observations. To account for why blocking the GABA transporter with tiagabine has an anticonvulsant effect, we propose that a balance normally exists between the excitatory and inhibitory roles of the transporter. Disruption of this balance may occur at different sites (in a seizure focus, propagation pathways, normal cortex) and at different times (interictally, at the onset of seizures, during or between ictal bursts) in patients with epilepsy. Depending upon which neurons are influenced (e.g., inhibitory or excitatory), their physiological state (i.e., whether GABA hyperpolarizes or depolarizes them), and the role they play in circuit function (e.g., whether they promote or destabilize neuronal synchronization), blocking the transporter could either lead to a proconvulsant or anticonvulsant effect. Further defining the physiology of the GABA transporter will help to understand its opposite roles. It is now clear that the GABA transporter is highly dynamic, and plays an active role in neuronal inhibition that goes far beyond simply GABA reuptake. In the future, it may be possible to develop novel interventions that target only specific aspects of GABA transporter function, such as enhancing GABA transporter reversal while also blocking reuptake. This may be a highly effective approach for developing novel anticonvulsants to treat epilepsy without producing major side effects.

Reference

1. Attwell D, Barbour B, Szatkowski M. Nonvesicular release of neurotransmitter. Neuron 1993; 11:401-407.
2. Belhage B, Hansen GH, Schousboe A. Depolarization by K^+ and glutamate activates different neurotransmitter release mechanisms in GABAergic neurons: vesicular versus nonvesicular release of GABA. Neurosci 1993; 54:1019-1034.
3. Bernath S, Zigmond MJ. Characterization of [^{3}H]GABA release from striatal slices: evidence for a calcium-independent process via the GABA uptake system. Neuroscience 1988; 27:563-570.
4. Bernstein EM, Quick MW. Regulation of γ-aminobutyric acid (GABA) transporters by extracellular GABA. J Biol Chem 1999; 274:889-895.
5. Borden LA. GABA transporter heterogeneity: pharmacology and cellular localization. Neurochem Int 1996; 29:335-356.
6. Brickley SG, Cull-Candy SG, Farrant M. Development of a tonic form of synaptic inhibition in rat cerebellar granule cells resulting from persistent activation of $GABA_A$ receptors. J Physiol Lond 1996; 497:753-759.
7. Cammack JN, Rakhilin SV, Schwartz EA. A GABA transporter operates asymmetrically and with variable stoichiometry. Neuron 1994; 13:949-960.
8. Cammack JN, Schwartz EA. Ions required for the electrogenic transport of GABA by horizontal cells of the catfish. J Physiol Lond 1993; 472:81-102.
9. Cherubini E, Gaiarsa JL, Ben-Ari Y. GABA: An excitatory transmitter in early postnatal life. Trends Neurosci 1991; 14:515-519.
10. Clark JA, Amara SG. Stable expression of a neuronal gamma-aminobutyric acid transporter, GAT-3, in mammalian cells demonstrates unique pharmacological properties and ion dependence. Mol Pharmacol 1994; 46:550-557.
11. Conti F, Melone M, De Biasi S. Neuronal and glial localization of GAT-1, a high-affinity gamma-aminobutyric acid plasma membrane transporter, in human cerebral cortex: with a note on its distribution in monkey cortex. J Comp Neurol 1998; 396:51-63.
12. Dingledine R, Korn SJ. Gamma-aminobutyric acid uptake and the termination of inhibitory synaptic potentials in the rat hippocampal slice. J Physiol Lond 1985; 366:387-409.
13. During MJ, Ryder KM, Spencer DD. Hippocampal GABA transporter function in temporal-lobe epilepsy. Nature 1995; 376:174-177.
14. Engel D, Pahner I, Schulze K et al. Plasticity of rat central inhibitory synapses through GABA metabolism. J Physiol Lond 2001; 535:473-482.
15. Gadea A, Lopez-Colome AM. Glial transporters for glutamate, glycine, and GABA: II. GABA transporters. J Neurosci Res 2001; 63:461-468.
16. Gaspary HL, Wang W, Richerson GB. Carrier-mediated GABA release activates GABA receptors on hippocampal neurons. J Neurophysiol 1998; 80:270-281.
17. Honmou O, Kocsis JD, Richerson GB. Gabapentin potentiates the conductance increase induced by nipecotic acid in CA1 pyramidal neurons in vitro. Epilepsy Res 1995; 20:193-202.
18. Isaacson JS, Solis JM, Nicoll RA. Local and diffuse synaptic actions of GABA in the hippocampus. Neuron 1993; 10:165-175.
19. Jung MJ, Palfreyman MG. Vigabatrin mechanisms of action. In: Levy RH, Mattson RH, Meldrum BS, eds. Antiepileptic drugs. New York: Raven Press, 1995:903-913.
20. Kanner BI, Schuldiner S. Mechanism of transport and storage of neurotransmitters. CRC Crit Rev Biochem 1987; 22:1-38.
21. Krnjevic K, Morris ME, Reiffenstein RJ. Changes in extracellular Ca^{2+} and K^+ activity accompanying hippocampal discharges. Can J Physiol Pharmacol 1980; 58:579-582.
22. Lerma J, Herranz AS, Herreras O et al. In vivo determination of extracellular concentration of amino acids in the rat hippocampus. A method based on brain dialysis and computerized analysis. Brain Res 1986; 384:145-155.
23. Levi G, Raiteri M. Carrier-mediated release of neurotransmitters. Trends Neurosci 1993; 16:415-419.
24. Levitan ES, Schofield PR, Burt DR et al. Structural and functional basis for $GABA_A$ receptor heterogeneity. Nature 1988; 335:76-79.

25. Lu CC, Hilgemann DW. GAT1 (GABA:Na$^+$:Cl$^-$) cotransport function. Steady state studies in giant Xenopus oocyte membrane patches. J Gen Physiol 1999; 114:429-444.
26. Matskevitch I, Wagner CA, Stegen C et al. Functional characterization of the Betaine/gamma-aminobutyric acid transporter BGT-1 expressed in Xenopus oocytes. J Biol Chem 1999; 274:16709-16716.
27. Michelson HB, Wong RK. Excitatory synaptic responses mediated by GABA$_A$ receptors in the hippocampus. Science 1991; 253:1420-1423.
28. Minelli A, Brecha NC, Karschin C et al. GAT-1, a high-affinity GABA plasma membrane transporter, is localized to neurons and astroglia in the cerebral cortex. J Neurosci 1995; 15:7734-7746.
29. Minelli A, DeBiasi S, Brecha NC et al. GAT-3, a high-affinity GABA plasma membrane transporter, is localized to astrocytic processes, and it is not confined to the vicinity of GABAergic synapses in the cerebral cortex. J Neurosci 1996; 16:6255-6264.
30. Nusser Z, Mody I. Selective modulation of tonic and phasic inhibitions in dentate gyrus granule cells. J Neurophysiol 2002; 87:2624-2628.
31. Otis TS, Staley KJ, Mody I. Perpetual inhibitory activity in mammalian brain slices generated by spontaneous GABA release. Brain Res 1991; 545:142-150.
32. Otsuka M, Obata K, Miyata Y et al. Measurement of gamma-aminobutyric acid in isolated nerve cells of cat central nervous system. J Neurochem 1971; 18:287-295.
33. Overstreet LS, Westbrook GL. Paradoxical reduction of synaptic inhibition by vigabatrin. J Neurophysiol 2001; 86:596-603.
34. Overstreet LS, Westbrook GL. Synapse density regulates independence at unitary inhibitory synapses. J Neurosci 2003; 23:2618-2626.
35. Petroff OA, Behar KL, Mattson RH et al. Human brain gamma-aminobutyric acid levels and seizure control following initiation of vigabatrin therapy. J Neurochem 1996a; 67:2399-2404.
36. Petroff OA, Rothman DL, Behar KL et al. The effect of gabapentin on brain gamma-aminobutyric acid in patients with epilepsy. Ann Neurol 1996b; 39:95-99.
37. Pietrini G, Suh YJ, Edelmann L et al. The axonal gamma-aminobutyric acid transporter GAT-1 is sorted to the apical membranes of polarized epithelial cells. J Biol Chem 1994; 269:4668-4674.
38. Pin JP, Bockaert J. Two distinct mechanisms, differentially affected by excitatory amino acids, trigger GABA release from fetal mouse striatal neurons in primary culture. J Neurosci 1989; 9:648-656.
39. Radian R, Ottersen OP, Storm-Mathisen J et al. Immunocytochemical localization of the GABA transporter in rat brain. J Neurosci 1990; 10:1319-1330.
40. Ribak CE, Tong WM, Brecha NC. GABA plasma membrane transporters, GAT-1 and GAT-3, display different distributions in the rat hippocampus. J Comp Neurol 1996; 367:595-606.
41. Roepstorff A, Lambert JD. Comparison of the effect of the GABA uptake blockers, tiagabine and nipecotic acid, on inhibitory synaptic efficacy in hippocampal CA1 neurones. Neurosci Lett 1992; 146:131-134.
42. Rossi DJ, Hamann M. Spillover-mediated transmission at inhibitory synapses promoted by high affinity α_6 subunit GABA$_A$ receptors and glomerular geometry. Neuron 1998; 20:783-795.
43. Saransaari P, Oja SS. Release of GABA and taurine from brain slices. Prog Neurobiol 1992; 38:455-482.
44. Saxena NC, Macdonald RL. Properties of putative cerebellar gamma-aminobutyric acid A receptor isoforms. Mol Pharmacol 1996; 49:567-579.
45. Schwartz EA. Calcium-independent release of GABA from isolated horizontal cells of the toad retina. J Physiol Lond 1982; 323:211-227.
46. Schwartz EA. Depolarization without calcium can release gamma-aminobutyric acid from a retinal neuron. Science 1987; 238:350-355.
47. Somjen G, Giacchino JL. Potassium and calcium concentrations in interstitial fluid of hippocampal formation during paroxysmal responses. J Neurophysiol 1985; 53:1098-1108.
48. Staley KJ, Soldo BL, Proctor WR. Ionic mechanisms of neuronal excitation by inhibitory GABA$_A$ receptors. Science 1995; 269:977-981.
49. Stell BM, Mody I. Receptors with different affinities mediate phasic and tonic GABA(A) conductances in hippocampal neurons. J Neurosci 2002; 22:RC223.
50. Taylor CP, Gee NS, Su TZ et al. A summary of mechanistic hypotheses of gabapentin pharmacology. Epilepsy Res 1998; 29:233-249.
51. Thompson SM, Gahwiler BH. Effects of the GABA uptake inhibitor tiagabine on inhibitory synaptic potentials in rat hippocampal slice cultures. J Neurophysiol 1992; 67:1698-1701.
52. Tossman U, Jonsson G, Ungerstedt U. Regional distribution and extracellular levels of amino acids in rat central nervous system. Acta Physiol Scand 1986; 127:533-545.

53. Turner TJ, Goldin SM. Multiple components of synaptosomal [^{3}H]-gamma-aminobutyric acid release resolved by a rapid superfusion system. Biochem 1989; 28:586-593.
54. Welty DF, Schielke GP, Vartanian MG et al. Gabapentin anticonvulsant action in rats: disequilibrium with peak drug concentrations in plasma and brain microdialysate. Epilepsy Res 1993; 16:175-181.
55. Whitworth TL, Quick MW. Upregulation of gamma-aminobutyric acid transporter expression: role of alkylated gamma-aminobutyric acid derivatives. Biochem Soc Trans 2001; 29:736-741.
56. Williamson A, Telfeian AE, Spencer DD. Prolonged GABA responses in dentate granule cells in slices isolated from patients with temporal lobe sclerosis. J Neurophysiol 1995; 74:378-387.
57. Wu Y, Wang W, Richerson GB. GABA transaminase inhibition induces spontaneous and enhances depolarization-evoked GABA efflux via reversal of the GABA transporter. J Neurosci 2001; 21:2630-2639.
58. Wu Y, Wang W, Richerson GB. Vigabatrin induces tonic inhibition via GABA transporter reversal without increasing vesicular GABA release. J Neurophysiol 2003; 89:2021-2034.
59. Zerangue N, Kavanaugh MP. Flux coupling in a neuronal glutamate transporter. Nature 1996; 383:634-637.

CHAPTER 7

GABA and Its Receptors in Epilepsy

Günther Sperk, Sabine Furtinger, Christoph Schwarzer and Susanne Pirker

Abstract

γ-aminobutyric acid (GABA) is the principal inhibitory neurotransmitter in the mammalian brain. It acts through 2 classes of receptors, GABA$_A$ receptors that are ligand-operated ion channels and the G-protein-coupled metabotropic GABA$_B$ receptors. Impairment of GABAergic transmission by genetic mutations or application of GABA receptor antagonists induces epileptic seizures, whereas drugs augmenting GABAergic transmission are used for antiepileptic therapy. In animal epilepsy models and in tissue from patients with temporal lobe epilepsy, loss in subsets of hippocampal GABA neurons is observed. On the other hand, electrophysiological and neurochemical studies indicate a compensatory increase in GABAergic transmission at certain synapses. Also, at the level of the GABA$_A$ receptor, neurodegeneration-induced loss in receptors is accompanied by markedly altered expression of receptor subunits in the dentate gyrus and other parts of the hippocampal formation, indicating altered physiology and pharmacology of GABA$_A$ receptors. Such mechanisms may be highly relevant for seizure induction, augmentation of endogenous protective mechanisms, and resistance to antiepileptic drug therapy. Other studies suggest a role of GABA$_B$ receptors in absence seizures. Presynaptic GABA$_B$ receptors suppress neurotransmitter release. Depending on whether this action is exerted in GABAergic or glutamatergic neurons, there may be anticonvulsant or proconvulsant actions.

γ–aminobutyric acid (GABA) is the principal inhibitory neurotransmitter in the mammalian brain.[1] It acts through 2 classes of receptors, GABA$_A$ receptors that are ligand-operated ion channels and the G-protein-coupled metabotropic GABA$_B$ receptors (for review see ref. 2). GABAergic neurons are ubiquitously distributed and encompass a fundamental role in processing and integration of all neuronal functions. It is therefore not surprising that blockade of the fast inhibitory GABA$_A$ receptors by bicuculline, pentylenetetrazol or picrotoxin causes severe motor seizures in experimental animals.[3,4] It has therefore been suggested that dysfunction of the GABAergic system may have a fundamental role in the propagation of acute seizures and in the manifestation of epilepsy syndromes. Indeed, mutant mice lacking the enzyme glutamate decarboxylase (GAD) or certain subunits of GABA$_A$ receptors are prone to spontaneous epileptic seizures.[5-7] In the same way, patients with auto-antibodies to the enzyme GAD-67 suffer from the so called Stiff-man-syndrome, and often develop also epilepsy.[8,9]

One of the most serious and frequent epilepsy syndromes is temporal lobe epilepsy (TLE). It is initiated by prolonged febrile seizures or status epilepticus, and takes years or even more than a decade until it is manifested.[10,11] In the clinic, TLE is difficult to treat, and patients frequently become resistant to drug therapy.[12] Repeated and prolonged seizures may also contribute to the severe neuronal damage observed in the temporal lobe, notably in the hippocampus, entorhinal cortex, amygdala and other brain areas.[13,14] One of the most typical features is the severe loss of principal neurons in the hippocampus proper, notably in sectors CA1 and

Recent Advances in Epilepsy Research, edited by Devin K. Binder and Helen E. Scharfman.
©2004 Eurekah.com and Kluwer Academic / Plenum Publishers.

CA3, whereas granule cells of the dentate gyrus, and pyramidal neurons of the sector CA2 and the subiculum, are relatively spared.[10,15]

Because of its clinical relevance and the feature that TLE develops over a prolonged "silent" period, considerable effort has been made through the past decades to investigate its pathophysiology. Animal models mimicking different aspects of TLE, like the induction by severe status epilepticus, subsequent epileptogenesis, and manifestation of spontaneous seizure activity, have been developed.[16-19] These models, and hippocampal tissue specimens obtained at surgery from patients with drug-resistant TLE, became valuable objects for studying the pathophysiology of the disease (see refs. 20-23).

Changes in the Function of GABA Neurons

In spite of considerable evidence for involvement of the GABA system in the pathophysiology of epilepsy, neurochemical and neurophysiological results do not allow decisive conclusions on alterations in GABAergic functions. There is considerable evidence for a loss in GABAergic neurons or function, as well as, in contrast, indications for compensatory activation of GABAergic neurons. Thus, in animal models of TLE involving an initial status epilepticus (induced by convulsant compounds such as kainic acid or pilocarpine, or by paradigms using sustained electrical stimulation), considerable loss of interneurons and GABAergic markers were found in the hippocampus, entorhinal cortex and other limbic brain areas.[17-19] Cell loss included both glutamatergic mossy cells and pyramidal cells, and inhibitory GABAergic neurons.[22,24-26]

Evidence for a loss in GABAergic neurons came from neurochemical studies in the kainic acid model of TLE and from epileptic human tissue showing decreases in the activity of the GABA-synthesizing enzyme glutamate decarboxylase.[27,28] Using immunocytochemistry, Ribak et al also demonstrated decreases in GABAergic terminals in a model of focal epilepsy,[29,30] and loss of GABA neurons were observed in kindled rats.[31,32] Variable degeneration of GABA- or GAD-positive neurons was reported in the hippocampal formation in animal models with Ammon's horn sclerosis. Notably, in the hilus of the dentate gyrus, a subpopulation of presumed GABAergic neurons containing the neuropeptide somatostatin as a co-transmitter is vulnerable to epilepsy, as well as to ischemia-induced brain damage.[26,33] On the other hand, basket cells containing the neurotransmitters cholecystokinin octapeptide, neurokinin B or the calcium-binding protein parvalbumin are resistant to this excitotoxic brain damage.[33-36] Considering the extensive loss of pyramidal cells and that of glutamatergic mossy cells in the dentate hilus, GABA neurons are apparently not preferentially lesioned and subpopulations of them may even be spared from epilepsy-induced brain damage.[26,35-38]

Electrophysiological experiments, however, suggested reduced GABA-mediated inhibition in the CA1 sector during and shortly after experimentally-induced repeated epileptic seizures in the rat.[39-41] It therefore had to be considered that surviving GABA neurons may exert reduced neuronal activity. This idea lead Sloviter to propose the so called "dormant basket cell hypothesis".[15,42] He suggested that GABAergic basket cells, normally driven by mossy cells in the dentate gyrus, lose their excitatory input and may therefore be "silent" in epilepsy. Considering the extensive loss in mossy cells and also in excitatory pyramidal cells, this seems to be a likely explanation for the transient reduction in inhibition during sustained epileptic seizures.[43,44] In chronically epileptic rats, however, increased GAD activity and intensified staining for GABA and GAD has been observed in the hippocampus and in hippocampal interneurons respectively.[28,37,45-48] At the same time, mRNAs for the neuropeptides somatostatin, neurokinin B and neuropeptide Y (all co-transmitters of GABA) are similarly up-regulated in local circuit neurons.[26,34,35] This indicates that GABAergic basket cells (at least in chronically epileptic rats) are metabolically highly active and may have increased turnover rates for GABA and its co-transmitters. This neurochemical evidence is clearly supported by electrophysiological data.[49,50] It becomes also more and more clear that GABAergic synapses may be heterogeneously affected at different sites of the epileptic brain. Thus, in the CA1 sector of pilocarpine-treated rats, GABAergic inhibition is enhanced in somata of pyramidal cells, but decreased in their dendrites.[51] This difference may be due to different functioning of presynaptic GABA neurons, enhanced activity of neurons innervating pyramidal cell somata, and a loss in neurons projecting to the dendrites, respectively.[51]

In terms of an overall response, it has to be considered that increased GABAergic activity may not only suppress convulsions but may also cause inhibition of inhibitory neurons and thus provoke convulsions.[52] The often dramatic increases in mRNA levels of surviving neurons also show the difficulty of performing unbiased cell counts using GABA and GAD immunocytochemistry or *in situ* hybridization. The marked increases in the expression of these parameters[26,46,47,53] make it likely that in epileptic rats also neurons become positively labeled which express these parameters below the detection limit in control rats. Such a bias may sometimes obscure actual loss of GABA neurons.

Also, in the hippocampus of TLE patients, the loss of GABAergic cells is limited and accompanied by enhanced labeling using immunocytochemical markers for GABA and its neuropeptide co-transmitters.[37,54,55] In the epileptic human hippocampus, sprouting of neuropeptide Y- or secretoneurin-containing GABAergic interneurons into the terminal field of mossy fibers and of the perforant path has been detected.[23,54-56] At these sites, neuropeptide Y or GABA (through GABA$_B$ receptors) may suppress glutamate release.[56]

Enhanced Expression of GABA in Dentate Granule Cells

Although dentate granule cells give rise to a main excitatory pathway, they also express GAD and synthesize GABA.[53,57,58] Notably, the expression of the 67 kD isoform of the enzyme becomes strongly up-regulated after sustained epileptic seizures.[53,57,59] Although the up-regulation of GAD67 mRNA occurs only transiently, the increase in GAD immunoreactivity appears to be long-lasting in epileptic rats.[53,58,60] A functional role for GABA formed in granule cells of normal and notably of epileptic rats has been suspected. A likely explanation may be that GABA formed in granule cells and their axons, the mossy fibers, becomes released by exocytosis onto dendrites of CA3 pyramidal neurons. This should result in augmented inhibition. Evidence for such a mechanism has been reported by two different groups.[61,62] Stimulation of granule cells results in monosynaptic GABA$_A$ receptor-mediated synaptic signals in CA3 pyramidal neurons of normal guinea pigs[62] and in kindled rats.[61] These signals have properties typical for mossy fibers. They are NMDA receptor-independent and induce long-term potentiation upon repeated stimulation.[62] Both research groups are in favor of mossy fibers being the source of GABA, but suggest that a small population of yet not demonstrated GABA neurons arising from the dentate gyrus could also explain the effects on CA3 pyramidal neurons.[61,62]

On the other hand, storage of GABA in synaptic vesicles and its exocytotic release from them implies the presence of a vesicular GABA transporter in mossy fiber vesicles. Although a vesicular glutamate transporter is highly expressed in granule cells,[63] the vesicular GABA transporter has been not found there.[60,64] It is thus unlikely that GABA may be transported by the vesicular glutamate transporter.[65] We therefore suggested that in addition or alternatively the newly formed GABA may be released from granule cells by reversal of its transport through the cytoplasmic membrane.[60] Such a mechanism would be favored by enhanced cytoplasmic Na$^+$ ions[66] or GABA concentration.[67] It is likely that both intracellular Na$^+$ concentration and GABA levels become elevated after sustained epileptic seizures.[53,58] The observed decrease in the expression of the GABA metabolizing enzyme GABA transaminase in granule cells of epileptic rats would also favor enhanced intracellular GABA levels.[60,68] Interestingly, expression of the plasma membrane transporters GAT-1 and GAT-3 has been observed in granule cells.[60,69,70] Taken together, these results indicate that alternatively or in addition to exocytotic release of GABA from mossy fiber terminals, the transmitter may be released by reversed transport through the plasma membrane during epileptic seizures. If this happens at granule cell dendrites, the liberated GABA may exert an inhibitory action at extrasynaptic GABA receptors that are highly concentrated there.[71]

In human TLE, as in experimental animals, it has been also shown that GABA reuptake into GABA neurons may be impaired by reversal of transport during acute seizures.[72,73] This mechanism has been proposed to promote epileptic activity. Furthermore, Mathern and co-workers[74] observed significant changes in the expression of cytoplasmic membrane GABA transporters in tissue from TLE patients, also indicating altered GABAergic transmission.

Altered Expression of GABA$_A$ Receptor and Its Subunits in TLE

Binding Studies

From the pronounced functional and degenerative changes affecting GABAergic neurons, it appears likely that pre- and postsynaptic GABA receptor functions may be altered in epilepsy. There are two types of GABA receptors present in the brain: GABA$_A$ receptors, which are ligand-operated ion channels;[2] and the G-protein coupled GABA$_B$ receptors (GABA$_B$R).[75,76] Like for the presynaptic markers GABA and GAD, both a loss in receptors (due to degeneration in receptor-carrying structures) or adaptive changes in receptors are observed in epilepsy. Overall decreases in GABA$_A$ receptor binding in the hippocampus proper using [^{3}H]-muscimol or [^{3}H]-labeled benzodiazepines as ligands are seen in animal models of TLE, such as lithium/pilocarpine- or kainate-induced status epilepticus, producing severe hippocampal damage[77,78]. However, also in the kindling model which seems to produce less severe, or no neurodegeneration, loss in binding sites is seen in the CA1 sector of the hippocampus.[79-81] Interestingly, also a lasting reduction in GABA$_A$ receptor sites was reported in the cerebellum by one group.[82]

The picture is quite different in the molecular layer of the dentate gyrus containing the dendrites of granule cells. It is rich in GABA$_A$ receptor subunits,[71] less vulnerable to neurodegeneration[17] and has a fundamental function in gating the excitatory input to the hippocampus. In rat models of TLE, granule cells are actually resistant to epilepsy-induced damage and express high concentrations of GABA$_A$ receptors. Neurochemical and electrophysiological studies in animal models suggest that the function of GABA$_A$ receptors is fundamentally altered.[83-85] Using ligands for GABA, benzodiazepine, and cage convulsant sites of the GABA$_A$ receptor, increases in binding sites were observed after electrical kindling or repeated electroshock.[79-81,86-89] In some studies, such changes persist for up to 4 weeks.[79-81]

Autoradiographic studies in the human brain are considerably more difficult to interpret. Extensive neuronal cell loss is seen in sectors CA1 and CA3 of the hippocampus proper and in the dentate hilus.[22,23] These are less severe in sector CA2 and in the granule cell layer, although they may comprise cell loss of more than 50%. The subiculum appears to be more resistant.[23,56] It is therefore difficult to judge whether an overall loss in binding sites is merely due to a loss in cells or also a down-regulation of receptors. Cell loss, however, results also in a shrinkage of the entire area. This could artificially even enhance the apparent binding signal.

Viewing the entire hippocampus, a loss in binding sites is observed in TLE patients, both in receptor autoradiography[90-93] and in positron emission tomography (PET) studies using [^{11}C]flumazenil accumulation.[92,94,95] Clinically, a decrease in [^{11}C]flumazenil binding with PET is even used as a diagnostic indicator for Ammon's horn sclerosis. In the dentate gyrus, GABA$_A$ receptor binding seems to be unchanged or less severely affected than in other subfields of the hippocampus as revealed by receptor autoradiographies.[91,96,97] Considering the severe cell loss, a lack of a decrease in GABA$_A$ binding sites, as observed in the study of McDonald et al,[97] points to an actually increased binding in surviving granule cells. Interestingly, in one PET study, a decrease in [^{11}C]flumazenil binding was reported in the insular cortex,[98] presumably reflecting neuropsychological changes in these TLE patients.

Expression of GABA$_A$ Receptor Subunits

GABA$_A$ receptors are constituted of five subunits enclosing the chloride channel. Various subunits deriving from multiple gene families (α1–α6, β1–β3, γ1–γ3, δ, ϵ, θ) have been identified as components of differently constituted GABA$_A$ receptor complexes in the mammalian brain (for review see ref. 2). While there are a huge number of theoretically possible differently constituted GABA$_A$ receptors, the actual number of differently assembled receptor pentamers, however, may be not as extensive. Preferred subunit combinations comprise combinations of two α–, two β– and one γ–subunit.[2,99,100]

From genetic studies, there is strong evidence that alteration of GABA$_A$ receptor function can be the cause of epilepsy syndromes. Thus, an altered composition of GABA$_A$ receptors, such as a deficiency in the β3-subunit, is related to epilepsy,[7,101,102] and in patients with mutations of the γ2-subunit epilepsy was diagnosed.[103-105]

Changes in the expression of GABA$_A$ receptor subunits have been extensively studied in animal models of TLE and more recently in human TLE specimens (Table 1). In animal models with sclerosis of Ammon's horn, losses in GABA$_A$ receptor subunits parallel those in pyramidal cells at least in the chronic situation.[106-108] More conclusive data were obtained in these animal models for the dentate gyrus. Since granule cells are resistant to seizure-induced brain damage in the rat, changes in GABA$_A$ receptor subunit expression may be more closely related to receptor adaptation. Lasting increases in subunit α1, α2, α4, and β1-3 (and to a limited extent γ2) mRNA levels and immunoreactivities were observed in the granule cell and molecular layer following kainic acid-induced seizures in the rat.[106,107] Consistent with this, increased levels of mRNAs encoding subunits α3, α4, β3, δ and ϵ were seen in granule cells also in the lithium/pilocarpine model of TLE, although subunit α1 and β1 mRNA levels were decreased.[109] As in the kainate model, mRNA levels of subunits α1, α2, α4, β1-3 and γ2 were increased in the granule cell layer in kindled rats.[110,111] Effects seen acutely during kindling are partially reversed at prolonged intervals.[110] This and the fact that little neurodegeneration is seen in these rats indicate a link of the neurochemical effects to the stimulation of neuronal pathways rather than to brain damage.

Changes observed in the expression of GABA$_A$ receptor subunits in the dentate gyrus of TLE patients are quite similar to those seen in rat models (Table 1).[112-114] Increases were observed for subunit α2–, α3–, β1-, β2-, β3- and γ2–, but not for α1-immunoreactivity.[112-114] On the other hand, marked reduction of subunit α1, α3, β1, β3 and γ2 immunoreactivities was observed in area CA1 of sclerotic specimens.[112-114] The data obtained for the dentate molecular layer are consistent with the idea of an over-expression of GABA$_A$ receptors in the dentate gyrus and with a possible shift of GABA$_A$ receptors containing the α1-subunit to receptors containing other α–subunits. Such a change in the receptor composition may also alter its physiological and pharmacological properties.[2] Evidence for receptor overactivity comes from work by Nusser et al.[115] They found a marked increase in subunit β2/3-IR at the electron microscopic level in somatic synapses of granule cells which was associated with increased synaptic GABA-mediated currents.[115]

In electrophysiological studies, the efficacy of clonazepam to stimulate GABA-mediated currents in cortical cells from TLE patients was comparable to that from adult rats.[116] In epileptic rats, not only an enhanced efficacy of GABA transmission[83,115] but also changes in the allosteric modulation of the GABA$_A$ receptor by drugs or zinc ions have been found in the dentate gyrus.[83,109] During acute lithium/pilocarpine-induced status epilepticus, a reduction of the sensitivity of GABA-evoked currents to benzodiazepines and zinc ions but not to barbiturates was observed in isolated dentate granule cells. Several weeks later, when these rats exerted spontaneous epileptic seizures, the effect of the benzodiazepine clonazepam on GABA currents became augmented.[85,109] In both situations, stimulation of GABA currents by zolpidem acting only on type I benzodiazepine receptors was reduced.[85,109] These diverse effects of classical benzodiazepines and of zolpidem point to a shift from α1- to α2-, α4- or α5-subunit-mediated GABA$_A$ receptor functions.[117] Finally, the augmented inhibitory action in chroni-

Table 1. Expression of GABA$_A$ receptor subunit mRNAs and immunoreactivities in the dentate gyrus in different rat models of TLE and patients with TLE and Ammon's horn sclerosis

	Animal Models of TLE													Human TLE	
	Kindling Kokaia et al[111]		Kindling Kamphuis et al[110]		Kindling Nishimura et al[107a]		SE Nishimura et al[107a]		KA Tsunashima et al[107]		KA (IR) Schwarzer et al[106]	Li^{2+}/Pilocarpine Brooks-Kayal et al[109]		Loup et al[112]	Pirker et al[113]
	Time after Termination of Kindling						Time after Status epilepticus								
	48h	5d	24h	28d	24h	7d	24h	30d	12h	7-30d	30d	24h	>30d		
α1	++	=	=	=	(-)	(+)	++	+	++	+	++	-	-	(+)	=*
α2	nd	nd	=	=	=	+	(-)	(+)	-	=	++	nd	nd	+	nd
α3	nd	nd	(+)	+	-	nd	nd	nd	-	(+)	=	++	+	(-)	(+)
α4	nd	nd	=	(+)	(+)	(+)	+	++	+	(+)	+	+	nd	nd	nd
α5	nd	nd	=	(+)	(+)	+	-	(-)	--	-	±	nd	nd	nd	nd
β1	nd	nd	=	(+)	+	+	+	+	+	+	±	--	-	nd	++
β2	nd	nd	+	=	+	+	++	++	±	++	++	nd	++	+	++
β3	+	=	+	=	+	+	+	(+)	(-)	(+)	++	++	nd	+	++
γ2	+	=	+	+	(+)	(+)	(+)	(+)	(-)	=	+	nd	nd	+	+
δ	nd	nd	=	=	--	(-)	(-)	(+)	(-)	(-)	-	++	++	nd	nd

Symbols are used for the following ranges (% of control): ++, >150 %; +, 115-150 %; =, 96-104 %; (+) 105-114 %; (-) 91 – 95 %; -, 50 – 80 %; --, <50 %; nd, not determined. For the animal models, data represent changes in mRNA levels in the granule cell layer (except for the data derived from Schwarzer et al.[106] in the kainate model). Data for the human tissue[112,113] and those derived from Schwarzer et al.[106] in the kainate model represent changes in the immunoreactivity in the dentate molecular layer. * Wolf et al.[114] reported a decrease in α1-IR.

cally epileptic rats of zinc could lead to a collapse of GABAergic function when released from sprouted mossy fibers onto $GABA_A$ receptors in the dentate molecular layer.[85]

Notably, in isolated CA1 pyramidal cells, different mechanisms for GABA-mediated transmission were observed. GABAergic transmission as well as clonazepam and zolpidem-mediated augmentation of GABA signals are reduced,[83] pointing again to reduced GABAergic transmission in sector CA1 of epileptic rats. [39,41,118-120]

$GABA_B$ Receptors

There is also strong evidence for an involvement of $GABA_B$ receptor ($GABA_BR$)-mediated neurotransmission in the generation of epileptic seizures. $GABA_BR$ are G-protein coupled receptors[121,122] and functionally active when inserted as a dimer of the two subforms $GABA_BR$-1 and $GABA_BR$-2 into the receptor.[75,76,123] $GABA_BRs$ can act postsynaptically or presynaptically.[75] Their main presynaptic action involves the inhibition of neurotransmitter release, notably that of glutamate or GABA. Depending on this, one would expect a suppression of excitatory and of inhibitory neurotransmitter action. In patients, baclofen, a $GABA_BR$ agonist, exerts a spasmolytic action but can also provoke convulsions.[124]

Strong evidence for a role of $GABA_BR$-mediated actions, presumably in thalamic nuclei, exists in absence seizures. $GABA_BR$ agonists promote and $GABA_BR$ antagonists suppress absence seizures in lethargic (lh/lh) mice, a model for absence seizures.[125] In these mice, $GABA_BR$ binding and $GABA_BR$-mediated inhibition of N-methyl-D-aspartate effects are augmented.[125] Similarly, injection of baclofen into thalamic relay nuclei and into the reticular nucleus increased, and injection of an $GABA_BR$ antagonist decreased (dose-dependently), spontaneous spike and wave discharges in rats with inborn absence seizures.[126]

Whereas stimulation of $GABA_BR$ appears to mediate proconvulsant effects in audiogenic (thalamus-mediated) seizures, baclofen has anticonvulsant effects in limbic (hippocampus-mediated) seizures. Baclofen prevents[127] and $GABA_BR$ antagonists accelerate amygdala kindling in the rat.[128] Mice deficient in $GABA_BR$-1 experience seizures and develop epilepsy.[129,130] Haas and coworkers[131] demonstrated that down-regulation of $GABA_BR$ enhanced inhibitory feedback in the dentate gyrus after kainic acid-induced epilepsy. These electrophysiological data are consistent with our neurochemical findings that the expression of $GABA_BR$-1 and $GABA_BR$-2 mRNAs and binding of ^3H-CGP54626 to $GABA_BR$ receptors is reduced in granule cells and pyramidal cells in kainic acid-treated rats.[132] In human TLE, $GABA_BR$ binding and mRNA levels are reduced in the dentate gyrus and pyramidal layer and enhanced in the subiculum.[132a,133,134] When corrected for cell loss, both parameters ($GABA_BR$ binding and mRNA expression) in surviving neurons appear to be enhanced.

In spite of solid evidence of a direct role of the GABA system on the *seizure threshold*, the role of the GABA system in *epileptogenesis* is still poorly understood. Activation of GABA turnover and modified expression of GABA (-A and -B) receptors may result in augmentation of GABA transmission and therefore in protection from subsequent seizures. Such a mechanism, however, may crucially depend on the anatomical site at which it is exerted and may also promote epileptogenesis by suppressing GABAergic inhibition.

Acknowledgements

We thank the Austrian Federal Ministry for Science and Transport (GZ 70.039/2-Pr/4/98) and the Dr. Legerlotz-Foundation.

References

1. Cooper JR, Bloom FE, Roth RH. The biochemical basis of neuropharmacology. New York: Oxford University Press, 1996.
2. Sieghart W, Sperk G. Subunit composition, distribution and function of $GABA_A$ receptor subtypes. Curr Top Med Chem 2002; 2:795-816.
3. Fisher RS. Animal models of the epilepsies. Brain Res Brain Res Rev 1989; 14:245-278.

4. Loscher W, Schmidt D. Which animal models should be used in the search for new antiepileptic drugs? A proposal based on experimental and clinical considerations. Epilepsy Res 1988; 2:145-181.
5. Asada H, Kawamura Y, Maruyama K, et al. Mice lacking the 65 kDa isoform of glutamic acid decarboxylase (GAD65) maintain normal levels of GAD67 and GABA in their brains but are susceptible to seizures. Biochem Biophys Res Commun 1996; 229:891-895.
6. Kash SF, Johnson RS, Tecott LH, et al. Epilepsy in mice deficient in the 65-kDa isoform of glutamic acid decarboxylase. Proc Natl Acad Sci USA 1997; 94:14060-14065.
7. DeLorey TM, Handforth A, Anagnostaras SG, et al. Mice lacking the beta3 subunit of the GABA$_A$ receptor have the epilepsy phenotype and many of the behavioral characteristics of Angelman syndrome. J Neurosci 1998; 18:8505-8514.
8. Peltola J, Kulmala P, Isojarvi J, et al. Autoantibodies to glutamic acid decarboxylase in patients with therapy-resistant epilepsy. Neurology 2000; 55:46-50.
9. Solimena M, Folli F, Denis-Donini S, et al. Autoantibodies to glutamic acid decarboxylase in a patient with stiff-man syndrome, epilepsy, and type I diabetes mellitus. N Engl J Med 1988; 318:1012-1020.
10. Gloor P. Mesial temporal sclerosis: Historical background and an overview from a modern perspective. In: Lüders H, eds. Epilepsy Surgery. New York: Raven Press, 1991:689-703
11. Mathern GW, Babb TL, Mischel PS, et al. Childhood generalized and mesial temporal epilepsies demonstrate different amounts and patterns of hippocampal neuron loss and mossy fibre synaptic reorganization. Brain 1996; 119:965-987.
12. Engel J, Jr. Finally, a randomized, controlled trial of epilepsy surgery. N Engl J Med 2001; 345:365-367.
13. Sutula TP, Hermann B. Progression in mesial temporal lobe epilepsy. Ann Neurol 1999; 45:553-556.
14. Van Paesschen W, Revesz T, Duncan JS, et al. Quantitative neuropathology and quantitative magnetic resonance imaging of the hippocampus in temporal lobe epilepsy. Ann Neurol 1997; 42:756-766.
15. Sloviter RS. The functional organization of the hippocampal dentate gyrus and its relevance to the pathogenesis of temporal lobe epilepsy. Ann Neurol 1994; 35:640-654.
16. Sloviter RS, Dempster DW. "Epileptic" brain damage is replicated qualitatively in the rat hippocampus by central injection of glutamate or aspartate but not by GABA or acetylcholine. Brain Res Bull 1985; 15:39-60.
17. Sperk G. Kainic acid seizures in the rat. Prog Neurobiol 1994; 42:1-32.
18. Turski L, Ikonomidou C, Turski WA, et al. Review: cholinergic mechanisms and epileptogenesis. The seizures induced by pilocarpine: a novel experimental model of intractable epilepsy. Synapse 1989; 3:154-171
19. VanLandingham KE, Lothman EW. Self-sustaining limbic status epilepticus. I. Acute and chronic cerebral metabolic studies: Limbic hypermetabolism and neocortical hypometabolism. Neurology 1991; 41:1942-1949.
20. Houser CR. Granule cell dispersion in the dentate gyrus of humans with temporal lobe epilepsy. Brain Res 1990; 535:195-204.
21. Houser CR, Miyashiro JE, Swartz BE, et al. Altered patterns of dynorphin immunoreactivity suggest mossy fiber reorganization in human hippocampal epilepsy. J Neurosci 1990; 10:267-282.
22. Mathern GW, Pretorius JK, Babb TL. Quantified patterns of mossy fiber sprouting and neuron densities in hippocampal and lesional seizures. J Neurosurg 1995; 82:211-219.
23. Pirker S, Czech T, Baumgartner C, et al. Chromogranins as markers of altered hippocampal circuitry in temporal lobe epilepsy. Ann Neurol 2001; 50:216-226.
24. Scharfman HE, Schwartzkroin PA. Consequences of prolonged afferent stimulation of the rat fascia dentata: Epileptiform activity in area CA3 of hippocampus. Neuroscience 1990; 35:505-517
25. Sloviter RS. A selective loss of hippocampal mossy fiber Timm stain accompanies granule cell seizure activity induced by perforant path stimulation. Brain Res 1985; 330:150-153.
26. Sperk G, Marksteiner J, Gruber B, et al. Functional changes in neuropeptide Y- and somatostatin-containing neurons induced by limbic seizures in the rat. Neuroscience 1992; 50:831-846.
27. Lloyd KG, Bossi L, Morselli PL, et al. Alterations of GABA-mediated synaptic transmission in human epilepsy. Adv Neurol 1986; 44:1033-1044
28. Sperk G, Lassmann H, Baran H, et al. Kainic acid induced seizures: neurochemical and histopathological changes. Neuroscience 1983; 10:1301-1315.
29. Ribak CE, Harris AB, Vaughn JE, et al. Inhibitory, GABAergic nerve terminals decrease at sites of focal epilepsy. Science 1979; 205:211-214.
30. Ribak CE, Bradurne RM, Harris AB. A preferential loss of GABAergic, symmetric synapses in epileptic foci: a quantitative ultrastructural analysis of monkey neocortex. J Neurosci 1982; 2:1725-1735.

31. Kamphuis W, Wadman WJ, Buijs RM, et al. Decrease in number of hippocampal gamma-aminobutyric acid (GABA) immunoreactive cells in the rat kindling model of epilepsy. Exp Brain Res 1986; 64:491-495.
32. Kamphuis W, Huisman E, Wadman WJ, et al. Decrease in GABA immunoreactivity and alteration of GABA metabolism after kindling in the rat hippocampus. Exp Brain Res 1989; 74:375-386.
33. Sloviter RS. Decreased hippocampal inhibition and a selective loss of interneurons in experimental epilepsy. Science 1987; 235:73-76.
34. Gruber B, Greber S, Sperk G. Kainic acid seizures cause enhanced expression of cholecystokinin-octapeptide in the cortex and hippocampus of the rat. Synapse 1993; 15:221-228.
35. Marksteiner J, Wahler R, Bellmann R, et al. Limbic seizures cause pronounced changes in the expression of neurokinin B in the hippocampus of the rat. Neuroscience 1992; 49:383-395.
36. Sloviter RS. Calcium-binding protein (calbindin-D28k) and parvalbumin immunocytochemistry: Localization in the rat hippocampus with specific reference to the selective vulnerability of hippocampal neurons to seizure activity. J Comp Neurol 1989; 280:183-196.
37. Babb TL, Pretorius JK, Kupfer WR, et al. Glutamate decarboxylase-immunoreactive neurons are preserved in human epileptic hippocampus. J Neurosci 1989; 9:2562-2574.
38. Davenport CJ, Brown WJ, Babb TL. Sprouting of GABAergic and mossy fiber axons in dentate gyrus following intrahippocampal kainate in the rat. Exp Neurol 1990; 109:180-190.
39. Franck JE, Kunkel DD, Baskin DG, et al. Inhibition in kainate-lesioned hyperexcitable hippocampi: physiologic, autoradiographic, and immunocytochemical observations. J Neurosci 1988; 8:1991-2002.
40. Sloviter RS, Damiano BP. On the relationship between kainic acid-induced epileptiform activity and hippocampal neuronal damage. Neuropharmacology 1981; 20:1003-1011.
41. Fisher RS, Alger BE. Electrophysiological mechanisms of kainic acid-induced epileptiform activity in the rat hippocampal slice. J Neurosci 1984; 4:1312-1323.
42. Sloviter RS. Permanently altered hippocampal structure, excitability, and inhibition after experimental status epilepticus in the rat: the "dormant basket cell" hypothesis and its possible relevance to temporal lobe epilepsy. Hippocampus 1991; 1:41-66.
43. Bekenstein JW, Lothman EW. Dormancy of inhibitory interneurons in a model of temporal lobe epilepsy. Science 1993; 259:97-100.
44. Sloviter RS. On the relationship between neuropathology and pathophysiology in the epileptic hippocampus of humans and experimental animals. Hippocampus 1994; 4:250-253.
45. Babb TL, Pretorius JK, Mello LE, et al. Synaptic reorganizations in epileptic human and rat kainate hippocampus may contribute to feedback and feedforward excitation. Epilepsy Res Suppl 1992; 9:193-202
46. Esclapez M, Houser CR. Up-regulation of GAD65 and GAD67 in remaining hippocampal GABA neurons in a model of temporal lobe epilepsy. J Comp Neurol 1999; 412:488-505.
47. Feldblum S, Ackermann RF, Tobin AJ. Long-term increase of glutamate decarboxylase mRNA in a rat model of temporal lobe epilepsy. Neuron 1990; 5:361-371.
48. Marksteiner J, Sperk G. Concomitant increase of somatostatin, neuropeptide Y and glutamate decarboxylase in the frontal cortex of rats with decreased seizure threshold. Neuroscience 1988; 26:379-385.
49. Bernard C, Esclapez M, Hirsch JC, et al. Interneurones are not so dormant in temporal lobe epilepsy: a critical reappraisal of the dormant basket cell hypothesis. Epilepsy Res 1998; 32:93-103.
50. Esclapez M, Hirsch JC, Khazipov R, et al. Operative GABAergic inhibition in hippocampal CA1 pyramidal neurons in experimental epilepsy. Proc Natl Acad Sci U S A 1997; 94:12151-12156.
51. Cossart R, Dinocourt C, Hirsch JC, et al. Dendritic but not somatic GABAergic inhibition is decreased in experimental epilepsy. Nat Neurosci 2001; 4:52-62.
52. Peterson GM, Ribak CE. Hippocampus of the seizure-sensitive gerbil is a specific site for anatomical changes in the GABAergic system. J Comp Neurol 1987; 261:405-422.
53. Schwarzer C, Sperk G. Hippocampal granule cells express glutamic acid decarboxylase-67 after limbic seizures in the rat. Neuroscience 1995; 69:705-709.
54. de Lanerolle NC, Kim JH, Robbins RJ, et al. Hippocampal interneuron loss and plasticity in human temporal lobe epilepsy. Brain Res 1989; 495:387-395.
55. Mathern GW, Babb TL, Pretorius JK, et al. Reactive synaptogenesis and neuron densities for neuropeptide Y, somatostatin, and glutamate decarboxylase immunoreactivity in the epileptogenic human fascia dentata. J Neurosci 1995; 15:3990-4004.
56. Furtinger S, Pirker S, Czech T, et al. Plasticity of Y1 and Y2 receptors and neuropeptide Y fibers in patients with temporal lobe epilepsy. J Neurosci 2001; 21:5804-5812.
57. Sandler R, Smith AD. Coexistence of GABA and glutamate in mossy fiber terminals of the primate hippocampus: an ultrastructural study. J Comp Neurol 1991; 303:177-192.

58. Sloviter RS, Dichter MA, Rachinsky TL, et al. Basal expression and induction of glutamate decarboxylase and GABA in excitatory granule cells of the rat and monkey hippocampal dentate gyrus. J Comp Neurol 1996; 373:593-618.
59. Lehmann H, Ebert U, Loscher W. Immunocytochemical localization of GABA immunoreactivity in dentate granule cells of normal and kindled rats. Neurosci Lett 1996; 212:41-44.
60. Sperk G, Schwarzer C, Heilman J, et al. Expression of plasma membrane but not vesicular GABA transporter indicates inverse GABA transport in dentate granule cells after kainic acid induced seizures. Hippocampus 2003; in press.
61. Gutierrez R, Heinemann U. Kindling induces transient fast inhibition in the dentate gyrus-CA3 projection. Eur J Neurosci 2001; 13:1371-1379.
62. Walker MC, Ruiz A, Kullmann DM. Monosynaptic GABAergic signaling from dentate to CA3 with a pharmacological and physiological profile typical of mossy fiber synapses. Neuron 2001; 29:703-715.
63. Takamori S, Rhee JS, Rosenmund C, et al. Identification of a vesicular glutamate transporter that defines a glutamatergic phenotype in neurons. Nature 2000; 407:189-194.
64. Chaudhry FA, Reimer RJ, Bellocchio EE, et al. The vesicular GABA transporter, VGAT, localizes to synaptic vesicles in sets of glycinergic as well as GABAergic neurons. J Neurosci 1998; 18:9733-9750.
65. Fykse EM, Iversen EG, Fonnum F. Inhibition of L-glutamate uptake into synaptic vesicles. Neurosci Lett 1992; 135:125-128.
66. Attwell D, Barbour B, Szatkowski M. Nonvesicular release of neurotransmitter. Neuron 1993; 11:401-407.
67. Wu Y, Wang W, Richerson GB. GABA transaminase inhibition induces spontaneous and enhances depolarization-evoked GABA efflux via reversal of the GABA transporter. J Neurosci 2001; 21:2630-2639.
68. Kang TC, Park SK, Bahn JH, et al. The alteration of gamma-aminobutyric acid-transaminase expression in the gerbil hippocampus induced by seizure. Neurochem Int 2001; 38:609-614.
69. Frahm C, Engel D, Piechotta A, et al. Presence of gamma-aminobutyric acid transporter mRNA in interneurons and principal cells of rat hippocampus. Neurosci Lett 2000; 288:175-178.
70. Ribak CE, Tong WM, Brecha NC. GABA plasma membrane transporters, GAT-1 and GAT-3, display different distributions in the rat hippocampus. J Comp Neurol 1996; 367:595-606.
71. Sperk G, Schwarzer C, Tsunashima K, et al. GABA$_A$ receptor subunits in the rat hippocampus I: Immunocytochemical distribution of 13 subunits. Neuroscience 1997; 80:987-1000.
72. During MJ, Ryder KM, Spencer DD. Hippocampal GABA transporter function in temporal-lobe epilepsy. Nature 1995; 376:174-177.
73. Patrylo PR, Spencer DD, Williamson A. GABA uptake and heterotransport are impaired in the dentate gyrus of epileptic rats and humans with temporal lobe sclerosis. J Neurophysiol 2001; 85:1533-1542.
74. Mathern GW, Mendoza D, Lozada A, et al. Hippocampal GABA and glutamate transporter immunoreactivity in patients with temporal lobe epilepsy. Neurology 1999; 52:453-472.
75. Couve A, Moss SJ, Pangalos MN. GABA$_B$ receptors: a new paradigm in G protein signaling. Mol Cell Neurosci 2000; 16:296-312.
76. Kaupmann K, Malitschek B, Schuler V, et al. GABA$_B$-receptor subtypes assemble into functional heteromeric complexes. Nature 1998; 396:683-687.
77. Kish SJ, Sperk G, Hornykiewicz O. Alterations in benzodiazepine and GABA receptor binding in rat brain following systemic injection of kainic acid. Neuropharmacology 1983; 22:1303-1309.
78. Kapur J, Lothman EW, DeLorenzo RJ. Loss of GABA$_A$ receptors during partial status epilepticus. Neurology 1994; 44:2407-2408.
79. Titulaer MN, Kamphuis W, Pool CW, et al. Kindling induces time-dependent and regional specific changes in the [^{3}H]muscimol binding in the rat hippocampus: a quantitative autoradiographic study. Neuroscience 1994; 59:817-826.
80. Titulaer MN, Kamphuis W, Lopes da Silva FH. Long-term and regional specific changes in [^{3}H]flunitrazepam binding in kindled rat hippocampus. Neuroscience 1995; 68:399-406.
81. Titulaer MN, Kamphuis W, Lopes da Silva FH. Autoradiographic analysis of [^{35}S]t-butylbicyclophosphorothionate binding in kindled rat hippocampus shows different changes in CA1 area and fascia dentata. Neuroscience 1995; 66:547-554.
82. Bazyan AS, Zhulin VV, Karpova MN, et al. Long-term reduction of benzodiazepine receptor density in the rat cerebellum by acute seizures and kindling and its recovery 6 months later by a pentylenetetrazole challenge. Brain Res 2001; 888:212-220.

83. Gibbs JW, 3rd, Shumate MD, Coulter DA. Differential epilepsy-associated alterations in postsynaptic GABA$_A$ receptor function in dentate granule and CA1 neurons. J Neurophysiol 1997; 77:1924-1938.
84. Kapur J, Coulter DA. Experimental status epilepticus alters γ-aminobutyric acid type A receptor function in CA1 pyramidal neurons. Ann Neurol 1995; 38:893-900.
85. Kapur J, Macdonald RL. Rapid seizure-induced reduction of benzodiazepine and Zn^{2+} sensitivity of hippocampal dentate granule cell GABA$_A$ receptors. J Neurosci 1997; 17:7532-7540.
86. Nobrega JN, Kish SJ, Burnham WM. Regional brain [^{3}H]muscimol binding in kindled rat brain: a quantitative autoradiographic examination. Epilepsy Res 1990; 6:102-109.
87. Shin C, Pedersen HB, McNamara JO. γ-Aminobutyric acid and benzodiazepine receptors in the kindling model of epilepsy: a quantitative radiohistochemical study. J Neurosci 1985; 5:2696-2701.
88. Valdes F, Dasheiff RM, Birmingham F, et al. Benzodiazepine receptor increases after repeated seizures: Evidence for localization to dentate granule cells. Proc Natl Acad Sci USA 1982; 79:193-197.
89. Tuff LP, Racine RJ, Mishra RK. The effects of kindling on GABA-mediated inhibition in the dentate gyrus of the rat. II. Receptor binding. Brain Res 1983; 277:91-98.
90. Hand KS, Baird VH, Van Paesschen W, et al. Central benzodiazepine receptor autoradiography in hippocampal sclerosis. Br J Pharmacol 1997; 122:358-364.
91. Johnson EW, de Lanerolle NC, Kim JH, et al. "Central" and "peripheral" benzodiazepine receptors: Opposite changes in human epileptogenic tissue. Neurology 1992; 42:811-815.
92. Koepp MJ, Hand KS, Labbe C, et al. In vivo [^{11}C]flumazenil-PET correlates with ex vivo [^{3}H]flumazenil autoradiography in hippocampal sclerosis. Ann Neurol 1998; 43:618-626.
93. Olsen RW, Bureau M, Houser CR, et al. GABA/benzodiazepine receptors in human focal epilepsy. Epilepsy Res Suppl 1992; 8:383-391.
94. Koepp MJ, Labbe C, Richardson MP, et al. Regional hippocampal [^{11}C]flumazenil PET in temporal lobe epilepsy with unilateral and bilateral hippocampal sclerosis. Brain 1997; 120:1865-1876.
95. Savic I, Persson A, Roland P, et al. In-vivo demonstration of reduced benzodiazepine receptor binding in human epileptic foci. Lancet 1988; 2:863-866.
96. Burdette DE, Sakurai SY, Henry TR, et al. Temporal lobe central benzodiazepine binding in unilateral mesial temporal lobe epilepsy. Neurology 1995; 45:934-941.
97. McDonald JW, Garofalo EA, Hood T, et al. Altered excitatory and inhibitory amino acid receptor binding in hippocampus of patients with temporal lobe epilepsy. Ann Neurol 1991; 29:529-541.
98. Bouilleret V, Dupont S, Spelle L, et al. Insular cortex involvement in mesiotemporal lobe epilepsy: a positron emission tomography study. Ann Neurol 2002; 51:202-208.
99. Chang Y, Wang R, Barot S, et al. Stoichiometry of a recombinant GABA$_A$ receptor. J Neurosci 1996; 16:5415-5424.
100. Tretter V, Ehya N, Fuchs K, et al. Stoichiometry and assembly of a recombinant GABA$_A$ receptor subtype. J Neurosci 1997; 17:2728-2737.
101. Homanics GE, DeLorey TM, Firestone LL, et al. Mice devoid of gamma-aminobutyrate type A receptor beta3 subunit have epilepsy, cleft palate, and hypersensitive behavior. Proc Natl Acad Sci USA 1997; 94:4143-4148.
102. Sugimoto T, Yasuhara A, Ohta T et al. Angelman syndrome in three siblings: Characteristic epileptic seizures and EEG abnormalities. Epilepsia 1992; 33:1078-1082.
103. Baulac S, Huberfeld G, Gourfinkel-An I et al. First genetic evidence of GABA$_A$ receptor dysfunction in epilepsy: a mutation in the γ2-subunit gene. Nat Genet 2001; 28:46-48.
104. Harkin LA, Bowser DN, Dibbens LM et al. Truncation of the GABA$_A$-Receptor γ2 Subunit in a family with generalized epilepsy with febrile seizures plus. Am J Hum Genet 2002; 70:530-536.
105. Wallace RH, Marini C, Petrou S et al. Mutant GABA$_A$ receptor γ2-subunit in childhood absence epilepsy and febrile seizures. Nat Genet 2001; 28:49-52.
106. Schwarzer C, Tsunashima K, Wanzenbock C et al. GABA$_A$ receptor subunits in the rat hippocampus II: Altered distribution in kainic acid-induced temporal lobe epilepsy. Neuroscience 1997; 80:1001-1017.
107. Tsunashima K, Schwarzer C, Kirchmair E et al. GABA$_A$ receptor subunits in the rat hippocampus III: Altered messenger RNA expression in kainic acid-induced epilepsy. Neuroscience 1997; 80:1019-1032.
107a. Nishimura T, Schwarzer C, Furtinger S et al. GABA$_A$ receptor subunits after kindling and status epilepticus induced by electrical stimulation. 2003; submitted.
108. Friedman LK, Pellegrini-Giampietro DE, Sperber EF et al. Kainate-induced status epilepticus alters glutamate and GABA$_A$ receptor gene expression in adult rat hippocampus: an in situ hybridization study. J Neurosci 1994; 14:2697-2707.

109. Brooks-Kayal AR, Shumate MD, Jin H et al. Selective changes in single cell GABA$_A$ receptor subunit expression and function in temporal lobe epilepsy. Nat Med 1998; 4:1166-1172.

110. Kamphuis W, De Rijk TC, Lopes da Silva FH. Expression of GABA- receptor subunit mRNAs in hippocampal pyramidal and granular neurons in the kindling model of epileptogenesis: an in situ hybridization study. Brain Res Mol Brain Res 1995; 31:33-47.

111. Kokaia M, Pratt GD, Elmer E et al. Biphasic differential changes of GABA$_A$ receptor subunit mRNA levels in dentate gyrus granule cells following recurrent kindling-induced seizures. Brain Res Mol Brain Res 1994; 23:323-332.

112. Loup F, Wieser HG, Yonekawa Y et al. Selective alterations in GABA$_A$ receptor subtypes in human temporal lobe epilepsy. J Neurosci 2000; 20:5401-5419.

113. Pirker S, Schwarzer C, Czech T et al. Increased expression of GABA$_A$ receptor β-subunits in the hippocampus of patients with temporal lobe epilepsy. J Neuropath Exp Neurol 2003, in press.

114. Wolf HK, Spanle M, Muller MB, et al. Hippocampal loss of the GABA$_A$ receptor α1 subunit in patients with chronic pharmacoresistant epilepsies. Acta Neuropathol 1994; 88:313-319

115. Nusser Z, Hajos N, Somogyi P, et al. Increased number of synaptic GABA$_A$ receptors underlies potentiation at hippocampal inhibitory synapses. Nature 1998; 395:172-177.

116. Gibbs JW, 3rd, Zhang YF, Kao CQ, et al. Characterization of GABA$_A$ receptor function in human temporal cortical neurons. J Neurophysiol 1996; 75:1458-1471.

117. Sieghart W. Structure and pharmacology of γ-aminobutyric acid$_A$ receptor subtypes. Pharmacol Rev 1995; 47:181-234.

118. Kamphuis W, Gorter JA, da Silva FL. A long-lasting decrease in the inhibitory effect of GABA on glutamate responses of hippocampal pyramidal neurons induced by kindling epileptogenesis. Neuroscience 1991; 41:425-431.

119. Mangan PS, Bertram EH, 3rd. Shortened-duration GABA$_A$ receptor-mediated synaptic potentials underlie enhanced CA1 excitability in a chronic model of temporal lobe epilepsy. Neuroscience 1997; 80:1101-1111.

120. Kapur J, Stringer JL, Lothman EW. Evidence that repetitive seizures in the hippocampus cause a lasting reduction of GABAergic inhibition. J Neurophysiol 1989; 61:417-426.

121. Dutar P, Nicoll RA. A physiological role for GABA$_B$ receptors in the central nervous system. Nature 1988; 332:156-158.

122. Klapstein GJ, Colmers WF. 4-Aminopyridine and low Ca^{2+} differentiate presynaptic inhibition mediated by neuropeptide Y, baclofen and 2-chloroadenosine in rat hippocampal CA1 in vitro. Br J Pharmacol 1992; 105:470-474.

123. Kuner R, Kohr G, Grunewald S, et al. Role of heteromer formation in GABA$_B$ receptor function. Science 1999; 283:74-77.

124. Kofler M, Kronenberg MF, Rifici C, et al. Epileptic seizures associated with intrathecal baclofen application. Neurology 1994; 44:25-27.

125. Hosford DA, Clark S, Cao Z, et al. The role of GABA$_B$ receptor activation in absence seizures of lethargic (lh/lh) mice. Science 1992; 257:398-401.

126. Liu Z, Vergnes M, Depaulis A, et al. Involvement of intrathalamic GABA$_B$ neurotransmission in the control of absence seizures in the rat. Neuroscience 1992; 48:87-93.

127. Wurpel JN. Baclofen prevents rapid amygdala kindling in adult rats. Experientia 1994; 50:475-478.

128. Karlsson G, Klebs K, Hafner T, et al. Blockade of GABA$_B$ receptors accelerates amygdala kindling development. Experientia 1992; 48:748-751.

129. Prosser HM, Gill CH, Hirst WD, et al. Epileptogenesis and enhanced prepulse inhibition in GABA$_B$1-deficient mice. Mol Cell Neurosci 2001; 17:1059-1070.

130. Schuler V, Luscher C, Blanchet C, et al. Epilepsy, hyperalgesia, impaired memory, and loss of pre- and postsynaptic GABA$_B$ responses in mice lacking GABA$_B$1. Neuron 2001; 31:47-58.

131. Haas KZ, Sperber EF, Moshe SL, et al. Kainic acid-induced seizures enhance dentate gyrus inhibition by downregulation of GABA$_B$ receptors. J Neurosci 1996; 16:4250-4260.

132. Furtinger S, Bettler B, Sperk G. Altered expression of GABA$_B$ receptors in the hippocampus after kainic acid-induced seizures in rats. Mol Brain Res 2003; 113:107-115.

132a.Furtinger S, Pirker S, Czech T et al. GABA$_B$ receptors in the hippocampus temporal lobe epilepsy. 2003, submitted.

133. Billinton A, Baird VH, Thom M, et al. GABA$_B$1 mRNA expression in hippocampal sclerosis associated with human temporal lobe epilepsy. Brain Res Mol Brain Res 2001; 86:84-89.

134. Billinton A, Baird VH, Thom M, et al. GABA$_B$ receptor autoradiography in hippocampal sclerosis associated with human temporal lobe epilepsy. Br J Pharmacol 2001; 132:475-480.

135. Scharfman HE. Epilepsy as an example of neuronal plasticity. The Neuroscientist 2002; 8:155-174.

CHAPTER 8

Role of the Depolarizing GABA Response in Epilepsy

Kevin J. Staley

The term "seizure" underscores two fundamental characteristics of epileptic phenomena: they are sudden and unexpected deviations from the normal function of the nervous system. Thus 2 important criteria for a candidate convulsant mechanism are that the mechanism is compatible with normal neural functioning, and that the mechanism can produce rapid deviations from normal function. For example, the blockade of GABA-mediated inhibition is a robust and frequently studied experimental convulsant mechanism,[26,28] but is unlikely to underlie human epilepsy because it is not compatible with normal function nor is it capable of sudden transitions from normal function. Of course, we can modify the GABA-block hypothesis to incorporate the "sudden and unexpected" criterion by proposing that the sudden and unexpected features arise from the condition that imposes the reduction in GABA efficacy.

One such condition is a curious laboratory phenomenon that is frequently referred to as the depolarizing GABA response. This mechanism was first described in the late 1970s and early 1980s during focal application of GABA to hippocampal pyramidal cells: although brief applications of agonists produced the expected hyperpolarizing response, more prolonged applications resulted in a depolarizing response.[1,3,6] Similar results were obtained by synaptic activation of GABA receptors in hippocampal slices.[1,17,19] The depolarizing response could be elicited by focal activation of GABA receptors in the dendrites but not the somata of cultured pyramidal cells.[6,21] These first experiments elucidated the salient characteristics of the depolarizing GABA response: there were both temporal and spatial features of GABA receptor activation that seemed necessary to elicit the response.

At about this time, the $GABA_B$ response was discovered, which clarified some aspects of the temporal pattern of GABA response. Activation of the $GABA_B$ receptor produced a slow hyperpolarization, so that application of GABA could produce 3 responses: a hyperpolarizing response, a subsequent depolarizing response, and a late, small hyperpolarizing response. In 1984, Newberry and Nicoll used the selective agonist baclofen to show that the late hyperpolarizing response reflected activation of potassium-permeable $GABA_B$ receptors. The subsequent development of specific $GABA_B$ antagonists supported the idea that the neither the early hyperpolarizing nor the depolarizing GABA response was related to $GABA_B$ receptors: in the presence of the $GABA_B$ antagonist phaclofen, the prolonged activation of $GABA_A$ receptors produced a late depolarizing response in the dendrites but not the soma of pyramidal cells.[22,25] The prolonged and dendritic features of GABA application were not absolute, however. Alger and Nicoll demonstrated that prolonged dendritic GABA application produced only a small hyperpolarizing response in the presence of low concentrations of $GABA_A$ receptor antagonists.[1]

What was the mechanism responsible for this depolarizing GABA response? Certainly a unique GABA receptor could be present in the dendrites. When the depolarizing GABA response was elicited by high-frequency activation of GABAergic synapses, the depolarizing re-

Recent Advances in Epilepsy Research, edited by Devin K. Binder and Helen E. Scharfman.
©2004 Eurekah.com and Kluwer Academic / Plenum Publishers.

sponse had a slower onset and occurred later than the hyperpolarizing response, suggesting diffusion to a more distant receptor.[2,9,17] Thus, an extrasynaptic GABA$_A$ receptor with a unique ionic permeability was a particularly appealing way to explain this phenomenon. However, such a receptor was difficult to reconcile with the observation that when GABA$_B$ receptors were blocked, no matter where the GABA was applied on the dendrite, the depolarizing response never preceded the hyperpolarizing response, as might be expected if some dendritic regions had a "depolarizing" GABA$_A$ receptor and some had a "hyperpolarizing" receptor.

A second possibility was suggested by studies that demonstrated that prolonged applications of GABA could actually change the GABA$_A$ reversal potential. This was first demonstrated in peripheral sympathetic ganglion neurons using Cl$^-$-sensitive electrodes and very prolonged somatic GABA applications.[5] Several groups subsequently demonstrated that shorter applications of GABA could shift the GABA reversal potential at the soma of hippocampal pyramidal cells.[10,26,27] Using the Nernst equation and the change in the GABA reversal potential, these groups calculated that the GABA applications actually resulted in an increase in intracellular chloride of several millimolar. Although these GABA-induced chloride concentration changes may seem large, they are in line with the micromolar increases in intracellular calcium concentration that have been measured during proportionately smaller and briefer calcium conductance increases, and have been confirmed using intracellular Cl$^-$-sensitive dyes.[8]

These calculated increases in intracellular chloride during prolonged GABA applications presumed that chloride was the only anion that permeated the activated GABA$_A$ channel. However, in 1987 it was conclusively demonstrated that the bicarbonate anion also permeates the GABA$_A$ channel.[7,12] The curious aspect of the GABA HCO$_3^-$ current is that it flows in the opposite direction from Cl$^-$. The intracellular Cl$^-$ concentration is much lower than the extracellular concentration, so despite the fact that the intracellular membrane potential is more negative than the extracellular potential, negatively charged Cl$^-$ anions flow into the neuron when the GABA$_A$ receptor channel is opened. This increases the negative charge in the neuron and further hyperpolarizes the neuronal membrane. In contrast to Cl$^-$, the concentration of bicarbonate anion is about the same on both sides of the neuronal membrane. This occurs for three reasons: first, bicarbonate is in equilibrium with CO$_2$ and protons: HCO$_3^-$ + H$^+$ $\leftrightarrow$ H$_2$CO$_3$ $\leftrightarrow$ H$_2$O + CO$_2$. Second, the proton concentration is about the same on both sides (pH 7.2 intracellular and 7.4 extracellular). Third, CO$_2$ is freely permeable across cell membranes, so any accumulation of HCO$_3^-$ on one side of the membrane will simply drive CO$_2$ the other way across the membrane until the HCO$_3^-$ concentrations reach equilibrium.[13] If the HCO$_3^-$ concentration is symmetric across the membrane, then the only force driving HCO$_3^-$ through the open GABA channel is the negative intracellular potential. Thus, HCO$_3^-$ anions leave the neuron through the open GABA channel. This loss of negatively charged anions depolarizes the neuronal membrane.

One might think that the actions of HCO$_3^-$ and Cl$^-$ should cancel each other out, so that nothing happens when the GABA channel is open. However, the concentration of Cl$^-$ is about 5 times higher than HCO$_3^-$, and Cl$^-$ permeates the GABA channel about 5 times more rapidly than HCO$_3^-$.[7] Thus, the GABA$_A$ current is predominantly Cl$^-$, and the net effect of GABA$_A$ receptor activation is accumulation of anions inside the neuron, with a consequent hyperpolarization of the neuronal membrane potential.

This accumulation of Cl$^-$ inside the neuron is what leads to the depolarizing response. As described above, the HCO$_3^-$ concentration never goes out of equilibrium: no matter how much HCO$_3^-$ leaves, more sneaks back into the neuron as CO$_2$. Thus the HCO$_3^-$ concentration remains stable as long as the neuron can buffer the H$^+$ that accompanies the HCO$_3^-$. However, as demonstrated by the somatic GABA application experiments, during prolonged GABA applications the intracellular Cl$^-$ concentration can increase substantially. This results in a change in the ratio of intracellular to extracellular Cl$^-$, which changes the Cl$^-$ reversal potential. If the intracellular Cl$^-$ increases a few mM, the Cl$^-$ reversal potential can equal the membrane potential. At this point, the Cl$^-$ flux stops and the only net anion flux is due to the egress of HCO$_3^-$.

Thus during prolonged application of GABA, the driving force for Cl$^-$ is degraded, allowing the smaller but more persistent HCO$_3^-$ flux to dominate. The result is that during prolonged GABA applications, the GABA response is initially hyperpolarizing (Cl$^-$ flux dominates) but then becomes depolarizing (Cl$^-$ gradient collapses and HCO$_3^-$ flux dominates).[23]

The above mechanism explains why the depolarizing response requires prolonged GABA application, but why is the response only observed in the dendrites? The answer involves the relative volumes of the dendrites vs. the soma. During prolonged GABA application, the rise in intracellular Cl$^-$ concentration will be directly proportional to the volume of the intracellular compartment into which the Cl$^-$ flows.[24] Because the volume of a square cylinder or sphere is proportional to the cube of the radius, the same size GABA current will cause Cl$^-$ to accumulate 1000 times faster in a 1 μm section of dendrite with a radius of 1 μm compared to a soma with a radius of 10 μm. This does not seem consistent with the early studies that demonstrated an increase in intracellular Cl$^-$ during prolonged GABA applications.[5,10] However, the Cl$^-$ accumulation has been demonstrated in voltage clamp experiments in which the membrane potential was clamped at very positive potentials; this increases the driving force for Cl$^-$ and greatly increases the rate of Cl$^-$ accumulation. Near resting membrane potential, the Cl$^-$ driving force and consequent Cl$^-$ flux are much smaller, and not sufficient to increase intracellular Cl$^-$ to the point of stopping Cl$^-$ flux.[29]

After a large GABA current, how does the Cl$^-$ ever get back out of the cell? KCC2 is a neuronal Cl$^-$ export protein that derives its energy from the transmembrane K$^+$ gradient.[16] Given enough time (several seconds at body temperature), KCC2 can export the Cl$^-$ that enters through the GABA$_A$ receptor.[11] Thus, the Cl$^-$ that enters via low-amplitude or infrequent GABA$_A$ currents can be removed in an electroneutral manner (one K$^+$ accompanies each Cl$^-$ out of the cell). This is the best explanation for the finding that low concentrations of GABA antagonists prevented the depolarizing response:[2] by diminishing the frequency of GABA$_A$ channel openings, the antagonist limited the Cl$^-$ influx to a range that could be compensated by KCC2. It also explains an apparent paradox in the literature: some laboratories think that the depolarizing response is due to extracellular K$^+$ accumulation rather than intracellular Cl$^-$ accumulation.[14] However, because KCC2 is a KCl cotransporter, both phenomena should occur nearly simultaneously: large increases in intracellular Cl$^-$ due to GABA$_A$ receptor-mediated Cl$^-$ flux will result in high KCl transport rates which may increase extracellular K$^+$.

There are many findings that on first glance do not seem compatible with the Cl$^-$ accumulation mechanism of the depolarizing GABA response. For example, in the GABA antagonist experiment,[2] why didn't the hyperpolarizing response get smaller in the presence of the antagonist? The Cl$^-$ accumulation theory would predict that the size of the depolarizing response was limited not by the amount of GABA channel opening but rather the Cl$^-$ driving force. Thus, low concentrations of antagonist did not affect the rate-limiting factor in the hyperpolarizing response (although higher concentrations of antagonist or lower amounts of applied GABA would have reduced the hyperpolarizing component of the response).

Another puzzling set of observations were made using the convulsant 4-aminopyridine (4AP). This agent blocks potassium currents in the terminals of pyramidal cells and interneurons, resulting in large increases in both glutamate and GABA release. The large amounts of synaptically released GABA produce very impressive depolarizing GABA responses.[18] However, the depolarizing responses sometimes occur before or even during hyperpolarizing responses:[17] this does not seem to fit with the slow accumulation of Cl$^-$. However, if we consider that the hyperpolarizing and depolarizing responses may occur at different cellular locations in the presence of 4AP, the paradox can be resolved. For example, if a small somatic GABA synaptic conductance occurs near the end of a large dendritic GABA synaptic conductance, there would be an initial hyperpolarization (dendritic Cl$^-$ influx) followed by a depolarization (dendritic HCO$_3^-$ efflux) followed by a hyperpolarization (new somatic Cl$^-$ influx). The different kinetics of the responses observed in 4AP are also consistent with the idea that the hyperpolarizing and depolarizing responses are not occurring at the same subsynaptic locations.

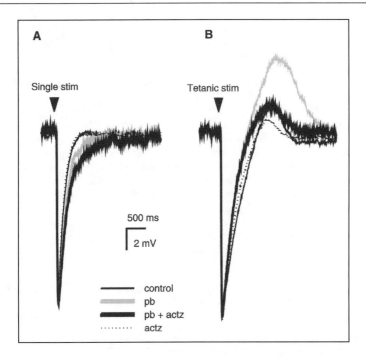

Figure 1. The GABA$_A$ depolarizing response is enhanced by pentobarbital but diminished by acetazolamide. Whole-cell recordings from CA1 pyramidal cell in which glutamate responses were blocked by 1 mM kynurenate and GABA$_B$ responses were blocked by 1μM CGP55845A demonstrates that at RMP of -65 mV, single electrical stimulation of s. moleculare evokes a hyperpolarizing GABA response (left panel). The peak amplitude of the response is not increased by 60 μM pentobarbital, suggesting that the Cl$^-$ gradient is the factor that limits the amplitude of the GABA$_A$ postsynaptic response. Addition of 10 μM acetazolamide inhibits the HCO$_3^-$ recycling described in the text, which stabilizes the GABA reversal potential and uncovers a more prolonged hyperpolarizing response. This action of acetazolamide indicates that in pentobarbital, the return of the membrane potential to RMP is not due to closure of GABA$_A$ channels but rather collapse of the Cl$^-$ gradient. Tetanic stimulation (10 stimuli at 100 Hz) produces a marginally higher amplitude of the response, though it dramatically increases the size of the depolarizing response, which is also consistent with the idea that the Cl$^-$ gradient is the limiting factor in determining the amplitude and direction of the dendritic GABA$_A$ response. Addition of 10 μM acetazolamide selectively inhibits the depolarizing response by destabilizing the HCO$_3^-$ gradient. Thus, acetazolamide prevents the pentobarbital-induced prolonged opening of the GABA$_A$ channel from depolarizing the neuronal membrane, which may limit the anticovulsant efficacy of pentobarbital.

So what does the depolarizing GABA response have to do with epilepsy? Going back to the term "seizure", epileptologists are always looking for mechanisms to generate sudden alterations in the balance between excitation and inhibition. The depolarizing GABA response is attractive in this regard because it is so activity-dependent: a circuit that works perfectly under normal circumstances may become epileptic when driven too hard by its inputs if the GABA signaling in the dendrites overwhelms the capacity of KCC2 to pump out Cl$^-$, resulting in a depolarizing rather than hyperpolarizing response. The depolarizing GABA response can act synergistically with the NMDA receptor to excite neurons by removing the voltage-dependent magnesium block of the NMDA receptor,[23] or by activating voltage-dependent calcium conductances.[9]

If the depolarizing GABA response were important in epileptogenesis, one would predict that an agent that inhibited the response should be a good anticonvulsant. One way to decrease the depolarizing response is to limit the ease with which HCO_3^- can sneak back into the cell as CO_2. Although it is not possible to alter CO_2 permeability, it is possible to decrease the rate at which HCO_3^- is dehydrated to H_2O and CO_2 by blocking the enzyme carbonic anhydrase. This is the primary action of the anticonvulsant acetazolamide. Acetazolamide blocks the depolarizing GABA response,[23] and this is the best-characterized anticonvulsant mechanism for this drug discovered to date.[15]

Another prediction regarding the role of the depolarizing GABA response and seizures is that anticonvulsants that act by increasing the open time of the GABA receptor should make the depolarizing response more prominent. Pentobarbital prolongs the average open time of the GABA receptor, and has been demonstrated to increase the depolarizing GABA response.[1,25] If acetazolamide is added to pentobarbital, the increased GABA effect of pentobarbital can be observed without an increase in the depolarizing response (Fig. 1). Thus, one prediction is that these two clinically-used anticonvulsants should have synergistic actions. This has been experimentally verified in vivo.[4,20] Benzodiazepines also increase the mean open time of the GABA channel, so similar synergy should exist between acetazolamide and benzodiazepines. Acetazolamide is available for intravenous as well as oral administration, so it may be time to explore this synergy clinically, perhaps in a status epilepticus therapeutic trial.

References

1. Alger BE, Nicoll RA. Pharmacological evidence for two kinds of GABA receptor on rat hippocampal pyramidal cells studied in vitro. J Physiol 1982; 328:125-141.
2. Alger BE, Nicoll RA. Feed-forward dendritic inhibition in rat hippocampal pyramidal cells studied in vitro. J Physiol 1982; 328:105-123.
3. Andersen P, Dingledine R, Gjerstad L. Two different responses of hippocampal pyramidal cells to application of gamma-amino butyric acid. J Physiol 1980; 305:279-296.
4. Archer DP, Samanani N, Roth SH. Small-dose pentobarbital enhances synaptic transmission in rat hippocampus. Anesth Analg 2001; 93:1521-1525
5. Ballanyi K, Grafe P. An intracellular analysis of gamma-aminobutyric-acid-associated ion movements in rat sympathetic neurones. J Physiol 1985; 365:41-58.
6. Barker JL, Ransom BR. Amino acid pharmacology of mammalian central neurones grown in tissue culture. J Physiol 1978; 280:331-54.
7. Bormann J, Hamill OP, Sakmann B. Mechanism of anion permeation through channels gated by glycine and gamma-aminobutyric acid in mouse cultured spinal neurones. J Physiol 1987; 385:243-286.
8. Dallwig R, Deitmer JW, Backus KH. On the mechanism of GABA-induced currents in cultured rat cortical neurons. Pflugers Arch 1999; 437:289-97.
9. Grover LM, Lambert NA, Schwartzkroin PA et al. Role of HCO_3^- ions in depolarizing $GABA_A$ receptor-mediated responses in pyramidal cells of rat hippocampus. J Neurophysiol 1993; 69:1541-1555.
10. Huguenard JR, Alger BE. Whole-cell voltage-clamp study of the fading of GABA-activated currents in acutely dissociated hippocampal neurons. J Neurophysiol 1986; 56:1-18.
11. Jarolimek W, Lewen A, Misgeld U. A furosemide-sensitive K^+-Cl^- cotransporter counteracts intracellular Cl^- accumulation and depletion in cultured rat midbrain neurons. J Neurosci 1999; 19:4695-4704.
12. Kaila K, Voipio J. Postsynaptic fall in intracellular pH induced by GABA-activated bicarbonate conductance. Nature 1987; 330:163-165.
13. Kaila K. Ionic basis of $GABA_A$ receptor channel function in the nervous system. Prog Neurobiol 1994; 42:489-537.
14. Kaila K, Lamsa K, Smirnov S. Long-lasting GABA-mediated depolarization evoked by high-frequency stimulation in pyramidal neurons of rat hippocampal slice is attributable to a network-driven, bicarbonate-dependent K^+ transient. J Neurosci 1997; 17:7662-7672.
15. Meldrum B. Update on the mechanism of action of antiepileptic drugs. Epilepsia 1996; 37 Suppl 6:S4-11.
16. Payne JA. Functional characterization of the neuronal-specific K-Cl cotransporter: implications for $[K^+]_o$ regulation. Am J Physiol 1997; 273(5 Pt 1):C1516-1525.
17. Perkins KL, Wong RKS. Ionic basis of the postsynaptic depolarizing GABA response in hippocampal pyramidal cells. J Neurophysiol 1996; 76:3886-3894.

18. Perreault P, Avoli M. A GABAergic depolarizing potential in the hippocampus disclosed by the convulsant 4-aminopyridine. Brain Res 1987; 400:191-195.
19. Perreault P, Avoli M. A depolarizing inhibitory postsynaptic potential activated by synaptically released gamma-aminobutyric acid under physiological conditions in rat hippocampal pyramidal cells. Can J Physiol Pharmacol 1988; 66:1100-1102.
20. Sato J, Nioka M, Owada E. Effect of acetazolamide on the anticonvulsant potency of phenobarbital in mice. J Pharmacobiodyn 1981; 4:952-960.
21. Scharfman HE, Sarvey JM. Responses to gamma-aminobutyric acid applied to cell bodies and dendrites of rat visual cortical neurons. Brain Res 1985; 358:385-389.
22. Scharfman HE, Sarvey JM. Responses to GABA recorded from identified rat visual cortical neurons. Neuroscience 1987; 23:407-422.
23. Staley KJ, Soldo BL, Proctor WR. Ionic mechanisms of neuronal excitation by inhibitory GABA$_A$ receptors. Science 1995; 269:977-981.
24. Staley KJ, Proctor WR. Modulation of mammalian dendritic GABA(A) receptor function by the kinetics of Cl$^-$ and HCO$_3^-$ transport. J Physiol 1999; 519 Pt 3:693-712.
25. Thalmann RH. Blockade of a late inhibitory postsynaptic potential in hippocampal CA3 neurons in vitro reveals a late depolarizing potential that is augmented by pentobarbital. Neurosci Lett 1988; 95:155-60.
26. Thompson SM, Deisz RA, Prince DA. Relative contributions of passive equilibrium and active transport to the distribution of chloride in mammalian cortical neurons. J Neurophysiol 1988; 60:105-124.
27. Thompson SM, Gähwhiler BH. Activity-dependent disinhibition. I. repetitive stimulation reduces IPSP driving force and conductance in the hippocampus in vitro. J Neurophysiol 1989a; 61:501-511.
28. Treiman DM. GABAergic mechanisms in epilepsy. Epilepsia 2001; 42(Suppl 3):8-12.
29. Staley K, Smith R. A new form of feedback at the GABA(A) receptor. Nat Neurosci 2001; 4:674-676.

CHAPTER 9

Gap Junctions, Fast Oscillations and the Initiation of Seizures

Roger D. Traub, Hillary Michelson-Law, Andrea E.J. Bibbig, Eberhard H. Buhl and Miles A. Whittington

I n this chapter, we shall review evidence that gap junctions can contribute to epileptogenesis in the hippocampus and cortex—but not just any gap junctions. Rather, we shall argue for a role for a newly described sort of gap junction, located between the proximal axons of principal neurons. Such axon-axon gap junctions promote epileptogenesis not so much by enhancing synchrony, as by providing pathways for the direct spread of action potentials between neurons. A by-product of such spread is the ability of axonally-coupled neurons to generate oscillations at very high frequencies (>~70 Hz). It is of note that seizure activity, both in vivo and in vitro, has been observed to begin with very high-frequency oscillations. If such oscillations can be shown to initiate the seizure discharge, and not just be an epiphenomenon, then targeting gap junction conductances may prove useful as an anticonvulsant strategy.

In Vitro Population Activities Paroxysmal to Greater or Lesser Extent Which Depend on Gap Junctions

Field Bursts

Field bursts are a form of paroxysmal activity, most readily induced in CA1 but also present in other areas (particularly those areas where the neurons are packed tightly in a laminar fashion), that may be noted when individual neurons are hyperexcitable and chemical synaptic transmission is suppressed.[1-5] Thus, field bursts are favored by very low Ca^{2+} media, perhaps with Mn^{2+} added, and by elevations of extracellular $[K^+]$. Field bursts are characterized both by slow extracellular potential shifts and by large population spikes. The amplitude of such extracellular potentials tends to be enhanced by tight laminar packing of the neurons. Given the suppression of chemical synaptic transmission, the large population spikes suggested the existence of "nonsynaptic" mechanisms for the tight temporal synchronization of action potentials. One such mechanism that was proposed was so-called field effects, produced by the flow of current loops in the extracellular space, across neuronal membranes, and through neuronal interiors.[3,6] Given that extracellular currents are produced by electrogenesis in the entire local neuronal population, field effects can, in principle, couple together nearby neurons. Indeed, the extracellular potential gradients occurring during field bursts can be large enough that effects on spike-firing timings could take place;[7] and small transmembrane depolarizations are sometimes seen just prior to action potential firing, with such depolarizations being expected with field effects.[3]

Nevertheless, experiments after the discovery of field bursts indicate that gap junctions are probably required for field bursts to occur;[8] the paroxysmal events are also influenced by pH changes (i.e., with increased pH enhancing, and reduced pH suppressing), in the way one would expect for gap junctions[9] (see also Spray et al[10]). It is noteworthy as well that low $[Ca^{2+}]_o$

Recent Advances in Epilepsy Research, edited by Devin K. Binder and Helen E. Scharfman. ©2004 Eurekah.com and Kluwer Academic / Plenum Publishers.

media, which promote the occurrence of field bursts, alkalinize the interior of cells, thereby presumably opening gap junctions.[11]

Fast (~200 Hz) Ripples in Vitro

Two distinctive (though probably related) types of "ripples" have been described in the hippocampus. The first to be discovered were in vivo ripples superimposed on the depolarizing potentials associated with physiological sharp waves;[12] intracellular recordings[13] indicate that such ripples consist predominantly of population IPSPs, and we shall refer to such events as "i-ripples", where "i" stands for IPSP. Similar very fast rippling is now known to occur in cortex in vivo in anesthetized cats, where it is also superimposed on depolarizing potentials in principal neurons—specifically, the depolarizing portion of the <1 Hz slow rhythm,[14,15] and such cortical ripples may be similar in mechanism to hippocampal i-ripples.

In addition, however, there are in vitro "p-ripples", where the "p" stands for "population spike"; the designation is applied because the ripple appears as a brief series of ~200 Hz small population spikes (usually 100 μV or less), even when synaptic transmission is blocked in low [Ca^{2+}] media.[16,17] P-ripples occur spontaneously in principal cell regions, changing from sporadic brief trains in normal media, to almost continuous oscillation at similar frequency in low [Ca^{2+}] media.[16] The extracellular potentials—typically tens of μV—seem far too small for field effects to be relevant, nor are intracellular recordings suggestive of this mechanism. On the other hand, a variety of gap-junction-blocking compounds suppress p-ripples, and p-ripples are greatly enhanced by alkalinizing the medium—thus, p-ripples depend on gap junctions. Intracellular recordings during p-ripples show spikes with notches and inflections on their rising phase (that is, appearing to be antidromic), as well as spikelets,[16] also called fast prepotentials[18] or d-spikes.[19] The shape of these latter potentials suggested that the requisite gap junctions responsible for p-ripples were in axons, a notion for which experimental support was subsequently adduced (see below). Furthermore, it was shown in simulations that axonal gap junctions—having a density corresponding to observed dye-coupling densities[20]—would indeed be expected to generate very fast (>100 Hz) network oscillations[21] (see also below). The necessary (and observed) density of gap junctions turns out to be surprisingly low, with each cell coupled to an average of about 1.5 to about 3.5 other cells. [In contrast, synaptic connectivity between CA1 pyramidal cells is about 1%;[22] in a slice with, say, 3,000 pyramidal cells, this means that each pyramidal cell would synaptically excite about 30 others.]

As noted above, field bursts seem to depend upon gap junctions.[8] Furthermore, action potentials recorded intracellularly appear to be antidromic, arising directly from baseline, sometimes with a notch on the rising phase; and small spikes are interspersed with large ones.[2] We would guess, therefore, that field bursts are actually runs of p-ripples, associated with especially large population spikes, due to hyperexcitability caused by the ionic conditions (or other nonsynaptic effects). If our hypothesis is correct, field bursts then represent a form of epileptogenesis that is a direct consequence of axonal gap junctions between principal neurons.

P-ripples may be related to in vivo i-ripples also, by the following mechanism: a burst of very fast oscillation, generated by a network of electrically coupled principal cell axons, can synaptically excite a collection of interneurons to generate a coherent fast oscillation as well. Recall that EPSPs in interneurons tend to be large and fast,[23,24] and that many interneurons can fire at high frequency. A fast network oscillation in a population of interneurons can, in turn, elicit a coherent train of IPSPs in pyramidal neurons, as observed in vivo.[13] Simulations based on this concept replicate many of the in vivo experiments.[25]

Very Fast Oscillations and Seizure Discharge in TMA (Tetramethylamine)

A more complex network behavior occurs in hippocampal slices bathed in TMA, an alkalinizing compound that is presumed to open gap junctions. It has previously been shown[26] that tetanic stimulation of the CA1 region leads to a ~1 second epoch of γ oscillation (~30-70 Hz), that is dependent upon activation of metabotropic glutamate receptors. When such a γ-inducing stimulus is given in the presence of TMA, the γ oscillation still occurs, but it can be followed by a period of low amplitude, very fast (~100 Hz) field potential oscillation, and then by a seizurelike discharge, lasting > 10 seconds, and consisting of a series of synchronized bursts.[27] Interestingly, the very fast oscillation continues during the seizure discharge (including being superimposed on the burst complexes), and also continues afterwards—one is reminded of the very fast "afterdischarge termination oscillation" observed in the limbic system of rats, in vivo, after tetanically-induced seizures,[28] a phenomenon that could be nonsynaptically mediated. [One is also reminded of the older observation of a ~200 Hz rhythm superimposed on single, temporally isolated, synchronized bursts in various in vitro experimental epilepsy models, including disinhibition[19,29] and low $[Mg^{2+}]_o$.[30] Simulations indicate that recurrent excitatory synapses, acting in concert with axonal gap junctions, could together lead to a synchronized burst with superimposed very fast oscillation.[21] While it is known that AMPA receptors are critically involved in these experimental epilepsies, the role of gap junctions in producing the superimposed oscillation remains to be investigated pharmacologically.]

That gap junctions are involved in the in vitro VFO γ seizure discharge phenomenon is suggested not only by the bathing medium (i.e., alkaline conditions); it is also suggested by the observation that the gap-junction-blocking compound carbenoxolone in the bath (along with TMA) does not qualitatively influence the evoked γ oscillation, but it aborts both the subsequent VFO and the seizure discharge.[27] For these reasons—and also because of strikingly similar electrographic phenomenology in some patients with focal seizures (see below)—we have proposed that the (putatively gap-junction-mediated) VFO might actually initiate the succeeding seizure discharge[27]—an hypothesis which, if true, could have practical consequences. Possible mechanisms linking the VFO to the seizure itself will be discussed below.

Giant IPSPs in 4AP

4AP (50 - 100 μM) applied to hippocampal and neocortical slices leads to a variety of population phenomena, including: a) epileptiform depolarizing bursts, dependent on AMPA and depolarizing $GABA_A$ receptors;[31,32] and b) giant GABA-dependent potentials, with hyperpolarizing and depolarizing $GABA_A$, as well as $GABA_B$, components.[33-36] We shall consider here, briefly, 4AP-induced $GABA_B$-dependent "giant" potentials (in principal cells), that occur when ionotropic glutamate receptors and $GABA_A$ receptors are pharmacologically blocked.[35] Not surprisingly, subpopulations of interneurons fire (at least approximately) in phase with each other, so as to elicit the $GABA_B$ potential in principal cells. Furthermore, at least some interneurons[35]—as well as some principal cells[37]—simultaneously generate spikelets, or small action potentials, during artificially induced hyperpolarizations, or during the giant IPSP itself. 4AP-induced giant $GABA_B$-dependent IPSPs, and synchronous interneuronal firing, are suppressed by the gap-junction-blocking compound carbenoxolone,[38,39] presumably via electrical uncoupling of interneurons.

The question, then, is how might a population of interneurons fire trains of approximately synchronous action potentials, when ionotropic receptors are blocked. DC electrical coupling, and ultrastructural dendrodendritic (or dendrosomatic) gap junctions, are known to occur between selected pairs of nearby interneurons.[40-45] Nevertheless, in simulations of interneuronal networks, electrically coupled between dendrites,[46] network bursts occurred only under conditions when action potentials could cross from dendrite to dendrite—and this appears unlikely, in view of the experimentally observed properties of interneuron gap junctions. If it is the case, however, that interneurons are also electrically coupled between their axons (at least to a functionally significant extent in 4AP), then network bursts are readily simulated.[38] As to the spikelets in principal cells—presumed to be antidromic—which occur during the giant IPSP: these spikelets can be explained if GABA released during the giant IPSP were to excite axons. This is not as paradoxical as it sounds: bicuculline has been observed to suppress ectopic spikes generated in Schaffer collaterals,[47,48] and some presynaptic terminals are depolarized by GABA.[49]

The mechanisms of 4AP-induced interneuronal network bursting are of possible relevance to epileptogenesis, given the contributions that excessive GABA release can make to neuronal hyperexcitability, at least under certain conditions.[50-52]

γ *Oscillations Generated by Underlying VFO*

Just as seizure discharges and gap-junction mediated VFO may be interrelated, so may certain types of γ oscillations and VFO also be interrelated. Such a relationship is important to understand: γ oscillations—at least some types of them—occur under physiological conditions, and in vivo γ oscillations are probably important for perception[53,54] and memory.[55] Nevertheless, certain in vitro γ oscillations are built, as it were, on a platform of VFO. (It is not known if any of the multiple in vivo γ oscillations are related to VFO.) The sorts of in vitro γ oscillations where this appears to be true are the ones evoked by carbachol,[56] kainate,[57] or by pressure ejection (so-called "puff") of hypertonic K^+ solution.[58,59] These sorts of γ share several properties:[60]

    a. AMPA and $GABA_A$ receptors are both required.

    b. Gap-junction-blocking agents, particularly octanol, suppress the γ.

    c. Principal cells are not strongly depolarized, and fire infrequently; but their firing, when it occurs, is still (on average) time-locked to the population rhythm.

    d. Interneurons fire at higher rates than principal cells, and are also time-locked to the population rhythm.

    e. The power spectrum of the field potential contains not only a γ peak, but also a higher frequency—albeit smaller—peak, typically 80 Hz or above.[27,56] Blockade of synaptic transmission, at least for γ produced by pressure ejection of potassium, suppresses the γ peak and "unmasks" the higher frequency oscillation.[27]

The above experimental observations, and others as well, can be explained by the following type of model[60] (see also Fig. 2): a combination of spontaneous "ectopic" axonal spikes, along with axonal gap junctions, leads to firing of principal cell axons that is organized into a high-frequency oscillation.[21] This activity synaptically (via AMPA receptors) forces the interneurons to fire. The interneurons, however, inhibit not only each other, but also the somata and axon initial segments of principal cells. Provided the gap junctions are not located too far from somata (and available data—see below—suggest that the distance is < ~150 μm), and provided the IPSCs are large enough,[25] then such synaptic inhibition "chops up" the VFO into segments whose duration is determined by the IPSC decay time constants, and hence gives a γ-frequency rhythm (cf. Whittington et al;[26] Traub et al[61])—but leaving some VFO component as well. This scheme shows why AMPA receptors, $GABA_A$ receptors, and gap junctions are all necessary for the rhythm.

Understanding the mechanisms of carbachol/kainate/pressure ejection γ oscillations is, we believe, clinically relevant, particularly if similar mechanisms are at work in vivo—for example during the γ and very fast oscillations that occur during the slow (<1 Hz) rhythms of sleep.[14,15] Thus, suppose one targets gap junctions pharmacologically in order to suppress the VFO that appear to initiate certain focal seizures.[27] Such targeting could well interfere with normal brain rhythms. One type of experimental evidence that could be useful in analyzing this possibility is based on a prediction of network simulations:[60] many principal cell somatic action potentials, during the oscillation, are expected to be antidromic. It would help to know if this occurs in vivo. Very fast oscillations also occur as an apparently normal phenomenon during the cortical sensory evoked responses, in rats, produced by whisker stimulation[62,63]—it would help to know if these oscillations depend on gap junctions and involve (at least partly) antidromic spikes.

What Effects Do Interneuronal Gap Junctions Have on Population Activities in Vitro?

Up to this point, we have been discussing the effects on population behavior of axonal gap junctions, primarily between principal cells, deferring for the moment a summary of the evidence that such axonal gap junctions even exist. Presentation in this way is motivated by our belief that the axonal gap junctions are the ones making a "primary" contribution to epileptogenesis. Nevertheless, the reader may well feel confused: what about dendritic gap junctions, between interneurons? These are the ones for which there is abundant ultrastructural evidence,[44,45] and whose properties—DC coupling ratio, prejunctional spike → postjunctional spikelet transduction, entraining behavior—have been studied directly with simultaneous dual intracellular recordings, in neocortex and hippocampus.[40-43,64] So as not to leave a gap in the reader's understanding, it is necessary to consider interneuron (putatively dendritic) electrical coupling as well: indeed, this coupling is also relevant to epileptogenesis, at least in one experimental model (kainate application in connexin36 knock-out mice)—although the connection betwen gap-junction-mediated effects and seizure phenomenology appears to be indirect. Our present (limited) understanding can be summarized with 3 experimental observations, paired with insights gleaned from network simulations:

1. "Stabilization" of Interneuron Network Gamma (ING). ING can be evoked experimentally in hippocampal CA1 among other places, by blockade of ionotropic glutamate receptors, combined with tonic excitation of interneurons, for example by metabotropic glutamate receptor agonists,[65,66] or by pressure ejection of hypertonic K^+ solution.[27,59] A synchronous γ-frequency oscillation then occurs that is critically dependent on mutual synaptic inhibition, with frequency being altered in predictable ways by application of drugs that act on $GABA_A$ receptors.[66,67] Simulations had predicted,[66,68,69] however, that this form of oscillation is readily "broken up" by "heterogeneity", that is, by dispersion in some parameter such as the tonic excitatory conductance "felt" by each interneuron. Further experiments and modeling[61] indicated that spatial extension of the cell array—which introduces both axon conduction delays, and also localization in synaptic connectivity—destabilizes the ING oscillation even further. Nevertheless, in experimental conditions—in which precise homogeneity is unlikely to occur—ING does indeed exist, at least transiently.

   A solution to this apparent paradox comes from the observation that carbenoxolone significantly reduces the degree of synchronization of ING.[70] Furthermore, spatially distributed network models that incorporate dendritic interneuron-interneuron gap junctions—at low density (average 2 per neuron) and reasonable conductance (~0.5 - 1.0 nS)—have greatly enhanced stability of ING, compared to the case where gap junctions are not present. These data suggest, therefore, one possible physiological role for the anatomically and physiologically observed dendritic gap junctions between interneurons: increasing coherence in a certain type of γ oscillation.

2. "Stabilization" of γ oscillations produced in vitro by kainate or carbachol. The γ oscillations induced by these drugs (as noted above) require ionotropic glutamate (specifically, AMPA) receptors: they are not "ING". Contributions of interneuron gap junctions to such γ oscillations have been studied by comparing hippocampal slices from wild-type mice with hippocampal slices from mice in which connexin36 has been knocked out.[57] [Connexin36 is a major neuronal gap junction protein that is expressed primarily in neurons, especially interneurons.[57,71-73]] In the connexin36 knockout mice, very fast oscillations observed in low $[Ca^{2+}]$ media[16] are indistinguishable from wild-type; it therefore appears likely that principal cell axon gap junctions do not depend critically upon connexin36. On the other hand, electrical coupling between nearby interneurons is far less common in the knockout mice than in the wild-type mice. Thus, it is reasonable to conclude that effects on oscillations, observed in the knock-out, are caused by the loss of gap junctions between interneurons. What is seen, in examining power spectra of oscillating field potentials in the presence of kainate or carbachol, is that γ power is reduced by almost 1/2 in the knockout. In preliminary studies, we have refined our models of these types of oscillations,[60] so as to examine this issue; specifically, we used longer model axons, so that ectopic spike generation was not directly influenced by synaptic inhibition, and we included interneuron gap junctions as in a recent study.[70] Principal cell axonal gap junctions were located about 180 μm from the soma, somewhat farther than what has been observed with dye coupling (65 - 117 μm in 4 instances[11]). In such a case, the model agrees with experiment, in that interneuron gap junctions influence the mean power of the gamma oscillation.[98]

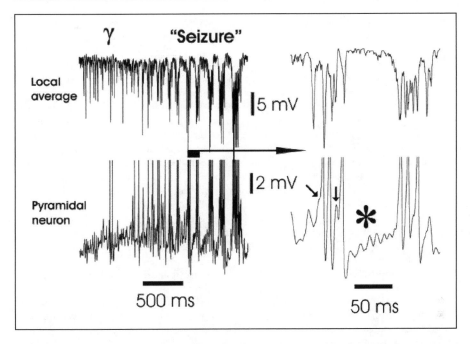

Figure 1. Simulated transition from γ-frequency (~40 Hz) population oscillation to rhythmic series of synchronized bursts ("seizure"), as the conductance of recurrent pyramidal/pyramidal EPSCs is gradually increased. The γ oscillation is generated cooperatively through an interplay of axonal gap junctions, and excitatory and inhibitory synapses (see Traub et al, ref. 60). Network structure was as in that paper, but with gap junctions here located ~187 μm from the soma, where they "feel" the influence of axon initial segment and somatic IPSCs. Intrinsic properties of the model neurons were as in Traub and Bibbig.[25] Two of the burst complexes are expanded on the right: note the VFO superimposed on the burst complexes themselves, and on the interburst interval (*), similar to what occurs in some epileptic patients, as well as in slices bathed in TMA.[27] (RD Traub, A Bibbig and MA Whittington, unpublished data.)

3. Epileptogenesis in connexin36 knockout. Hippocampal slices from the connexin36 knockout, when bathed in kainate, produce not only γ oscillations, but also epileptiform bursts; slices from wild-type mice, bathed in the same concentrations of kainate, produce only γ oscillations, without epileptiform bursts.[74] Evidently, at least under particular experimental conditions, interneuron gap junctions are protective against epileptiform activity. Why might this be? Here, we can so far only offer hypotheses. For example, 1) suppose that short-term plasticity of pyramidal cell-to-interneuron synapses is different in wild-type and knockout, a possibility suggested by the different amplitudes of population EPSPs in the two cases.[57] Alternatively, 2) suppose that short episodes of very fast oscillation lead to facilitation of recurrent excitatory synapses, something which might plausibly occur in the presence of kainate.[75] Then, slices from knockout mice would have more stretches of VFO (their γ being somewhat unstable) than do slices from wild-type mice, and—as a result—experience greater enhancement of recurrent synaptic connections. We know that, at least in simulations, an imposed enhancement of recurrent excitation can lead to a seizure discharge (Fig. 1).

If this latter hypothesis proves correct, then the link between interneuron gap junctions and epileptogenesis is an indirect one, being mediated through a predisposition to VFO, itself mediated by principal cell axonal gap junctions. Note also that in this scheme, principal cell axonal and interneuron dendritic gap junctions act with opposite "sign": the axonal gap junctions promoting, and the dendritic gap junctions (apparently) protecting against, epileptogenesis. Undoubtedly, this scheme represents a major oversimplification.

What Is the Evidence That Principal Cell Gap Junctions Exist?

Much of the preceding discussion has assumed the existence of gap junctions between principal cell axons. The hypothesis that such gap junctions exist arose from study of the shape of spikelets—presumed electrotonic coupling potentials—which were recorded during a very fast (~200 Hz) population oscillation that depended on gap junctions.[16] Here, we very briefly summarize the experimental evidence from hippocampus[11] that axon gap junctions exist, and that (at least some) spikelets represent decrementally conducted antidromic spikes. (See also a recent review.[76]) The experimental data[11] include these findings:

1. Spikelets can be evoked in low $[Ca^{2+}]$ media by stimulating s. oriens at some distance from the recorded neuron. Spikelets evoked in this way can follow, faithfully, stimulation frequencies as fast as 500 Hz, characteristic of axonal firing behavior.
2. Spikelets quickly disappear when the Na^+-channel blocker QX314 is in the whole-cell recording pipette; additionally, spikelets increase in amplitude after a hyperpolarizing prepulse, such as would diminish Na^+ channel inactivation. Hence, spikelets are not passively conducted electrotonic potentials.
3. The amplitude and rate-of-rise of spikelets are reduced by carbenoxolone; this gap-junction-blocking drug, however, does not affect the shape of action potentials directly evoked and recorded in axons or in mossy fiber presynaptic terminals. Thus, a gap junction lies in the path between the stimulating electrode and the soma of the recorded neuron.
4. Dual simultaneous recordings at soma and proximal axon of a single neuron show that the spikelet is, indeed, conducted antidromically. Additionally, the spikelet appears in soma before it appears in the apical dendrite.
5. Time-lapse confocal imaging shows that fluorescent dye injected into one pyramidal neuron through a recording pipette passes along the axon and into the axon of a "dye-coupled" second pyramidal neuron, whose soma and dendrites fill subsequently. Ultrastructural confirmation of the existence of axonal gap junctions is not yet available, to our knowledge.

How Do Axonal Gap Junctions Lead to Very Fast Oscillations?

The means by which axonal gap junctions generate very fast oscillations (> ~70 Hz) are not intuitive. We use the word "generate" here deliberately and with reason, drawing a contrast with a previously known role for gap junctions in rapidly oscillating systems, for example in the medullary pacemaker nucleus of the electric fish.[77,78] In this pacemaker nucleus, gap junctions act to couple together, to entrain, individual intrinsically oscillating neurons, each of which has a natural frequency (were it to be isolated) that is close to the population frequency. This is not what happens in a hippocampal network, where the axons do not "naturally" and in isolation oscillate at 100-200 Hz. Indeed, so far as is known, the period of the hippocampal very fast oscillation is not directly related to a 5 or 10 ms time constant in the axonal membrane biophysics, but rather is related most closely to the sparse connectivity of the gap junctional network. In principal cell networks, the gap junctions generate VFO, rather than entraining them. The reasons underlying this assertion are laid out elsewhere.[21,76]

What we can present here are the basic conditions under which VFO appear able to occur in an axonal network:

1. There is a low rate of spontaneous "ectopic" action potentials in the axons (or presynaptic terminals), 0.05 Hz/axon being sufficient in simulations.
2. There are enough axoaxonal gap junctions (each axon coupled to more than one other on average), but not too many (each axon coupled to less than about 3 or 4, on average).
3. Action potentials can cross from axon to coupled axon.

From these considerations, which are based on extensive network simulations, we see that VFO should be favored by factors that increase axonal excitability, such as kainate in the bath:[79-81]

there should be more ectopic spikes and greater ease of spikes "crossing" the gap junctions; and VFO should be favored by measures (e.g., alkalinization) that increase gap junctional conductance. On the other hand, synaptic inhibition on the axonal initial segment might shunt the axonal membrane near the gap junction, and make it less likely for an antidromic spike to reach the axonal gap junction, or—if the spike does reach there—then it is less likely for the spike to cross the junction and evoke a spike in the coupled fiber.

Additional in Vitro Hippocampal Epilepsy Models

Besides the in vitro models discussed so far, there are other studies as well suggesting that gap-junction-blocking compounds suppress epileptogenesis, including Ross et al,[82] Köhling et al,[83] and (Q Yang and HB Michelson, personal comm.). The mechanisms by which these effects could be occurring are not yet worked out.

Gap Junctions and Epileptogenesis in Vivo

The link between gap junctions and epileptogenesis in vivo is indirect: there are associations, temporal correlations, between very fast oscillations and seizures. In experimental epileptogenesis, for example using spontaneous seizures occurring in rats after kainic acid injection, very fast field potential oscillations (even up to 500 Hz) have been observed prior to seizure discharges,[84,85] with the sites of VFO generation being discrete and patchy. In clinical observations, fast oscillations have been observed leading into focal seizures, in both adults and children.[27,86,87] In one study,[27] a series of children with focal cortical dysplasias was briefly described, with such a pattern of seizure initiation (i.e., with initial localized VFO) apparently being characteristic.

Another suggestive set of clinical/genetic observations is this: on the one hand, the neuronal connexin, connexin36, has been assigned to the gene band 15q14 in humans.[88] On the other hand, both familial rolandic epilepsy with centrotemporal spikes, and also juvenile myoclonic epilepsy, have an association with this same locus.[89,90] It will be instructive to know the genetic localization of the protein(s) that mediate electrical coupling between axons.

How Could the Transition from Very Fast Oscillations to Seizure Discharge Occur?

We have noted that fast oscillations can evolve into seizure discharges, both in experimental (in vitro and in vivo), as well as clinical contexts.[27,91] In addition, we showed (Fig. 1) that one possible way for such an evolution to take place is through enhancement of recurrent synaptic excitation. It is important to emphasize, however, that the evolution could occur by several other means, including reduction in synaptic inhibition (Fig. 2), or by reduction in synaptic excitation of interneurons (R Traub, A Bibbig, MA Whittington, unpublished data). Undoubtedly, details of these mechanisms will be worked out in experimental models; determining the mechanisms in humans will likely prove difficult. There are, however, potentially significant awards in therapy.

Summary and Unifying Hypotheses

The most important points we have raised are these:

1. Both experimental (in vitro and in vivo) and human seizure discharges may begin with a very fast (>70 Hz) neuronal network oscillation.[27,76]
2. Very fast oscillations (VFO) can occur in vitro without chemical synapses, but depending on gap junctions.[16]
3. The gap junctions which generate VFO are located between the axons of principal neurons.[11,21,76]
4. Axonal gap junctions, working in concert with chemical synapses, can generate—depending on parameters—VFO or gamma oscillations.[25,60]

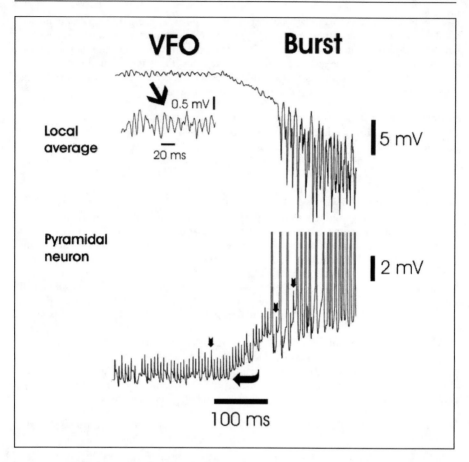

Figure 2. Simulated transition from continuous very fast oscillation (VFO) to synchronized burst, as synaptic inhibition of pyramidal cells is reduced (beginning at horizontal arrow, time constant for reducing maximal IPSC conductance = 250 ms). The VFO was generated by distally located (~560 µm from the soma) axonal gap junctions between pyramidal cells, with 1.6 junctions per neuron on average. At this distance, the gap junction site is little influenced by synaptic inhibition, so that VFO is continuous. Ectopic action potentials originate in the axon distal to the gap junction. Synaptic interactions (including especially recurrent excitation between pyramidal neurons) are present in this detailed network of 3,072 pyramidal and 384 inhibitory neurons, similar to Traub and Bibbig.[25] Some of the spikelets in the pyramidal cell are indicated by vertical arrows. Note the fast oscillations superimposed as well on the synchronized burst. (RD Traub, A Bibbig and MA Whittington, unpublished data).

   5. The transition from fast (gamma or VFO) oscillation to seizure may depend upon time-dependent alterations in one or more synaptic conductances, including recurrent excitation between principal neurons.
   6. The relation between interneuron dendritic gap junctions and epilepsy—at least in the kainate model[74]—appears to be an indirect one.

How might the presumed link between gap junctions and clinical epilepsy be further investigated? Two possible approaches come to mind: First, does pharmacological blockade of neuronal gap junctions, in vivo, suppress seizures? It is of note, for example, that carbenoxolone is approved for human use in Europe, for treatment of ulcer disease,[92] and it seems to cross the blood-brain barrier.[93] It is unclear, though, whether gap-junction-blocking compounds can attain CNS concentrations sufficient to block putative pathological gap junctions without causing visual loss.[71,94,95] Second, how might one investigate whether axonal gap junctions are increased in number, or have altered functional properties, in human epileptogenic foci? [There is reason to believe that neuronal injury leads both to axonal reorganization and to enhanced expression of gap junctions.[96,97]] This issue may require technically difficult experiments in resected brain tissue specimens, and also a means of visualizing axons, as well as gap junctions located on them, in neuropathological material. For this, it would be most helpful to identify the protein(s) constituting axonal gap junctions and to develop markers for them.

Acknowledgments

Supported by NIH, the Wellcome Trust, and the Medical Research Council (U.K.). We thank Dietmar Schmitz, Andreas Draguhn, Hannah Monyer, Nancy Kopell, Torsten Baldeweg, and Fiona E.N. LeBeau for helpful discussion, and Robert Walkup for invaluable help with computing.

References

1. Haas HL, Jefferys JGR. Low-calcium field burst discharges of CA1 pyramidal neurones in rat hippocampal slices. J Physiol 1984; 354:185-201.
2. Taylor CP, Dudek FE. Synchronous neural after discharges in rat hippocampal slices without active chemical synapses. Science 1982; 218:810-812.
3. Taylor CP, Dudek FE. Excitation of hippocampal pyramidal cells by an electrical field effect. J Neurophysiol 1984; 52:126-142.
4. Schweitzer JS, Patrylo PR, Dudek FE. Prolonged field bursts in the dentate gyrus: Dependence on low calcium, high potassium, and nonsynaptic mechanisms. J Neurophysiol 1992; 68:2016-2025.
5. Konnerth A, Heinemann U, Yaari Y. Slow transmission of neural activity in hippocampal area CA1 in absence of active chemical synapses. Nature 1984; 307:69-71.
6. Traub RD, Dudek FE, Taylor CP et al. Simulation of hippocampal afterdischarges synchronized by electrical interactions. Neuroscience 1985; 14:1033-1038.
7. Jefferys JGR. Influence of electric fields on the excitability of granule cells in guinea-pig hippocampal slices. J Physiol 1981; 319:143-152.
8. Perez-Velazquez JL, Valiante TA, Carlen PL. Modulation of gap junctional mechanisms during calcium-free induced field burst activity: a possible role for electrotonic coupling in epileptogenesis. J Neurosci 1994; 14:4308-4317.
9. Schweitzer JS, Wang H, Xiong ZQ et al. pH sensitivity of nonsynaptic field bursts in the dentate gyrus. J Neurophysiol 2000; 84:927-933.
10. Spray DC, Harris AL, Bennett MVL. Gap junctional conductance is a simple and sensitive function of intracellular pH. Science 1981; 211:712-715.
11. Schmitz D, Schuchmann S, Fisahn A et al. Axo-axonal coupling: a novel mechanism for ultrafast neuronal communication. Neuron 2001; 31:831-840.
12. Buzsáki G, Horváth Z, Urioste R et al. High-frequency network oscillation in the hippocampus. Science 1992; 256:025-1027.
13. Ylinen A, Bragin A, Nádasdy Z et al. Sharp wave-associated high frequency oscillation (200 Hz) in the intact hippocampus: network and intracellular mechanisms. J Neurosci 1995; 15:30-46.
14. Grenier F, Timofeev I, Steriade M. Focal synchronization of ripples (80-200 Hz) in neocortex and their neuronal correlates. J Neurophysiol 2001; 86:1884-1898.
15. Steriade M, Contreras D, Curró Dossi R et al. The slow (<1 Hz) oscillation in reticular thalamic and thalamocortical neurons: scenario of sleep rhythm generation in interacting thalamic and neocortical networks. J Neurosci 1993; 13:3284-3299.
16. Draguhn A, Traub RD, Schmitz D et al. Electrical coupling underlies high-frequency oscillations in the hippocampus in vitro. Nature 1998; 394:189-192.
17. Draguhn A, Traub RD, Bibbig A et al. Ripple (approximately 200-Hz) oscillations in temporal structures. J Clin Neurophysiol 2000; 17:361-376.
18. Spencer WA, Kandel ER. Electrophysiology of hippocampal neurons. IV. Fast prepotentials. J Neurophysiol 1961; 24:272-285.
19. Schwartzkroin PA, Prince DA. Penicillin-induced epileptiform activity in the hippocampal in vitro preparation. Ann Neurol 1977; 1:463-469.
20. Church J, Baimbridge KG. Exposure to high-pH medium increases the incidence and extent of dye coupling between rat hippocampal CA1 pyramidal neurons in vitro. J Neurosci 1991; 11:3289-3295.

21. Traub RD, Schmitz D, Jefferys JGR et al. High-frequency population oscillations are predicted to occur in hippocampal pyramidal neuronal networks interconnected by axoaxonal gap junctions. Neuroscience 1999; 92:407-426.
22. Deuchars J, Thomson AM. CA1 pyramid-pyramid connections in rat hippocampus in vitro: dual intracellular recordings with biocytin filling. Neuroscience 1996; 74:1009-1018.
23. Miles R. Synaptic excitation of inhibitory cells by single CA3 hippocampal pyramidal cells of the guinea-pig in vitro. J Physiol 1990; 428:61-77.
24. Geiger JRP, Lübke J, Roth A et al. Submillisecond AMPA receptor-mediated signaling at a principal neuron-interneuron synapse. Neuron 1997; 18:1009-1023.
25. Traub RD, Bibbig A. A model of high-frequency ripples in the hippocampus, based on synaptic coupling plus axon-axon gap junctions between pyramidal neurons. J Neurosci 2000; 20:2086-2093.
26. Whittington MA, Stanford IM, Colling SB et al. Spatiotemporal patterns of γ frequency oscillations tetanically induced in the rat hippocampal slice. J Physiol 1997; 502: 591-607.
27. Traub RD, Whittington MA, Buhl EH et al. A possible role for gap junctions in generation of very fast EEG oscillations preceding the onset of, and perhaps initiating, seizures. Epilepsia 2001; 42:153-170.
28. Bragin A, Penttonen M, Buzsáki G. Termination of epileptic afterdischarge in the hippocampus. J Neurosci 1997; 17:2567-2579.
29. Wong RKS, Traub RD. Synchronized burst discharge in disinhibited hippocampal slice. I. Initiation in CA2-CA3 region. J Neurophysiol 1983; 49:442-458.
30. Traub RD, Jefferys JGR, Whittington MA. Enhanced NMDA conductances can account for epileptiform activities induced by low Mg^{2+} in the rat hippocampal slice. J Physiol 1994; 478: 379-393.
31. Perreault P, Avoli M. Physiology and pharmacology of epileptiform activity induced by 4-aminopyridine in rat hippocampal slices. J Neurophysiol 1991; 65:771-785.
32. Traub RD, Colling SB, Jefferys JGR. Cellular mechanisms of 4-aminopyridine-induced synchronized after-discharges in the rat hippocampal slice. J Physiol 1995; 489:127-140.
33. Aram JA, Michelson HB, Wong RKS. Synchronized GABAergic IPSPs recorded in the neocortex after blockade of synaptic transmission mediated by excitatory amino acids. J Neurophysiol 1991; 65:1034-1041.
34. Michelson HB, Wong RKS. Excitatory synaptic responses mediated by $GABA_A$ receptors in the hippocampus. Science 1991; 253:1420-1423.
35. Michelson HB, Wong RKS. Synchronization of inhibitory neurones in the guinea pig hippocampus in vitro. J Physiol 1994; 477:35-45.
36. Müller W, Misgeld U. Picrotoxin- and 4-aminopyridine-induced activity in hilar neurons in the guinea pig hippocampal slice. J Neurophysiol 1991; 65:141-147.
37. Avoli M, Methot M, Kawasaki H. GABA-dependent generation of ectopic action potentials in the rat hippocampus. Eur J Neurosci 1998; 10:2714-2722.
38. Traub RD, Bibbig A, Piechotta A et al. Synaptic and nonsynaptic contributions to giant IPSPs and ectopic spikes induced by 4-aminopyridine in the hippocampus in vitro. J Neurophysiol 2001; 85: 1246-1256.
39. Yang Q, Michelson HB. Gap junctions synchronize the firing of inhibitory interneurons in guinea pig hippocampus. Brain Res 2001; 907:139-143.
40. Galarreta M, Hestrin S. A network of fast-spiking cells in the neocortex connected by electrical synapses. Nature 1999; 402:72-75.
41. Gibson JR, Beierlein M, Connors BW. Two networks of electrically coupled inhibitory neurons in neocortex. Nature 1999; 402:75-79.
42. Venance L, Rozov A, Blatow M et al. Connexin expression in electrically coupled postnatal rat brain neurons. Proc Natl Acad Sci USA 2000; 97:10260-10265.
43. Tamás G, Buhl EH, Lörincz A et al. Proximally targeted GABAergic synapses and gap junctions precisely synchronize cortical interneurons. Nat Neurosci 2000; 3:366-371.
44. Fukuda T, Kosaka T. Gap junctions linking the dendritic network of GABAergic interneurons in the hippocampus. J Neurosci 2000; 20:1519-1528.
45. Kosaka T. Gap junctions between nonpyramidal cell dendrites in the rat hippocampus (CA1 and CA3 regions). Brain Res 1983; 271:157-161.
46. Traub RD. Model of synchronized population bursts in electrically coupled interneurons containing active dendritic conductances. J Comput Neurosci 1995; 2:283-289.
47. Stasheff SF, Hines M, Wilson WA. Axon terminal hyperexcitability associated with epileptogenesis in vitro. I. Origin of ectopic spikes. J Neurophysiol 1993; 70:960-975.
48. Stasheff SF, Mott DD, Wilson WA. Axon terminal hyperexcitability associated with epileptogenesis in vitro. II. Pharmacological regulation by NMDA and $GABA_A$ receptors. J Neurophysiol 1993; 70:976-984.

49. Jang IS, Jeong HJ, Akaike N. Contribution of the Na-K-Cl cotransporter on GABA$_A$ receptor-mediated presynaptic depolarization in excitatory nerve terminals. J Neurosci 2001; 21:5962-5972.
50. Thalmann RH, Peck EJ, Ayala GF. Biphasic response of hippocampal pyramidal neurons to GABA. Neurosci Lett 1981; 21:319-324.
51. Wong RKS, Watkins DJ. Cellular factors influencing GABA response in hippocampal pyramidal cells. J Neurophysiol 1982; 48:938-951.
52. Taira T, Lamsa K, Kaila K. Posttetanic excitation mediated by GABA$_A$ receptors in rat CA1 pyramidal neurons. J Neurophysiol 1997; 77:2213-2218.
53. Roelfsema PR, König P, Engel AK et al. Reduced synchronization in the visual cortex of cats with strabismic amblyopia. Eur J Neurosci 1994; 6:1645-1655.
54. Singer W, Gray CM. Visual feature integration and the temporal correlation hypothesis. Ann Rev Neurosci 1995; 18:555-586.
55. Tallon-Baudry C, Bertrand O, Fischer C. Oscillatory synchrony between human extrastriate areas during visual short-term memory maintenance. J Neurosci 2001; 21:RC177:1-5.
56. Fisahn A, Pike FG, Buhl EH et al. Cholinergic induction of network oscillations at 40 Hz in the hippocampus in vitro. Nature 1998; 394:86-189.
57. Hormuzdi SG, Pais I, LeBeau FEN. Impaired electrical signaling disrupts gamma frequency oscillations in connexin 36-deficient mice. Neuron 2001; 31:487-495.
58. Towers SK, LeBeau FEN, Gloveli T et al. Fast network oscillations in the rat dentate gyrus in vitro. J Neurophysiol 2002; 87:1165-1168.
59. LeBeau FEN, Towers SK, Traub RD et al. Fast network oscillations induced by potassium transients in the rat hippocampus in vitro. J Physiol 2002; 542:167-179.
60. Traub RD, Bibbig A, Fisahn A et al. A model of gamma-frequency network oscillations induced in the rat CA3 region by carbachol in vitro. Eur J Neurosci 2000; 12:4093-4106.
61. Traub RD, Jefferys JGR, Whittington MA. Fast Oscillations in Cortical Circuits. Cambridge, MA: MIT Press, 1999.
62. Jones MS, Barth DS. Spatiotemporal organization of fast (>200 Hz) electrical oscillations in rat vibrissa/barrel cortex. J Neurophysiol 1999; 82:1599-1609.
63. Jones MS, MacDonald KD, Choi B et al. Intracellular correlates of fast (>200 Hz) electrical oscillations in rat somatosensory cortex. J Neurophysiol 2000; 84:1505-1518.
64. Bartos M, Vida I, Frotscher M et al. Rapid signaling at inhibitory synapses in a dentate gyrus interneuron network. J Neurosci 2001; 21:2687-2698.
65. Whittington MA, Traub RD, Jefferys JGR. Synchronized oscillations in interneuron networks driven by metabotropic glutamate receptor activation. Nature 1995; 373:612-615.
66. Traub RD, Whittington MA, Colling SB et al. Analysis of gamma rhythms in the rat hippocampus in vitro and in vivo. J Physiol 1996; 493:471-484.
67. Faulkner HJ, Traub RD, Whittington MA. Disruption of synchronous gamma oscillations in the rat hippocampal slice: a common mechanism of anaesthetic drug action. Br J Pharmacol 1998; 125:483-492.
68. Wang X-J, Buzsáki G. Gamma oscillation by synaptic inhibition in a hippocampal interneuronal network model. J Neurosci 1996; 16:6402-6413.
69. White JA, Chow CC, Ritt J et al. Synchronization and oscillatory dynamics in heterogeneous, mutually inhibited neurons. J Comput Neurosci 1998; 5:5-16.
70. Traub RD, Kopell N, Bibbig A et al. Gap junctions between interneuron dendrites can enhance long-range synchrony of gamma oscillations. J Neurosci 2001; 21:9478-9486.
71. Teubner B, Degen J, Sohl G et al. Functional expression of the murine connexin 36 gene coding for a neuron-specific gap junctional protein. J Membr Biol 2000; 176:249-262.
72. Rash JE, Staines WA, Yasumura T et al. Immunogold evidence that neuronal gap junctions in adult rat brain and spinal cord contain connexin-36 but not connexin-32 or connexin-43. Proc Natl Acad Sci USA 2000; 97:7573-7578.
73. Condorelli DF, Belluardo N, Trovato-Salinaro A et al. Expression of Cx36 in mammalian neurons. Brain Res Brain Res Rev 2000; 32:72-85.
74. Pais I, Hormuzdi SG, Monyer H et al. Sharp wave-like activity in the hippocampus in vitro in mice lacking the gap junction protein connexin 36. J Neurophysiol 2003; 89:2046-2054.
75. Lauri SE, Delany C, Clarke VR et al. Synaptic activation of a presynaptic kainate receptor facilitates AMPA receptor-mediated synaptic transmission at hippocampal mossy fibre synapses. Neuropharm. 2001; 41:907-915.
76. Traub RD, Draguhn A, Whittington MA et al. Axonal gap junctions between principal neurons: a novel source of network oscillations, and perhaps epileptogenesis. Rev Neurosci 2002; 13:1-30.

77. Moortgat KT, Bullock TH, Sejnowski TJ. Gap junction effects on precision and frequency of a model pacemaker network. J Neurophysiol 2000; 83:984-97.
78. Moortgat KT, Bullock TH, Sejnowski TJ. Precision of the pacemaker nucleus in a weakly electric fish: network versus cellular influences. J Neurophysiol 2000; 83:971-83.
79. Schmitz D, Frerking M, Nicoll RA. Synaptic activation of presynaptic kainate receptors on hippocampal mossy fiber synapses. Neuron 2000; 27:327-338.
80. Schmitz D, Mellor J, Frerking M et al. Presynaptic kainate receptors at hippocampal mossy fiber synapses. Proc Natl Acad Sci USA 2001; 98:11003-11008.
81. Semyanov A, Kullmann DM. Kainate receptor-dependent axonal depolarization and action potential initiation in interneurons. Nat Neurosci 2001; 4:718-723.
82. Ross FM, Gwyn P, Spanswick D et al. Carbenoxolone depresses spontaneous epileptiform activity in the CA1 region of rat hippocampal slices. Neuroscience 2000; 100:789-796.
83. Köhling R, Gladwell SJ, Bracci E et al. Prolonged epileptiform bursting induced by 0-Mg^{2+} in rat hippocampal slices depends on gap junctional coupling. Neuroscience 2001; 105:579-587.
84. Bragin A, Engel Jr J, Wilson CL et al. Hippocampal and entorhinal cortex high-frequency oscillations (100-500 Hz) in human epileptic brain and in kainic acid-treated rats with chronic seizures. Epilepsia 1999; 40:127-137.
85. Bragin A, Mody I, Wilson CL et al. A local generation of fast ripples in epileptic brain. J Neurosci 2002; 22:2012-2021.
86. Alarcon G, Binnie CD, Elwes RDC et al. Power spectrum and intracranial EEG patterns at seizure onset in partial epilepsy. Electroencephalogr Clin Neurophysiol 1995; 94:326-337.
87. Fisher RS, Webber WRS, Lesser RP et al. High-frequency EEG activity at the start of seizures. J Clin Neurophysiol 1992; 9:441-448.
88. Belluardo N, Trovato-Salinaro A, Mudo G et al. Structure, chromosomal localization, and brain expression of human Cx36 gene. J Neurosci Res 1999; 57:740-752.
89. Neubauer BA, Fiedler B, Himmelein B et al. Centrotemporal spikes in families with rolandic epilepsy: linkage to chromosome 15q14. Neurology 1998; 51:1608-1612.
90. Sander T, Schulz H, Vieira-Saeker AM et al. Evaluation of a putative major susceptibility locus for juvenile myoclonic epilepsy on chromosome 15q14. Am J Med Genet 1999; 88:182-187.
91. Bragin A, Engel Jr J, Wilson CL et al. High-frequency oscillations in the human brain. Hippocampus 1999; 9:137-142.
92. Nagy GS. Evaluation of carbenoxolone sodium in the treatment of duodenal ulcer. Gastroenterology 1978; 74:7-10.
93. Jellinck PH, Monder C, McEwen BS et al. Differential inhibition of 11 beta-hydroxysteroid dehydrogenase by carbenoxolone in rat brain regions and peripheral tissues. J Steroid Biochem Molec Biol 1993; 46:209-213.
94. Dobbins KR, Saul RF. Transient visual loss after licorice ingestion. J Neuroophthalmol 2000; 20:38-41.
95. Guldenagel M, Ammermuller J, Feigenspan A et al. Visual transmission deficits in mice with targeted disruption of the gap junction gene connexin36. J Neurosci 2001; 21:6036-6044.
96. Buckmaster PS, Dudek FE. In vivo intracellular analysis of granule cell axon reorganization in epileptic rats. J Neurophysiol 1999; 81:712-721.
97. Chang Q, Pereda A, Pinter MJ et al. Nerve injury induces gap junctional coupling among axotomized adult motor neurons. J Neurosci 2000; 20:674-684.
98. Traub RD, Pais I, Bibbig A et al. Contrasting roles of axonal (pyramidal cell) and dendritic (interneuron) electrical coupling in the generation of neuronal network oscillations. Proc Natl Acad Sci USA 2003; 100:1370-1374.

CHAPTER 10

Functional Role of Proinflammatory and Anti-Inflammatory Cytokines in Seizures

Annamaria Vezzani, Daniela Moneta, Cristina Richichi, Carlo Perego and Maria G. De Simoni

Abstract

Recent evidence has shown that proinflammatory and anti-inflammatory molecules are synthesized during epileptic activity in glial cells in CNS regions where seizures initiate and spread. These molecules are released and interact with specific receptors on neurons. Since various cytokines have been shown to affect neuronal excitability, this led to the hypothesis that they may have a role in altering synaptic transmission in epileptic conditions. Indeed, intracerebral application of IL-1β enhances epileptic activity in experimental models while its naturally occurring receptor antagonist (IL-1Ra) mediates anticonvulsant actions. Transgenic mice overexpressing IL-1Ra in astrocytes are less susceptible to seizures, indicating that endogenous IL-1 has proconvulsant activity. Several studies indicate a central role of IL-1β for the exacerbation of brain damage after ischemic, traumatic or excitotoxic insults, suggesting that it may also contribute to neuronal cell injury associated with seizures. Finally, a functional polymorphism in the IL-1β gene promoter, possibly associated with enhanced ability to produce this cytokine, has been specifically found in temporal lobe epilepsy patients with hippocampal sclerosis and in children with febrile seizures. Thus, the IL-1 system may represent a novel target for controlling seizure activity and/or the associated long-term sequelae. Furthermore, these studies suggest that other inflammatory and anti-inflammatory molecules produced in the CNS may have a role in the pathophysiology of seizure disorders.

Introduction

The concept of the brain as an immunologically isolated organ has been recently modified since various evidence has shown functional interactions between components of the nervous and immune systems.

In particular, many molecules and pathways associated with immune and inflammatory responses have been identified in the CNS, and are activated in various pathophysiological conditions. Among these molecules, pro- and anti-inflammatory cytokines are polypeptide hormones which mediate a functional cross-talk between neurons, glia and cells of the immune system (Fig. 1).

Cytokines have been implicated as mediators and modulators of various physiological CNS functions, including effects on the hypothalamo-pituitary-adrenal axis, fever responses and somnogenic effects.[1,2] Cytokines also affect various neurotransmitter systems,[3,4] and the expression of neuropeptides and neurotrophic factors in several forebrain areas.[5-7]

Cytokines have been suggested to play a role in diverse forms of neurodegeneration occurring in hypoxia/ischemia, brain trauma, apoptosis and excitotoxicity (for review see ref. 8) and in neuronal network excitability. In particular, electrophysiological findings have shown that

Recent Advances in Epilepsy Research, edited by Devin K. Binder and Helen E. Scharfman.

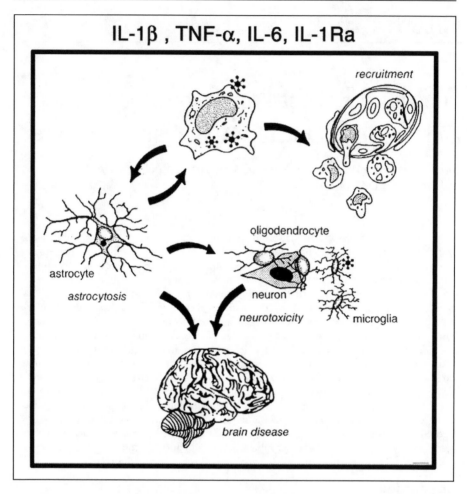

IL-1β , TNF-α, IL-6, IL-1Ra

Figure 1. Schematic drawing depicting resident glia, neurons and blood monocytes recruited from blood vessels. These cell types can synthesize and release proinflammatory cytokines (interleukin(IL)-1, tumor necrosis factor (TNF)-α, IL-6,IL-1 receptor antagonist (Ra)) in brain tissue following a variety of insults (e.g., ischemia, brain trauma, excitotoxicity, seizures).

relatively low concentrations of IL-1β, IL-6 and TNF-α inhibit long-term potentiation[9-11] and affect excitatory synaptic transmission[12-15] and ionic currents.[15-17] Most recently, biochemical studies in in vivo models of seizures indicate that various cytokines are induced in CNS by seizures.[17-22] Moreover, pharmacological evidence in rodents and in genetically-modified mice suggests that some cytokines have a significant role in modulating seizure activity.[23-25]

Cytokine Induction by Seizures

In vitro data have shown that several cell types, including microglia, astrocytes and neurons, produce and release cytokines under certain conditions.[2,26,27] Recruitment of blood monocytes may also contribute to the in vivo CNS production of cytokines in response to an inflammatory or neurotoxic insult.[28,29] Activated microglia have been suggested to represent an early inducible and major source of cytokines in epileptic tissue.[19,20,23,25,30] In particular, immunocytochemical analysis of cytokines (TNF-α, IL-1β, IL-6 and IL-1 receptor antagonist (Ra)) in

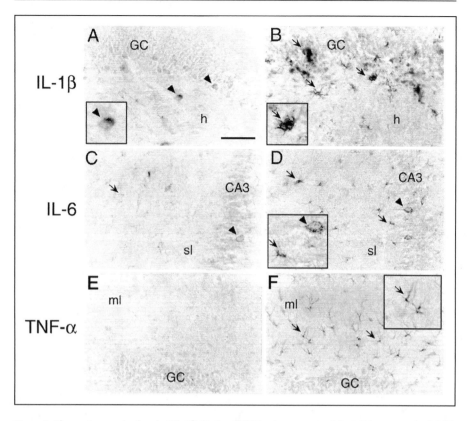

Figure 2. Photomicrographs showing IL-1β, IL-6 and TNF-α immunoreactivity in hippocampal subfields of control tissue (A,C,E) and after limbic seizures induced by electrical stimulation of the rat ventral hippocampus (B,D,F). Staining was enhanced by seizures in glia (arrows). Arrowheads depict immunopositive neurons. GC, granule cell layer; h, hilus; CA3, CA3 pyramidal cell layer; ml, molecular layer. Adapted from ref. 24.

rodent CNS has shown that constitutive expression of inflammatory cytokines in normal brain is low or barely detectable, whereas a marked increase both in microglia and astrocytes can be induced within 2 hours by an acute epileptic event (Fig. 2).[22-25] The increase in cytokines after seizures is transient. For example, after limbic status epilepticus induced by electrical stimulation of the ventral hippocampus, cytokines are rapidly synthesized (≤ 2h) and return to their basal level within 24 hours (Fig. 3).[24] Among the pro-inflammatory cytokines, only IL-1β was increased in the hippocampus of rats having spontaneous seizures 60 days after status epilepticus. Interestingly, immunocytochemistry showed IL-1β staining in hilar interneurons and CA3 cells in the dentate gyrus but not in glia where this cytokine was produced after the acute seizures were triggered.[24]

Cytokines are induced in forebrain regions where seizures originate and spread and their synthesis is higher when seizure activity is associated with nerve cell loss, as shown by comparing kainic acid- vs. bicuculline-induced seizures, the latter causing epileptic activity but not cell loss.[23]

The evidence that both mRNA and protein content increase in brain tissue indicates that seizures trigger de novo synthesis of cytokines. The mechanisms by which seizures induce the synthesis of cytokines are still unknown. Although specific effects triggered by seizures may play a role (e.g., protein extravasation from the blood-brain barrier[31] or stress-associated events

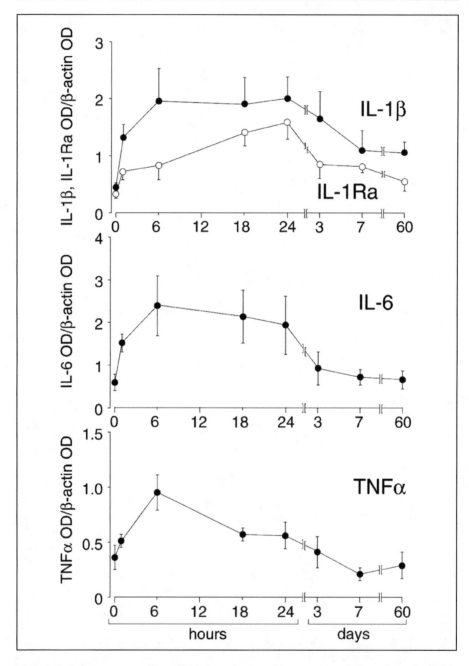

Figure 3. Proinflammatory cytokine mRNA expression in rat hippocampus at different times after limbic seizures induced by electrical stimulation of the ventral hippocampus. The level of mRNA is expressed as the ratio of densitometric measurement (OD) of the sample to the corresponding β-actin. Data are the mean±SE. Time 0 represents rats implanted with electrodes but not stimulated. Note the rapid and transient increase in cytokine mRNA. IL-1β transcript was still higher than control level 60 days after seizures (see text for details). Adapted from ref. 24.

priming glial cells), an attractive hypothesis is that glutamate released from neurons during seizures activates cytokine transcription in glial cells. This hypothesis is supported by the evidence that both ionotropic and metabotropic glutamate receptors exist in glia and that NMDA-receptor stimulation both in vitro and in vivo induces cytokine synthesis in glia.[32]

Pharmacological Effects: The Focus on IL-1

Two IL-1 receptors have been identified. The type I and type II receptors (IL-1RI and IL-1RII) are membrane proteins but only IL-1RI can induce a biological response (Fig. 4). Type II receptor functions as a decoy receptor to attenuate IL-1-induced responses. An important component of the IL-1 system is IL-1 receptor antagonist (Ra), a naturally occurring antagonist of IL-1β which binds IL-1RI with a lower affinity than the proinflammatory cytokine. Since only a few molecules of IL-1β are needed for cellular signaling, high molar excess (100- to 1000-fold) of IL-1Ra is produced during inflammatory reactions to counteract IL-1β effects. However, in contrast to septic shock or inflammatory diseases, seizures increase IL-1 and its receptor antagonist to a similar extent indicating that the brain is less effective in inducing a mechanism to rapidly terminate IL-1β actions than the periphery. This suggests that the balance between IL-1β and IL-1Ra during seizures may play a role in altering neuronal network excitability and/or the associated long-term events (e.g., epileptogenesis, neuronal cell loss).

Intracerebral injection of human recombinant IL-1β in rodents enhances seizures caused by intrahippocampal injection of kainic acid or bicuculline methiodide.[23,25] This action is receptor-mediated since it is blocked by co-administration of IL-1Ra and involves at some steps, not yet identified, an increase in glutamate neurotransmission. Thus, the proconvulsant activity of IL-1β is prevented by selective blockade of the NMDA-type of glutamate receptors (Table 1).[23]

It is noteworthy that IL-1Ra per se mediates anticonvulsant actions as shown by its ability to decrease seizures in rats injected intrahippocampally with kainic acid or electrically stimulated in the ventral hippocampus[23,24] and by reduced seizure susceptibility to bicuculline methiodide of transgenic mice overexpressing IL-1Ra in astrocytes (Table 2).[25] This supports the hypothesis that endogenous IL-1β has proepileptogenic effects.

Electrophysiological findings have shown both inhibitory[12,13] and facilitatory effects[14,15] of IL-1β on glutamatergic transmission. High ranges of IL-1 (1 nmol to 1 micromol) produce excitatory actions in brain, in accordance with evidence that the effects of IL-1β on neuronal activity (and viability) depends on its concentration.[3]

Table 1. *Effect of IL-1β and its receptor antagonist (IL-1Ra) on kainic acid-induced EEG seizures*

Drug	Dose	No Seizures	Time in Seizures (min)
Vehicle	-	15.0 ± 1.0	24.0 ± 2.5
IL-1β	10 µg	16.0 ± 1.0	52.3 ± 4.9**
IL-1Ra	1 µg	14.0 ± 2.0	22.0 ± 4.9
+IL-1β		15.0 ± 1.0	21.0 ± 3.0
IL-1Ra	6 µg	8.0 ± 2.0*	13.4 ± 2.6**
R-CPP	0.1 ng	16.0 ± 2.0	26.8 ± 3.3
+IL-1β		16.0 ± 3.0	27.0 ± 5.0

Data are the mean ± SE. Seizures were induced by intrahippocampal injection of 0.2 nmol kainic acid. Drugs were intrahippocampally applied 10 min before the convulsant. IL-1Ra, 6 µg, was injected intraventricularly 10 min before and 10 min after kainic acid. Seizure activity was recorded for 180 min. *p<0.05, **p<0.01 vs vehicle-injected rats. Adapted from refs. 23,25.

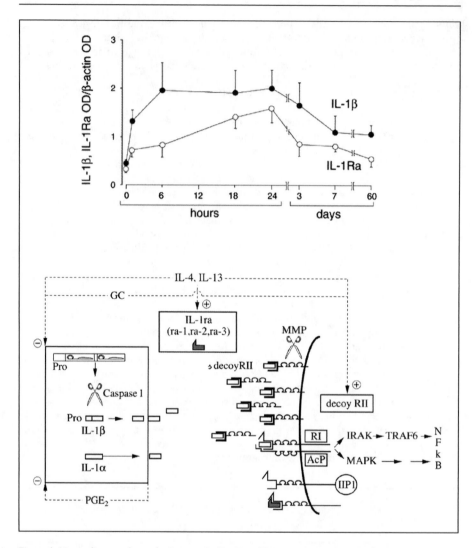

Figure 4. Upper diagram shows the increase in IL-1Ra mRNA as compared to IL-1β. IL-1Ra increase is delayed by several hours and returns to control levels faster that IL-1β. In addition, the molar ratio of IL-1Ra to IL-1 is approx. 1: 1 or lower. This indicates that the brain is ill equipped for rapidly terminating the actions of IL-1, differing from the periphery (see text for details). Lower diagram depicts the major components of the IL-1 system: mature IL-1 is cleaved from the prepropeptide by caspase 1 and acts on IL-1 type I receptors. IL-1 and receptor type I, in turn, couple to the accessory protein (AcP) to trigger the intracellular signal cascade leading to activation of NFkB. IL-1 R type II acts as a decoy receptor; thus, by binding the cytokine prevents its interaction with R type I. Both receptors exist in a membrane-bound and soluble (metalloproteinase-cleaved, MMP) form. IL-1Ra is produced by differential splicing of mRNA in three different forms: IL-1Ra is released extracellularly while type 2 and 3 remain intracellular. IL-1 levels are decreased by glucocorticoids (GC) or PGE2; IL-1Ra levels can be enhanced by GC or by anti-inflammatory cytokines (IL-4, IL-13). GC also increase the synthesis of R type II (decoy).

Other factors are important for determining the effects of this cytokine, e.g., the functional state of neurons (healthy or injured), the timing of cytokine release, the duration of tissue exposure, and the brain region involved.

Table 2. Transgenic mice overexpressing the human soluble form of IL-1Ra in astrocytes (GILRA2) have decreased susceptibility to seizures

	EEG Seizures		Behavioral Seizures			
	Onset	Duration	Onset		Duration	
			Clonic	Tonic-Clonic	Clonic	Tonic-Clonic
	(min)				(min)	
Wild-type	2.0 ± 0.8	8.4 ± 1.5	2.9± 0.4	6.9± 4.2	68.9± 4.2	2.9± 0.7
GIRLA2	7.7± 4.4	2.8 ± 1.6*	2.5± 0.5	44.7 ±16.0*	22.4 ± 7.9*	0.7 ± 0.3**

Data are the mean ± SE. Seizures were induced by intrahippocampal administration of 0.24 nmol bicuculline methiodide. Note that EEG and behavioral seizures were strongly inhibited in transgenic mice (*p<0.05, p<0.01 vs wild-type). GIRLA2 mice show an increase of about 15-fold in CSF IL-1Ra. Adapted from ref. 25.

IL-1, GABA and Glutamate

The interactions between IL-1β and these neurotransmitters are of particular relevance for epilepsy since it is well established that epileptic activity depends on the imbalance between inhibitory and excitatory neurotransmission. In particular, 0.6 to 6 nmol/ml of IL-1 enhances muscimol-induced Cl⁻ uptake in chick cortical synaptoneurosomes and potentiates GABA receptor-evoked currents in chick cortical neurons in culture.[16] In this study, systemic IL-1 raised the threshold to pentylenetetrazol seizures in mice. However, high concentrations of IL-1 (>60 nmol) have been shown to decrease the peak magnitude of currents elicited by GABA in hippocampal primary cultures and inhibitory postsynaptic potentials in CA3 neurons.[14,15]

In vivo microdialysis has shown that intracerebral injection of 0.6 pmol IL-1 enhances extracellular glutamate levels in the cortex[33] and 0.1 pmol/ml IL-1 inhibits astrocytic, high-affinity glutamate uptake by nitric oxide production.[34] However, sub-femtomolar concentrations of IL-1β increase adenosine release from hippocampal slices, which in turn inhibits glutamate release acting at A_1 receptors.[35] Thus, the concentration of IL-1β in the extracellular compartment appears to be a key factor for determining its functional actions in the brain.

TNF-α and IL-6

In addition to their effects on synaptic transmission (see Introduction), these cytokines appear to affect seizure susceptibility. Thus, mice specifically over-expressing IL-6 and TNF-α in glia develop neurological symptoms and random seizures.[36,37] Our recent evidence, however, suggests that TNF-α has anticonvulsant actions mediated by the p75 receptor.[38]

Proinflammatory Cytokines and Nerve Cell Injury

Several in vitro and in vivo studies indicate a central role of IL-1β in the exacerbation of brain damage after focal ischemia, brain trauma or excitotoxicity (for a review see ref. 8). The role of TNF-α is still controversial: protective effects are suggested by studies in TNF receptor knock-out mice,[39] since these mice showed increased neurodegeneration after intrahippocampal injection of kainic acid; however, transgenic mice overexpressing TNF-α or IL-6 in glia develop chronic inflammatory demyelination, reactive gliosis and neurodegeneration.[36,37]

The role of proinflammatory cytokines in seizure-induced cell damage is suggested by the evidence that glia produce them to a larger extent when seizures cause neuronal injury;[23] however, this aspect needs to be further investigated.

Human Epileptic Tissue

The expression of inflammatory cytokines has not been investigated in great detail in human epileptic tissue. Sheng et al[40] and Barres[41] have reported increased IL-1 and functional activation of microglia and astrocytes in brain tissue of humans affected by temporal lobe epilepsy. The mean concentration of IL-6 was significantly increased in CSF and plasma within 24 h from tonic-clonic seizures in humans, while no significant differences were found in IL-1β and TNF-α. IL-1Ra was increased as well by 2-fold.[42] In children with febrile seizures, the concentration of IL-1β in CSF was within the normal range.[43,44] However, in these studies CSF measurements were done on days 1-2 after seizures, which may be too late to detect cytokine changes. Thus, cytokine induction occurs within 2 h from seizure onset in experimental models and is transient. In addition, very low amounts of cytokines are enough to produce autocrine/paracrine effects and it is possible that CSF measurements do not detect these small changes.

Finally, a functional polymorphism in the IL-1β gene promoter, possibly associated with enhanced ability to produce this cytokine, has been specifically found in temporal lobe epilepsy patients with hippocampal sclerosis[45] and in children with febrile seizures.[46] Thus, a subtle anomaly in IL-1β to IL-1Ra ratio in the brain or developmental events during early life could set up a cascade of events leading to hippocampal sclerosis and epilepsy in individuals with this genetic predisposition.

Mechanism of Action

Cytokines are released from cells producing them: e.g., TNF-α is released from hippocampal slices at a larger extent after seizures.[47] Once released from glia, which are the major source of cytokines in the brain, they likely act upon their receptors on neurons (for review see ref. 8). Often these receptors show rapid up-regulation in response to injury, e.g., IL-1R type II was increased in the rat hippocampus and various forebrain areas after limbic seizures.[20]

In this scenario, while information about the functional role of TNF-α and IL-6 is still scarce, IL-1β will act on target neurons to enhance their excitability. This may occur directly by affecting ionic currents or indirectly by enhancing glutamatergic function and/or decreasing GABA actions (Fig. 5).[14,15,33] It is possible that IL-1β induces post-translational changes in NMDA or GABA receptor subunits.

Thus, IL-1R type I activation is associated with activation of signal transduction pathways (e.g., activation of protein kinase C, protein kinase A and nitric oxide production)[1] which are known to affect the responses of NMDA or GABA receptors to their endogenous ligands. IL-1Ra will afford protection from seizures by antagonizing the action of IL-1β on receptor type I. It is interesting to note that these receptors are densely concentrated on dendrites of granule neurons in the hippocampus.[20,48,49]

Thus, the balance between IL-1β and its receptor antagonist in brain may play a significant role in the maintenance and spread of seizures and this may be of pharmacological relevance.

IL-1R antagonist is highly inducible and appears to be a well-tolerated protein,[50] thus pharmacological means that increase IL-1Ra/IL-1β ratio may be useful for controlling seizure activity even when other factors act to increase brain excitability.

Acknowledgements

We gratefully acknowledge Prof K Iverfeldt (University of Stockholm, Sweden) and Prof T Bartfai (Scripps Res Inst, La Jolla, CA) for their invaluable contribution to part of these studies. Some of the work presented in this chapter has been supported by Telethon Onlus Foundation.

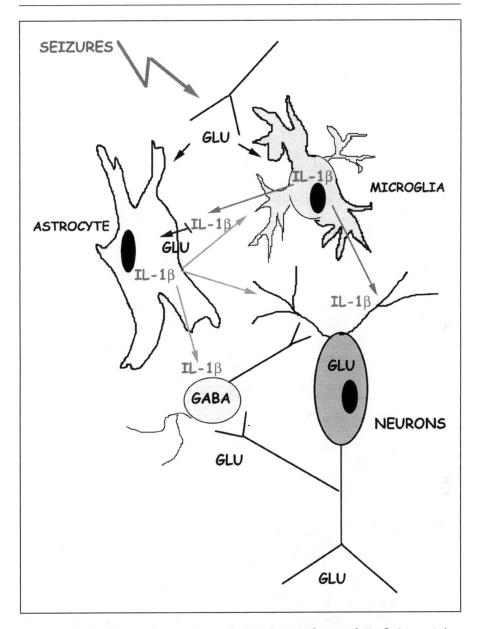

Figure 5. Schematic drawing depicting the proposed mechanism of action of IL-1β. Seizures induce glutamate release from excitatory synaptic terminals. Glutamate acting on ionotropic and/or metabotropic receptors on glia induces the transcription of the IL-1 gene and/or activates caspase 1. The mature cytokine is then released and acts on receptor (R) type I on neurons.[49] Electrophysiological evidence indicates that IL-1 inhibits GABA-mediated currents[15] and increases extracellular glutamate disposition (enhanced release and decreased reuptake[33,34]). These actions will mediate its proconvulsant activity in rodents. IL-1Ra antagonizes these effects by blocking IL-1 R type 1.

References

1. Schöbitz B, Ron De Kloet E, Holsboer F. Gene expression and function of interleukin-1, interleukin-6 and tumor necrosis factor in the brain. Prog Neurobiol 1994; 44:397-432.
2. Hopkins SJ, Rothwell NJ. Cytokines and the nervous system I. Expression and recognition. Trends Neurosci 1995; 18:83-88.
3. Rothwell NJ, Hopkins SJ. Cytokines and the nervous system II: Actions and mechanisms of action. Trends Neurosci 1995; 18:130-136.
4. De Simoni MG, Imeri L. Cytokine-neurotransmitter interactions in the brain. Biol Signals 1998; 7:33-44.
5. Scarborough DE, Lee SL, Dinarello CA et al. Interleukin-1β stimulates somatostatin biosynthesis in primary cultures of fetal rat brain. Endocrinology 1986; 124:549-551.
6. Spranger M, Lindholm D, Bandtlow C et al. Regulation of nerve growth factor (NGF) synthesis in the rat central nervous system: comparison between the effects of interleukin-1 and various growth factors in astrocyte cultures and in vivo. Eur J Neurosci 1990; 2:69-76.
7. Lapchak PA, Araujo DM. Hippocampal interleukin-2 regulates monoamine and opioid peptide release from the hippocampus. Neuroreport 1993; 4:303-306.
8. Allan SM, Rothwell NJ. Cytokines and acute neurodegeneration. Nature Rev Neurosci 2001; 2:734-744.
9. Katsuki H, Nakai S, Hirai Y et al. Interleukin-1β inhibits long-term potentiation in the CA3 region of the mouse hippocampal slice. Eur J Pharmacol 1990; 181:323-326.
10. Bellinger FP, Madamba S, Siggins GR. Interleukin 1β inhibits synaptic strength and long-term potentiation in the rat CA1 hippocampus. Brain Res 1993; 628:227-234.
11. Cunningham AJ, Murray CA, O'Neil LAJ et al. Interleukin 1β and tumor necrosis factor (TNF) inhibit long-term potentiation in the rat dentate gyrus in vitro. Neurosci Lett 1996; 203:17-20.
12. Coogan A, O'Connor JJ. Inhibition of NMDA receptor-mediated synaptic transmission in the rat dentate gyrus in vitro by IL-1β. Neuroreport 1997; 8:2107-2110.
13. D'Arcangelo G, Dodt H, Zieglgansberger W. Reduction of excitation by interleukin-1β in rat neocortical slices visualized using infrared darkfield videomicroscopy. Neuroreport 1997; 8:2079-2083.
14. Zeise ML, Espinoza J, Morales P et al. Interleukin-1β does not increase synaptic inhibition in hippocampal CA3 pyramidal and dentate gyrus granule cells of the rat in vitro. Brain Res 1997; 768:341-344.
15. Wang S, Cheng Q, Malik S. Interleukin-1β inhibits γ-amino butyric acid type A (GABA$_A$) receptor current in cultured hippocampal neurons. J Pharmacol Exp Ther 2000; 292:497-504.
16. Miller LG, Galpern WG, Lumpkin M et al. Interleukin-1 (IL-1) augments gamma-aminobutyric acid receptor function in the brain. Mol Pharmacol 1991; 39:105-108.
17. Plata-Salaman CR, French-Mullen JMH. Interleukin-1β depresses calcium current in CA1 hippocampal neurons at pathophysiological concentrations. Brain Res Bull 1992; 29:221-223.
18. Minami M, Kuraishi Y, Satoh M. Effects of kainic acid on messenger RNA levels of IL-1 beta, IL-6, TNF alpha and LIF in the rat brain. Biochem Biophys Res Commun 1991; 176:593-598.
19. Yabuuchi K, Minami M, Katsumata S et al. In situ hybridization study of interleukin-1β mRNA induced by kainic acid in the rat brain. Brain Res Mol Brain Res 1993; 20:153-161.
20. Nishiyori A, Minami M, Takami S et al. Type 2 interleukin-1 receptor mRNA is induced by kainic acid in the rat brain. Mol Brain Res 1997; 50:237-245.
21. Gahring LC, White SH, Skradski SL et al. Interleukin-1α in the brain is induced by audiogenic seizure. Neurobiol Dis 1997; 3:263-269.
22. Eriksson C, Winblad B, Schultzberg M. Immunohistochemical localization of interleukin-1beta, interleukin-1 receptor antagonist and interleukin-1beta converting enzyme/caspase-1 in the rat brain after peripheral administration of kainic acid. Neuroscience 1999; 93:915-930.
23. Vezzani A, Conti M, De Luigi A et al. Interleukin-1β immunoreactivity and microglia are enhanced in the rat hippocampus by focal kainate application: functional evidence for enhancement of electrographic seizures. J Neurosci 1999; 19:5054-5065.
24. De Simoni MG, Perego C, Ravizza T et al. Inflammatory cytokines and related genes are induced in the rat hippocampus by limbic status epilepticus. Eur J Neurosci 2000; 12:2623-2633.
25. Vezzani A, Moneta D, Conti M et al. Powerful anticonvulsant action of IL-1 receptor antagonist on intracerebral injection and astrocytic overexpression in mice. Proc Natl Acad Sci USA 2000; 97:11534-11539.
26. Benveniste EN. Inflammatory cytokines within the central nervous system: sources, function and mechanism of action. Am J Physiol 1992; 263:C1-C16.

27. Bartfai T, Shultzberg M. Cytokines in neuronal cell types. Neurochem Int 1993; 22:435-444.
28. Perry VH, Andersson PB, Gordon S. Macrophages and inflammation in the central nervous system. Trends Neurosci 1993; 16 (7):268-273.
29. Montero-Menei CN, Sindji L, Garcion E et al. Early events of the inflammatory reaction induced in rat brain by lipolysaccharide intracerebral injection: relative contribution of peripheral monocytes and activated microglia. Brain Res 1996; 724:55-66.
30. Zhao B, Schwartz JP. Involvement of cytokines in normal CNS development and neurological diseases: recent progress and perspective. J Neurosci Res 1998; 52:7-16.
31. Nitsch C, Goping G, Klatzo I. Pathophysiological aspects of blood brain barrier permeability in epileptic seizures. In: Schwarcz R, Ben-Ari Y, eds. Advances in Experimental Medicine and Biology. New York, London: 1986:175-184.
32. Jander S, Schroeter M, Stoll G. Role of NMDA receptor signaling in the regulation of inflammatory gene expression after focal brain ischemia. J Neuroimmunol 2000; 109:181-187.
33. Kamikawa H, Hori T, Nakane H et al. IL-1β increases norepinephrine level in the rat frontal cortex: involvement of prostanoids, NO, and glutamate: Am J Physiol 1998; 275:R806-R810.
34. Ye ZC, Sontheimer H. Cytokine modulation of glial glutamate uptake: a possible involvement of nitric oxide. Neuroreport 1996; 72:181-2185.
35. Luk WP, Zhang Y, White TD et al. Adenosine: a mediator of interleukin-1β induced hippocampal synaptic inhibition. J Neurosci 1999; 19:4238-4244.
36. Campbell IL, Abraham CR, Masliah E et al. Neurologic disease induced in transgenic mice by cerebral overexpression of interleukin 6. Proc Natl Acad Sci USA 1993; 90:10061-10065.
37. Akassoglou K, Probert L, Kontogeorgos G et al. Astrocyte-specific but not neuron specific transmembrane TNF triggers inflammation and degeneration in the central nervous system of transgenic mice. J Immunol 1997; 158:438-445.
38. Vezzani A, Moneta D, Richichi C et al. Functional role of TNF-α and IL-1β systems in seizure susceptibility and epileptogenesis. 31st Annual meeting Society for Neuroscience Nov 10-15 2001; 557:18.
39. Bruce AJ, Boling W, Kindy MS. Altered neuronal and microglial responses to excitotoxic and ischemic brain injury in mice lacking TNF receptors. Nature Med 1996; 2:788-794.
40. Sheng JG, Boop FA, Mrak RE et al. Increased neuronal β-amyloid precursor protein expression in human temporal lobe epilepsy: association with interleukin-1α immunoreactivity. J Neurochem 1994; 63:1872-1879.
41. Barres BA. New roles for microglia. J Neurosci 1991; 11:3685-3694.
42. Peltola J, Palmio J, Korhonen L. Interleukin-6 and interleukin-1 receptor antagonist in cerebrospinal fluid from patients with recent tonic-clonic seizures. Epilepsy Res 2000; 41:205-211.
43. Lahat E, Livine M, Barr J. Interleukin-1 levels in serum and cerebrospinal fluid of children with febrile seizures. Pediatr Neurol 1997; 17:34-36.
44. Ichiyama T, Nishikawa M, Yoshitomi T. Tumor necrosis factor-α, interleukin-1, and interleukin-6 in cerebrospinal fluid from children with prolonged seizures. Neurology 1998; 50:407-411.
45. Kanemoto K, Kawasaki J, Miyamoto T. Interleukin(IL)-1β, IL-1α, and IL-1 receptor antagonist gene polymorphisms in patients with temporal lobe epilepsy. Ann Neurol 2000; 47:571-574.
46. Hurme VM, Helminen M. Increased frequency of interleukin-1beta (-511) allele 2 in febrile seizures. Pediatr Neurol 2002; 26:192-195.
47. de Bock F, Dornand J, Rondouin G. Release of TNF in the rat hippocampus following epileptic seizures and excitotoxic neuronal damage. Neuroreport 1996; 7:1125-1129.
48. Takao T, Tracey DE, Mitchell WM. Interleukin-1 receptors in mouse brain: Characterization and neuronal localization. Endocrinology 1990; 127:3070-3078.
49. Ban E, Milon G, Prudhomme N et al. Receptors for interleukin-1 (alpha and beta) in mouse brain: Mapping and neuronal localization in hippocampus. Neuroscience 1991; 43:21-30.
50. Dinarello CA. Biological basis for interleukin-1 in disease. Blood 1996; 87:2095-2147.

CHAPTER 11

Using the Immune System to Target Epilepsy

Deborah Young and Matthew J. During

Introduction

The sudden and transient disruption from normal brain function by the disordered, synchronous and rhythmic firing of populations of neurons or seizures is the common feature of a diverse collection of disease syndromes collectively called the epilepsies. The epilepsies are estimated to affect 1-2% of the population, with the most common form being temporal lobe epilepsy, which accounts for 40% of all types. The underlying cause of this form of epilepsy is thought to involve an imbalance in excitatory and inhibitory neurotransmission due to altered circuitry in the hippocampus resulting from the selective degeneration of hilar, CA1 and CA3 pyramidal neurons but relative preservation of dentate granule neurons.[39] Hyperexcitability in the hippocampus then occurs following synaptic reorganization of the surviving dentate granule cell mossy fiber axons into the inner molecular layer of the dentate gyrus. While surgical resection is useful in alleviating seizures, particularly for intractable temporal lobe epilepsy,[38,24] pharmacotherapy is the mainstay of treatment for most forms. At the cellular level, anti-epileptic agents mediate their effects through three major pharmacological modes of actions: enhancement of inhibitory neurotransmission, attenuation of excitatory transmission or modulation of voltage-dependent ion channels. While the use of one or a combination of antiepileptic drugs can provide seizure control in most cases, there are also many cases refractory to treatment. Additionally, many of these compounds have been reported to negatively impact cognitive function,[11] suggesting the continuing requirement for better and more specific treatments.

Using the Immune System to Treat Neurological Disease— The Relationship between the Brain and the Immune System

While researchers have attempted to harness the power of the immune system for treating a diverse range of diseases including cancer, using the immune system to treat neurological disease in general has never been seriously considered a viable option. This is because the humoral effectors of the immune system, the immunoglobulins, are large proteins and have limited access into the brain due to the presence of the blood-brain barrier (BBB) which impedes their entry and thus subsequent binding to target antigens in the CNS. However, there is limited evidence that under certain conditions, antibodies can gain access into the CNS. One of the most compelling and intriguing lines of evidence are the autoimmune diseases that involve the CNS. Rasmussen's encephalitis (RE) is a rare form of pediatric epilepsy characterized by severe epileptic seizures that are relatively resistant to treatment with typical anti-seizure medication and inflammatory neuropathology restricted to one cerebral hemisphere.[50] Perivascular lymphocytic cuffs and scattered microglial nodules in the cortex, neuronal degeneration, and gliosis are key histopathological features of the affected hemisphere. The link between RE and autoimmune disease was established following the serendipitous discovery that a small subset of rabbits immunized with a recombinant GluR3 glutamate receptor fusion protein not only

Recent Advances in Epilepsy Research, edited by Devin K. Binder and Helen E. Scharfman. ©2004 Eurekah.com and Kluwer Academic / Plenum Publishers.

developed high-titer anti-GluR3 antibodies in the serum, but also a phenotype characterized by seizures and inflammatory neuropathology strikingly similar to that found in RE.[51] IgG subsequently isolated from the sera from 'epileptic' GluR3-immunized rabbits was found to have the ability to bind and evoke excitatory currents and kill mouse cortical neurons in culture. These cytotoxic effects could be reversed by the selective competititve antagonist CNQX, suggesting anti-GluR3 antibodies promote seizures and excitotoxic cell death by acting as an agonist at AMPA receptors.[56] Anti-GluR3 autoantibodies were subsequently demonstrated in human RE sera and shown to bind with moderate affinity to GluR3 overexpressed in HEK293 cells.[51] Removal of serum antibodies by plasmapheresis in RE patients transiently reduces the seizure frequency and improves neurologic function as serum concentrations of GluR3 antibodies are reduced, further corroborating humoral autoimmune processes in the pathogenesis of RE.[2,51] Similar improvements in neurologic function were found with the selective removal of IgG and IgM from the plasma of a RE patient by protein A affinity chromatography, with direct correlation between reduced GluR3 antibody titers and clinical benefit.[3] A more recent and independent study suggests that GluR3 autoantibodies can also destroy cells by a complement-dependent mechanism.[26] Rabbits immunized with a glutathione S-transferase-fused GluR3 protein developed anti-GluR3 antibodies that had no excitatory properties when applied to cortical neurons in whole cell patch-clamp studies. However, robust cytotoxic effects were found which were identified as being complement- dependent.[26] In corroboration of this finding, IgG and complement factors of the membrane attack complex were found on neurons and their processes in the cortex of brains of RE patients.[58] This has led to the proposal that these autoantibodies may therefore gain access into the CNS to trigger complement-mediated neuronal damage, events that may be important in the initial phases of the disease. Similar neuropathological findings have been reproduced in murine models of RE.[31,32]

While GluR3 may be the prototypic autoantigen, autoimmune attack on other brain antigens, some identified[61] and others as yet to be identified, may also be involved in some cases of RE. Autoantibodies to Munc-18, an intracellular protein found in presynaptic terminals with a role in synaptic vesicle release, has been reported in a RE patient also displaying serum anti-GluR3 autoantibodies.[61] Thus, Yang et al, propose that additive or synergistic effects of autoimmune attack on two different epitopes at two distinct sites of synaptic transmission could impair function, leading to seizures and neuronal death. The ratio of the autoantibodies in the cerebrospinal fluid and the serum relative to the total IgG concentration in the two compartments supports the passage of Munc-18 autoantibodies across the BBB.[61] However, the significance of anti-Munc-18 antibodies in RE pathogenesis is debatable as it has not been found in 14 other patients, suggesting it may be ancillary to the disease. Similarly, recent reports suggest prevalence of anti-GluR3 autoantibodies in other forms of epilepsy, and thus whether anti-GluR3 antibodies are specific for the diagnosis of RE is unclear.[59]

RE is but one of several autoimmune diseases involving the CNS that are associated with serum antibodies to brain antigens which appear to gain access into the brain and mediate the generation of complex neurological disease phenotypes.[57] Paraneoplastic disorders are autoimmune disorders that are a remote complication of some forms of cancer. These develop when the ectopic expression of brain antigens by the tumor leads to a T-cell mediated-immune response and/or autoantibodies that gain access to the brain to target neurons expressing those antigens and mediate a disease phenotype.[1,14] Stiff-man syndrome is characterized by progressive skeletal muscle rigidity, deficiency in GABA and circulating autoantibodies to GAD65, the key enzyme involved in GABA synthesis.[17,33,55] Gephyrin, a cytosolic protein required for clustering of glycine and $GABA_A$ receptors at the postsynaptic density of inhibitory synapses, has also been identified as an autoantigen for Stiff-man.[12] Rare cases of temporal lobe epilepsy and other forms of epilepsy are associated with GAD65 autoantibodies[25,43] and it has also been suggested that a subgroup of drug-refractory temporal lobe epilepsy patients may be associated with GAD65 autoantibodies, although further investigation is warranted.[30,46] Additionally, autoantibodies to other brain antigens have also been linked to other rare forms of epilepsy.[44]

Although many of these conditions are relatively rare, importantly these findings suggest that under certain circumstances antibodies do have the ability to pass the BBB, bind and modify the function of their target antigen. As antibodies to GluR3 are found in serum in the large majority of immunized animals without apparent disease,[51] it is clear that secondary factors are necessary for initiating the disease process. The identity of these factors has remained elusive, although it has been proposed that a focal lesion leading to disruption of the BBB due to trauma may be essential for immune-mediated attack. Indeed, trauma-induced breakdown of the BBB has been demonstrated in experimental epilepsy and excitotoxic lesions.[10,48,52]

Using the Immune System to Treat Neurological Disease— Therapeutic Agents for Neurological Disease

While these examples argue against exploiting the use of the immune system to treat neurological disease, there is good evidence that antibodies can potentially be used as therapeutic agents for altering the function of brain proteins. Monoclonal antibodies have been applied in experimental paradigms to disrupt signal transduction cascades mediated by neurotrophic factor receptors.[13] Following direct injection into the brain and binding to the extracellular domains of the receptor, the subsequent blockade of the TrkA signalling pathway has allowed researchers to elucidate the role of nerve growth factor in maintenance of long term potentiation in rat visual cortex.[47] More recently, the Alzheimer's disease (AD) vaccines have sparked a great deal of interest. One of the defining characteristics of AD pathology is the accumulation of amyloid β-peptide and its deposition as plaques within cortical and hippocampal brain regions. Immunization with amyloid β-peptide is associated with generation of high-titer amyloid β-peptide antibodies and the dramatic reduction of amyloid plaque burden and cognitive deficits in transgenic mouse models of AD.[42,53] While several mechanisms of action of the response to immunization have been proposed, the reduction in amyloid burden following passive transfer of amyloid β-peptide antibodies indicates that antibodies can cross the BBB directly to ameliorate the disease phenotype in an AD mouse model.[4] Importantly, this evidence supports the contention that the immune-privileged state of the brain as being relative, not absolute. Although anecdotal evidence has suggested the possibility of cerebritis associated with the early stage clinical trials of a β-amyloid vaccine,[8] some researchers believe that this might reflect the disease process, since AD is associated with a significant microglial component, cytokine release and inflammation, complicating the analysis of adverse events associated with the vaccine clinical trials.

Developing a Vaccine Strategy for Epilepsy

The NMDA receptor plays a key role in mediating excitotoxic-induced neuronal cell death in addition to other activity-dependent functions such as learning and memory.[9] Structurally, the NMDA receptor is a hetero-oligomeric complex composed of subunits from the NR1, NR2A-D and NR3A subunit families arranged to form a ligand-gated, voltage-dependent ion channel principally permeable to Ca^{2+} ions.[27] Although widely expressed throughout the brain, functional diversity of NMDA receptors in vivo is effected by incorporation of different NR2 and NR3 subunits with the essential NR1 subunit. The complete repertoire of subunit combinations and heterogeneity of NMDA receptors throughout the CNS is still unclear. The NR1 subunit is necessary for formation of a functional ion channel,[22] while the NR2 and NR3 subunits modulate channel properties.[15,54] Both the NR1 and NR2 proteins are transmembrane spanning proteins with an extracellular N-terminus and intracellular C-terminus and three complete transmembrane regions. An intramembrane loop forms between the first and second complete transmembrane regions. The glycine binding site is found on the NR1 subunit while the glutamate binding site is found on the NR2 subunit.[16] At resting membrane potential, the ion channel is blocked and activation of the receptor is dependent on binding of the coligands glutamate and glycine and simultaneous neuronal depolarization which relieves the Mg^{2+}-dependent block, allowing intracellular passage of Ca^{2+} ions.[16]

We have recently described a novel approach for epilepsy and stroke treatment involving manipulation of the immune system to generate a humoral response against a target brain antigen.[21] In an experimental situation analogous to the autoimmune diseases described above and in particular GluR3 immunization,[51] we hypothesized that ectopic expression of a specific brain protein that has been sequestered behind the BBB may lead to generation of autoantibodies specific to the target. These circulating autoantibodies would have limited access into the brain under normal basal conditions. However, following an insult and associated BBB compromise, events that occur following kainate-induced seizures[7,63] and excessive glutamate release,[10,37] increased passage of these autoantibodies would occur. These antibodies, instead of having excitotoxic or disease-inducing activity, would have the ability to bind and alter the function of the target protein to block processes involved in mediating insult-induced neuronal injury.[21]

Because the NR1 subunit is the common element in all NMDA receptors, we targeted the NR1 subunit in our vaccination strategy.[22] Thus, we proposed that vaccination against the NR1 subunit would generate serum autoantibodies to NR1 that would have restricted brain access under normal conditions. However, following an insult and associated BBB breakdown, increased passage of NR1 autoantibodies, which could bind and neutralize NMDA receptor function and thus block injury-induced neuronal death, would occur.

A Novel Genetic Vaccine for Stroke and Epilepsy

We utilized a novel genetic vaccination strategy based on the use of recombinant adeno-associated viral vectors (AAV). This idea stemmed from our previous work showing robust gene transfer and expression of β-galactosidase protein in gut lamina propria cells following oral administration of an AAV vector expressing a lacZ reporter gene.[20] Many of the transduced β-galactosidase-expressing cells were antigen-presenting cells involved in generating humoral immune responses, indicating their potential for use as vaccines. AAV vectors are derived from a nonpathogenic virus and genetically modified to undergo a single round of host infection to efficiently deliver a gene-of-interest to mammalian cells. Several desirable features including a broad host cell range and the ability to transduce both nondividing and dividing cells leading to high levels of stable and sustained gene expression, make AAV vectors arguably the vector of choice for human gene therapy.[41,45]

To test our vaccination hypothesis, a high-titer AAV vector expressing a full-length mouse NR1 cDNA under control of a CMV promoter was orally administered to fasted rats while control animals received an AAV vector expressing the lacZ reporter gene or the saline vehicle (Fig. 1). Robust NR1 or lacZ gene transfer and protein expression was found in gut lamina propria cells subsequently confirmed as antigen-presenting dendritic cells and macrophages by double-labelling immunohistochemistry using the corresponding cell markers (Fig. 1). PCR analysis confirmed that transfer of the NR1 or lacZ genes following oral immunization was confined to the gut, with no spread to the spleen, liver or testes. NR1 protein expression was associated with NR1 IgG autoantibodies in the serum by as detected by Western blot analysis, which increased in titer over the 1 to 5 month period. Similarly, β-galactosidase autoantibodies were found in the corresponding AAVlacZ-immunized animals. To further define the polyclonal response and the epitope regions to which the autoantibodies were generated against, a series of overlapping 16-mer peptides representing the entire 938 amino acids of the NR1 subunit was used to screen the NR1 serums. The most prevalent peptides that the NR1 autoantibodies bound to were those that corresponded to functional domains within the extracellular vestibule of the receptor. More specifically, NR1 autoantibodies were generated against epitopes that mapped to regions in the extracellular domains that are critical for glycine binding.[5] These results suggested that targeted immunization against a brain antigen could result in generation of a polyclonal antibody response to the target protein.

Vaccinated rats were screened for neuroprotective efficacy against the kainate model of experimental epilepsy. Peripheral administration of kainate to rodents leads to a stereotypical pro-

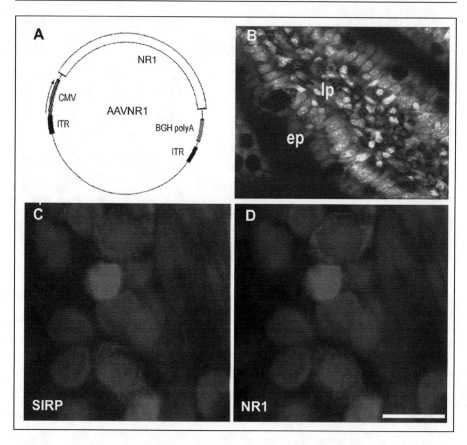

Figure 1. The AAVNR1 vector construct (A) was a full-length mouse NR1 cDNA under the control of a CMV immediate-early promoter and bovine growth hormone (BGH) polyadenylation site flanked by the AAV inverted terminal repeats (ITR). B. NR1 protein expression (blue) in the lamina propria (lp) as assessed by immunohistochemistry with propidium iodide counterstaining (orange) from a section from the proximal intestine at 4 weeks following NR1 immunization. The epithelial cell layer (ep) is also indicated. C. SIRP, a dendritic and macrophage cell marker (red) and (D) NR1 (blue) double immunofluorescent labeling with acridine orange counterstaining (green) in the gut lamina propria, showed NR1 protein expression in antigen-presenting cells. (reproduced from During et al, 2000; Science 287:1453-1460).

gression through various seizure behaviors culminating in prolonged seizures or status epilepticus.[6] We administered kainate at a dose of 10 mg/kg i.p. and monitored the appearance of electrographic seizures and found there was an overall reduction in the expected frequency of developing kainate-induced seizures from 68% in the control animals to 22% in the NR1-immunized animals. The anti-epileptic and neuroprotective efficacy of NR1 immunization was verified by TUNEL labelling and clusterin immunohistochemistry, a cell death marker[18] (Fig. 2). Of interest, one of the two NR1-immunized rats that had a robust NR1 autoantibody titer and developed status epilepticus did not show any TUNEL or clusterin labelling. The neuroprotective efficacy was further explored in the endothelin-1 model of ischemic stroke, and NR1-immunized animals were more protected against stroke-induced damage when compared to controls[21] (Fig. 3). Antibody passage across an intact BBB was confirmed by sampling cerebrospinal fluid from the cisterna magna from naïve and both lacZ- and NR1-immunized groups of animals, with only NR1 autoantibodies detectable in cerebrospinal fluid obtained from NR1 animals.

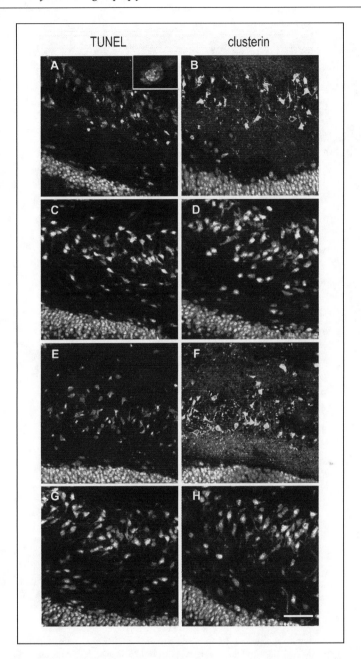

Figure 2. Hippocampal damage in the hilus assessed by fluorescent TUNEL (green) or clusterin (green) immunohistochemistry combined with immunohistochemistry to the neuronal marker NeuN (orange). After 45 min of status epilepticus, control lacZ animals (A,B) had extensive damage as indicated by these cell death markers. No injury was found in any animal not developing status epilepticus (C,D). Only 2 out of 9 NR1-immunized animals developed SE. One NR1-immunized rat showed extensive damage (E,F). However, subsequent serum analysis confirmed this animal had not developed an NR1 autoantibody titer. The remaining animal which developed a robust NR1 autoantibody titer showed no signs of injury despite having 45 min of status epilepticus (G,H). (reproduced from During et al, 2000; Science 287:1453-1460).

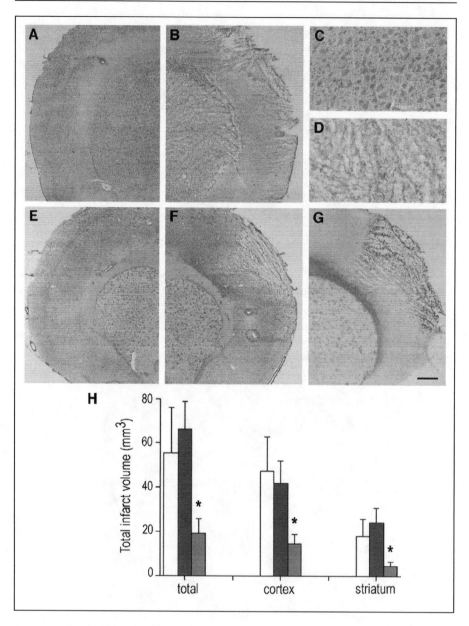

Figure 3. Reduced ischemic damage in the cortex and striatum of NR1-vaccinated animals after middle cerebral artery occlusion. H&E staining of the striatum and/or cortical regions contralateral (A,C,E) and ipsilateral (B,D,F) to the occlusion in lac (A-D) and NR1-vaccinated animals (E,F). Isolectin B4 staining (G) detected microglial infiltration in the ipsilateral hemisphere. H. Total infarct volume (mean ±SEM) in naïve (white bars), lacZ- (blue bars) and NR1-vaccinated groups (red bars). *P<0.01. (reproduced from During et al, 2000; Science 287:1453-1460).

As our results showed that NR1-immunization is able to manipulate NMDA receptor function, the effects on other NMDA receptor-mediated brain functions need also be considered. With these receptors playing major roles in mediating synaptic plasticity associated with learning and memory, previous reports have established that systemic administration of pharmacological NMDA receptor antagonists at neuroprotective doses to experimental animals is generally associated with impaired mobility and deficits in learning and memory function.[60,62] For this reason, we have conducted a preliminary assessment of the effects of NR1 immunization on mobility and performance of NR1-vaccinated and control rats in tests of cognitive function. Mobility as assessed by a line crossing and circular track test was similar for both NR1 and control animals.[21] Surprisingly, when these animals were challenged with more complex behavioral tasks that test learning and memory function, significant differences in performance in these tasks were observed between the NR1 and controls. NR1-immunized animals showed improved performance in the Barnes circular maze, a hippocampal-dependent spatial navigation task that requires rats to locate a dark escape tunnel hidden beneath one of 18 holes positioned around the perimeter of a brightly lit table using visual cues. In comparison to controls, NR1-animals showed reduced latency and required fewer trials to find the escape tunnel.[19] Similarly, NR1-immunized rats also exhibited a stronger freezing response than lacZ-immunized rats in a fear conditioning behavioral paradigm, a contextual association task.[28]

The results from our study suggest the neuroprotective effects found in the stroke and epilepsy models following NR1 vaccination are mediated by NR1 autoantibodies that have the ability to cross the BBB to bind and modulate NMDA receptor function. There is some evidence that suggests the NR1 vaccination approach has some clinical relevance. An analysis of serum from 270 stroke patients in Russia showed that presence of significant NMDA receptor autoantibody titers was associated with improved neurological recovery, while the absence of autoantibodies was associated with poor prognosis and death.[25] Increased penetration of NR1 autoantibodies was found after kainate administration, suggesting autoantibodies may directly mediate the neuroprotective effects at the time of insult. However, despite this observation, there is also the possibility that the neuroprotective and anti-epileptic effects observed with the vaccine may not be due to a direct effect of autoantibody antagonism at NMDA receptors at the time of insult but result from downstream modulation of neuroprotective genes. This is undergoing further investigation.

Concluding Remarks

One of the inherent problems with epilepsy and stroke is that the cascade of molecular and biochemical mechanisms mediating insult-induced neuronal death are switched on rapidly, leaving only a short therapeutic window for intervention against these events. The prophylactic nature of an epilepsy vaccine is a novel feature of this approach, providing the advantage that treatment could be targeted to patients considered at high risk for developing stroke or chronically epileptic patients that may be pharmacoresistant.

However, while these initial results look extremely promising, more research is required to determine the efficacy and safety of moving such a treatment approach into humans. The persistence of NR1 autoantibodies for many months and perhaps even years may have long-term detrimental effects on brain function. Further work is also required to determine the specificity of the effect and whether NR1 immunization modulates glutamatergic function at other glutamate receptor subtypes, particularly in light of the effects of GluR3 immunization.[51] Additional concerns relate to the cerebritis in the β-amyloid vaccinated Alzheimer's disease patients[8] suggesting that caution is needed in moving towards clinical trials. Furthermore, comparisons between an AAV approach and traditional protein vaccination strategies are also required to determine whether similar neuroprotective responses would occur or whether this is a unique feature of an AAV-based vaccine. Other considerations include determining the effect of the vaccination in animals with preexisting epileptic conditions modeling that found in human patients, particularly as altered NMDA receptor subunit composition in epileptic patients has been reported.[34,35] The enthusiasm for trials of NMDA receptor antagonists for epilepsy and stroke treatment has diminished because they have performed relatively poorly in clinical trials in comparison to the dramatic neuroprotective efficacy displayed in preclinical studies and they have been associated with side-effects that have limited their clinical utility. Several of the latest generation NMDA receptor antagonists have been reported to have fewer side-effects.[29] The NR1 vaccination is not associated with any obvious detrimental behavioral abnormalities, but in fact may improve cognitive performance, reminiscent to that found with partial glycine agonists.[36,40,49] One of the key issues for future work will be determining the precise site of action of these autoantibodies.

While we may be a long way from translating such technology to the clinic, it is clear that these approaches, if successful, could pave the way for alternative methods for epilepsy treatment. Active immunization against brain proteins and generation of autoantibodies that are capable of entering the brain and modulating function of target brain proteins may open new avenues for treatment of neurological disease.

References

1. Albert ML, Austin LM, Darnell RB. Detection and treatment of activated T cells in the cerebrospinal fluid of patients with paraneoplastic cerebellar degeneration. Ann Neurol 2000; 47:9-17.
2. Andrews PI, Dichter MA, Berkovic SF et al. Plasmapheresis in Rasmussen's encephalitis. Neurology 1996; 46:242-246.
3. Antozzi C, Granata T, Aurisano N et al. Long-term selective IgG immuno-adsorption improves Rasmussen's encephalitis. Neurology 1998; 51:302-305.
4. Bard F, Cannon C, Barbour R et al. Peripherally administered antibodies against amyloid beta-peptide enter the central nervous system and reduce pathology in a mouse model of Alzheimer disease. Nat Med 2000; 6:916-919.
5. Beck C, Wollmuth LP, Seeburg PH et al. NMDAR channel segments forming the extracellular vestibule inferred from the accessibility of substituted cysteines. Neuron 1999; 22:559-570.
6. Ben-Ari Y, Cossart R. Kainate, a double agent that generates seizures: two decades of progress. Trends Neurosci 2000; 23:580-587.
7. Bennett SA, Stevenson B, Staines WA et al. Periodic acid-Schiff (PAS)-positive deposits in brain following kainic acid-induced seizures: relationships to fos induction, neuronal necrosis, reactive gliosis and blood-brain barrier breakdown. Acta Neuropathol (Berl) 1995; 89:126-138.
8. Birmingham K, Frantz S. Set back to Alzheimer vaccine studies. Nature Med 2002; 8:199-200.
9. Bliss T, Collingridge GL. A synaptic model of memory: long-term potentiation in the hippocampus. Nature 1993; 361:31-39.
10. Bolton SJ, Perry VH. Differential blood-brain barrier breakdown and leucocyte recruitment following excitotoxic lesions in juvenile and adult rats. Exp Neurol 1998; 154:231-240.
11. Brunbech L, Sabers A. Effect of antiepileptic drugs on cognitive function in individuals with epilepsy: A comparative review of newer versus older agents. Drugs 2002; 62:593-604.
12. Butler MH, Hayashi A, Ohkoshi N et al. Autoimmunity to gephyrin in Stiff-man syndrome. Neuron 2000; 26:307-312.
13. Cattaneo A, Capsoni S, Margotti E et al. Functional blockade of tyrosine kinase A in the rat basal forebrain by a novel antagonistic anti-receptor monoclonal antibody. J Neurosci 1999; 19:9687-9697.
14. Darnell RB. Onconeural antigens and the paraneoplastic neurologic disorders: at the intersection of cancer, immunity and the brain. Proc Natl Acad Sci USA 1996; 93:4529-4536.
15. Das S, Sasaki YF, Rothe T et al. Increased NMDA current and spine density in mice lacking the NMDA receptor subunit NR3A. Nature 1998; 393:377-381.
16. Dingledine R, Borges K, Bowie D et al. The glutamate receptor ion channels. Pharmacol Rev 1999; 51:7-61.
17. Dinkel K, Meinck HM, Jury KM et al. Inhibition of gamma-aminobutyric acid synthesis by glutamic acid decarboxylase autoantibodies in stiff-man syndrome: acute behavioral effects of MK-801 in the mouse. Ann Neurol 1998; 44:194-201.
18. Dragunow M, Preston K, Dodd J et al. Clusterin accumulates in dying neurons following status epilepticus. Brain Res Mol Brain Res 1995; 32:279-290.
19. During MJ, Jiao XY, Lawlor PA et al. NRI vaccine increases NMDA receptor expression, neurogenesis and memory. Nature (submitted).
20. During MJ, Xu R, Young D et al. Peroral gene therapy of lactose intolerance using an adeno-associated virus vector. Nat Med 1998; 4:1131-1135.

21. During MJ, Symes CW, Lawlor PA et al. An oral vaccine against NMDAR1 with efficacy in experimental stroke and epilepsy. Science 2000; 287:1453-1460.
22. Forrest D, Yuzaki M, Soares HD et al. Targeted disruption of NMDA receptor 1 gene abolishes NMDA response and results in neonatal death. Neuron 1994; 13:325-338.
23. Giometto B, Nicolao P, Macucci M et al. Temporal-lobe epilepsy associated with glutamic-acid-decarboxylase autoantibodies. Lancet 1998; 352:457.
24. Gleissner U, Helmstaedter C, Schramm J et al. Memory outcome after selective amygdalo-hippocampectomy: a study in 140 patients with temporal lobe epilepsy. Epilepsia 2002; 43:87-95.
25. Gusev EI, Skvortsova VI, Izykenova GA et al. The level of autoantibodies to glutamate receptors in the blood serum of patients in the acute period of ischemic stroke. Zh Nevropatol Psikhiatr Im S S Korsakova 1996; 96:68-72.
26. He XP, Patel M, Whitney KD et al. Glutamate receptor GluR3 antibodies and death of cortical cells. Neuron 1998; 20:153-163.
27. Hollmann M, Heinemann S. Cloned glutamate receptors. Annu Rev Neurosci 1994; 17:31-108.
28. Kiyama Y, Manabe T, Sakimura K et al. Increased thresholds for long-term potentiation and contextual learning in mice lacking the NMDA-type glutamate receptor epsilon1 subunit. J Neurosci 1998; 18:6704-6712.
29. Kohl BK, Dannhardt G. The NMDA receptor complex: a promising target for novel antiepileptic strategies. Curr Med Chem 2001; 8:1275-1289.
30. Kwan P, Sills GJ, Kelly K et al. Glutamic acid decarboxylase autoantibodies in controlled and uncontrolled epilepsy: a pilot study. Epilepsy Res 2000; 42:191-195.
31. Levite M, Hermelin A. Autoimmunity to the glutamate receptor in mice—a model for Rasmussen's encephalitis? J Autoimmun 1999; 13:73-82.
32. Levite M, Fleidervish IA, Schwarz A et al. Autoantibodies to the glutamate receptor kill neurons via activation of the receptor ion channel. J Autoimmun 1999; 13:61-72.
33. Lohmann T, Hawa M, Leslie RD et al. Immune reactivity to glutamic acid decarboxylase 65 in stiffman syndrome and type 1 diabetes mellitus. Lancet 2000; 356:31-35.
34. Mathern GW, Pretorius JK, Mendoza D et al. Increased hippocampal AMPA and NMDA receptor subunit immunoreactivity in temporal lobe epilepsy patients. J Neuropathol Exp Neurol 1998; 57:615-634.
35. Mathern GW, Pretorius JK, Mendoza D et al. Hippocampal N-methyl-D-aspartate receptor subunit mRNA levels in temporal lobe epilepsy patients. Ann Neurol 1999; 46:343-358.
36. Matsuoka N, Aigner TG. D-cycloserine, a partial agonist at the glycine site coupled to N-methyl-D-aspartate receptors, improves visual recognition memory in rhesus monkeys. J Pharmacol Exp Ther 1996; 278:891-897.
37. Mayhan WG, Didion SP, Sheng M et al. Glutamate-induced disruption of the blood-brain barrier in rats. Role of nitric oxide. Stroke 1996; 27:965-969.
38. McIntosh AM, Wilson SJ, Berkovic SF. Seizure outcome after temporal lobectomy: current research practice and findings. Epilepsia 2001; 42:1288-1307.
39. McNamara JO. Emerging insights into the genesis of epilepsy. Nature 1999; 399:A15-22.
40. Monahan JB, Handelmann GE, Hood WF et al. D-cycloserine, a positive modulator of the N-methyl-D-aspartate receptor, enhances performance of learning tasks in rats. Pharmacol Biochem Behav 1989; 34:649-653.
41. Monahan PE, Samulski RJ. Adeno-associated virus vectors for gene therapy: more pros than cons? Mol Med Today 2000; 6:433-440.
42. Morgan D, Diamond DM, Gottschall PE et al. A beta peptide vaccination prevents memory loss in an animal model of Alzheimer's disease. Nature 2000; 408:982-985.
43. Nemni R, Braghi S, Natali-Sora MG et al. Autoantibodies to glutamic acid decarboxylase in palatal myoclonus and epilepsy. Ann Neurol 1994; 36:665-667.
44. Palace J, Lang B. Epilepsy: an autoimmune disease? J Neurol Neurosurg Psychiatry 2000; 69:711-714.
45. Peel AL, Klein RL. Adeno-associated virus vectors: activity and applications in the CNS. J Neurosci Methods 2000; 98:95-104.
46. Peltola J, Kulmala P, Isojarvi J et al. Autoantibodies to glutamic acid decarboxylase in patients with therapy-resistant epilepsy. Neurology 2000; 55:46-50.
47. Pesavento E, Margotti E, Righi M et al. Blocking the NGF-TrkA interaction rescues the developmental loss of LTP in the rat visual cortex: role of the cholinergic system. Neuron 2000; 25:165-175.
48. Pont F, Collet A, Lallement G. Early and transient increase of rat hippocampal blood-brain barrier permeability to amino acids during kainic acid-induced seizures. Neurosci Lett 1995; 184:52-54.
49. Popik P, Rygielska Z. A partial agonist at strychnine-insensitive glycine sites facilitates spatial learning in aged rats. J Physiol Pharmacol 1999; 50:139-151.

50. Rasmussen T, Olszewski J, Lloyd-Smith DK. Focal seizures due to chronic localized encephalitis. Neurology 1958; 8:435-455.
51. Rogers SW, Andrews PI, Gahring LC et al. Autoantibodies to glutamate receptor GluR3 in Rasmussen's encephalitis. Science 1994; 265:648-651.
52. Saija A, Princi P, Pisani A et al. Blood-brain barrier dysfunctions following systemic injection of kainic acid in the rat. Life Sci 1992; 51:467-477.
53. Schenk D, Barbour R, Dunn W et al. Immunization with amyloid-beta attenuates Alzheimer-disease-like pathology in the PDAPP mouse. Nature 1999; 400:173-177.
54. Sheng M, Cummings J, Roldan LA et al. Changing subunit composition of heteromeric NMDA receptors during development of rat cortex. Nature 1994; 368:144-147.
55. Solimena M, Folli F, Aparisi R et al. Autoantibodies to GABA-ergic neurons and pancreatic beta cells in stiff-man syndrome. N Engl J Med 1990; 322:1555-1560.
56. Twyman RE, Gahring LC, Spiess J et al. Glutamate receptor antibodies activate a subset of receptors and reveal an agonist binding site. Neuron 1995; 14:755-762.
57. Whitney KD, McNamara JO. Autoimmunity and neurological disease: antibody modulation of synaptic transmission. Annu Rev Neurosci 1999; 22:175-195.
58. Whitney KD, Andrews PI, McNamara JO. Immunoglobulin G and complement immunoreactivity in the cerebral cortex of patients with Rasmussen's encephalitis. Neurology 1999; 53:699-708.
59. Wiendl H, Bien CG, Bernasconi P et al. GluR3 antibodies: prevalence in focal epilepsy but no specificity for Rasmussen's encephalitis. Neurology 2001; 57:1511-1514.
60. Wozniak DF, Olney JW, Kettinger L et al. Behavioral effects of MK-801 in the rat. Psychopharmacology (Berl) 1990; 101:47-56.
61. Yang R, Puranam RS, Butler LS et al. Autoimmunity to munc-18 in Rasmussen's encephalitis. Neuron 2000; 28:375-383.
62. Ylinen A, Pitkanen M, Sirvio J et al. The effects of NMDA receptor antagonists at anticonvulsive doses on the performance of rats in the water maze task. Eur J Pharmacol 1995; 274:159-165.
63. Zucker DK, Wooten GF, Lothman EW. Blood-brain barrier changes with kainic acid-induced limbic seizures. Exp Neurol 1983; 79:422-433.

CHAPTER 12

Cortical Dysplasia and Epilepsy:
Animal Models

Philip A. Schwartzkroin, Steven N. Roper and H. Jurgen Wenzel

Abstract

Cortical dysplasia syndromes—those conditions of abnormal brain structure/organization that arise during aberrant brain development—frequently involve epileptic seizures. Neuropathological and neuroradiological analyses have provided descriptions and categorizations based on gross anatomical and cellular histological features (e.g., lissencephaly, heterotopia, giant cells), as well as on the developmental mechanisms likely to be involved in the abnormality (e.g., cell proliferation, migration). Recently, the genes responsible for several cortical dysplastic conditions have been identified and the underlying molecular processes investigated. However, it is still unclear how the various structural abnormalities associated with cortical dysplasia are related to (i.e., "cause") chronic seizures. To elucidate these relationships, a number of animal models of cortical dysplasia have been developed in rats and mice. Some models are based on laboratory manipulations that injure the brain (e.g., freeze, undercut, irradiation, teratogen exposure) of immature animals; others are based on spontaneous genetic mutations or on gene manipulations (knockouts/transgenics) that give rise to abnormal cortical structures. Such models of cortical dysplasia provide a means by which investigators can not only study the developmental mechanisms that give rise to these brain lesions, but also examine the cause-effect relationships between structural abnormalities and epileptogenesis.

What Is Cortical Dysplasia?

The term "cortical dysplasia" has been used by different authors/investigators to refer to a number of different histopathological brain phenomena. For the purposes of the present discussion, we will use the term in a relatively broad sense, to refer to a condition of abnormal brain structure/organization that, in most cases, is due to a deviation from the normal pattern of cortical development.[35,36,108] It is present in a wide variety of human diseases and syndromes. The structural abnormality may involve general/widespread conditions (such as lissencephaly) that are often associated with profound mental retardation; other cortical dysplastic patterns are restricted to discrete areas of focal cortical abnormality, and may be asymptomatic. Much of the recent interest in cortical dysplasia stems from its frequent association with seizure disorders (i.e., epilepsy)—although the cause-effect relationships remain to be elucidated. There have been numerous attempts to classify cortical dysplasia syndromes based on gross pathology, radiographic findings, histological features, genetic abnormalities, and stages of development most affected by the pathological process. None of these methods of categorization is entirely satisfactory or all-encompassing, but each one may be useful depending on the question at hand.

Recent Advances in Epilepsy Research, edited by Devin K. Binder and Helen E. Scharfman.
©2004 Eurekah.com and Kluwer Academic / Plenum Publishers.

Gross Pathology

Lissencephaly refers to a diffusely smooth-surfaced cerebral hemisphere without sulcation. This condition is often associated with severe mental retardation and seizures in affected individuals. Lissencephaly has generally been classified based on histological characteristics. Type I lissencephaly has a four-layered cortex: layer 1 corresponds to the remnants of the marginal zone; layer 2 contains a mixture of normal layer II, III, and V neurons; layer 3 is a cell-sparse zone; and layer 4 consists of a diffuse accumulation of heterotopic neurons.[16] Type I lissencephaly most commonly occurs in a sporadic form (isolated lissencephaly sequence) or as the Miller-Dieker syndrome (MDS), a haploinsufficiency disorder with characteristic dysmorphic facies, retardation, and seizures.[53] MDS is caused by a mutation in the *LIS1* gene,[126] which appears to be an important regulator of neuronal mictrotubule systems. Disruption of microtubule function may be the basis of impaired neuronal migration in MDS (see ref. 59 for review). Mutations of *LIS1* have also been identified in 40% of people with nonsyndromic lissencephaly.[116]

X-linked lissencephaly is another genetic disorder that produces lissencephaly in affected males, and gives rise to subcortical band heterotopia in females. Subcortical band heterotopia is characterized by a thick layer of cortex that extends diffusely throughout the white matter of the cerebral hemispheres. It is caused by mutation in a gene, *XLIS* (or *DCX*), which encodes for doublecortin,[49] a microtubule-associated protein that appears to be important for normal neuronal migration.

Lissencephaly with cerebellar hypoplasia (LCH) is a lissencephalic syndrome that includes malformations in the cerebellum and brainstem. Two families with LCH have shown mutations in the gene, *RELN*, which encodes for the protein reelin.[81] *RELN* was first identified in the mutant mouse, reeler. Reelin is produced by Cajal-Retzius cells in the marginal zone and is important in the late phases of neuronal migration.[45]

Type II lissencephaly (also called cobblestone lissencephaly) is characterized by a complete loss of normal horizontal lamination. Cortical neurons are arranged in vertical columns and clumps, separated by thick septations of glial and vascular tissue; glioneuronal heterotopias are present on the surface of the hemispheres. Type II lissencephaly is present in the Walker-Warburg syndrome, where lissencephaly is accompanied by hydrocephalus, retinal dysplasia, and (in some cases) encephalocele.[16] Type II lissencephaly is also seen in the cerebro-ocular muscle disorders.

Agyria and pachygyria also describe "smooth" areas of the cortex. These terms usually refer, however, to focal or regional areas without sulcation (as opposed to lissencephaly, where the process affects both hemispheres diffusely).

Hemimegalencephaly is a rare disorder with enlargement of one cerebral hemisphere, and may be accompanied by enlargement of the ipsilateral face and body as well. The involved hemisphere may show several types of cortical dysplasia, and dysmorphic neurons have also been described.[39,130] Hemimegalencephaly may be seen in a sporadic neurocutaneous disorder, the epidermal nevus syndrome, where it is associated with skin lesions and abnormalities of the eyes and muscloskeletal system.[114] Mental retardation, hemiparesis, and seizures (especially infantile spasms) are the most common clinical manifestations of hemimegalencephaly.

Neuronal heterotopia is the occurrence of neurons anywhere in the brain outside of their normal position. In the cerebral hemispheres, it may be classified as periventricular or subcortical, and diffuse or nodular. Diffuse periventricular heterotopia may have relatively mild clinical manifestations or may be asymptomatic. X-linked periventricular heterotopia is a genetic disorder characterized by embryonic lethality in males and bilateral, nodular, periventricular heterotopia in females. Affected females may also have patent ductus arteriosus and coagulopathies. The gene responsible for X-linked periventricular heterotopia is *FLN1*. This gene encodes for filamin-1, an actin cross-linking protein.[62]

Focal cortical dysplasia (FCD) (also called Taylor's-type dysplasia, forme fruste of tuberous sclerosis, and type III cortical dysplasia in the Montreal classification system as described by ref.

112) refers to localized malformations of the cortex that occur in their normal location (i.e., at the outer surface of the hemisphere). Epilepsy is a common manifestation of FCD and, because it is localized, this disorder comprises the major disorder in patients who undergo surgical resection for epilepsy associated with cortical dysplasia.[109,120] Taylor et al[151] gave one of the first detailed descriptions of FCD in surgical specimens from patients with intractable epilepsy, and proposed a causal role for FCD in seizure generation. Gross descriptions of FCD often include gyral abnormalities, thickening of the cortical mantle, and blurring of the interface between the gray and white matter. Cortical areas exhibiting FCD show loss of lamination, loss of spatial orientation of neurons, enlarged neurons with abnormal processes, and giant balloon cells.[154] Areas of FCD may also be associated with benign, cortically-based brain tumors (such as ganglioglioma and dysembryoplastic neuroepithelial tumor).[120] Because of their association with FCD and their indolent nature, these tumors are suspected to be developmental in origin.

Tuberous sclerosis (Bourneville disease) is an autosomal dominant neurocutaneous syndrome characterized by cortical tubers, subependymal nodules, giant cell astrocytoma, and hamartomatous growths and tumors in the retina, skin, kidneys and lungs. Histologically, cortical tubers are indistinguishable from some types of FCD. Tuberous sclerosis is associated with defects in two genes, *TSC1* and *TSC2*, which encode for hamartin and tuberin, respectively.[155,163] The relationship between cortical dysplasia and the protein defects are still under investigation.

Schizencephaly refers to an abnormal cleft in the cerebral hemisphere that extends from the pial surface to the ependymal surface.[161,162] This cleft is lined by malformed cortex that often shows polymicrogyria. Schizencephalic clefts are often bilateral or associated with an area of cortical dysplasia in the homologous contralateral hemisphere.[12] Because it involves loss of tissue, schizencephaly has often been described as the result of an in utero injury. However, recent reports of mutations in the homeobox gene, EMX2, suggest a genetic etiology in some cases of schizencephaly.[19,75]

Polymicrogyria is a gyral abnormality characterized by shallow, poorly formed, closely spaced sulci. Histologically, polymicrogyria is classified as layered and unlayered. Layered polymicrogyria shows a four-layered arrangement similar to that described for type I lissencephaly, with a cell-sparse zone separating two neuronal layers. However, in polymicrogyria, the cell-sparse zone appears to result from laminar necrosis of layer V, whereas in type I lissencephaly the cell-sparse layer represents a zone of demarcation between the normotopic and heterotopic neuronal layers.[10,12] Unlayered polymicrogyria does not have this cell-sparse zone, and consists of the remnant of the marginal zone overlying a disorganized neuronal layer.[60] In many cases, polymicrogyria is thought to result from ischemic injury to the cortex in the late prenatal or perinatal period. This hypothesis is based on the appearance of laminar necrosis in four-layered polymicrogyria, its vascular distribution, its common association with encephaloclastic lesions, and the fact that perinatal cortical injuries can produce a similar histology in animals.[55]

Histology

Histological classification schemes are potentially very important for our understanding of cortical dysplasia, since microscopic analysis can give a much better indication of the cellular constituents of the dysplastic cortex. Cortical dysplasia can be broadly described by loss of lamination, loss of spatial orientation of the neurons, abnormalities of neuronal morphology, and abnormalities of cellular commitment. Loss of normal lamination is an almost ubiquitous finding in cortical dysplasia, since proper lamination is dependent upon successful completion of all of the major stages of cortical development.[105,120] Loss of spatial orientation is very common in cortical dysplasia for the same reason. The incidence/frequency of neuronal cytomegaly and/or cytoskeletal abnormalities is less clear. Enlarged, dysmorphic neurons have been the focus of pathological descriptions of surgically resected dysplastic tissue since the early report of Taylor et al.[151] These neurons are usually described as "giant," with abnormal neuritic processes, containing large amounts of clumped Nissl substance surrounding the nucleus and increased amounts of high and medium molecular mass neurofilament proteins.[54] In a large surgical series from children with cortical dysplasia and intractable epilepsy, 56% of specimens showed neuronal cytomegaly.[105] A particularly dramatic example of such a dysmorphic cellular condition is the "balloon cell"—large, ovoid cells with eccentric nuclei and abundant opalescent cytoplasm.[105] They are found in some forms of FCD, and are also a hallmark of the cortical tubers of tuberous sclerosis. A report that showed simultaneous staining for both GFAP (a glial marker) and synaptophysin (a neuronal marker) in balloon cells pointed toward an abnormality of cell commitment in these cells.[156] The actual incidence of balloon cells in cortical dysplasia is not well known; they were present in 22% of specimens in the pediatric surgical series of Mischel et al.[105]

There have been several proposed classification systems based on histology. Mischel et al listed nine histological features of cortical dysplasia, and grouped them according to presumed early, intermediate, and late developmental lesions. These authors hypothesized that earlier lesions would result in more severe clinical phenotypes and later lesions would produce milder problems for the patient. Using this system, they classified histological changes as severe (balloon cells, neuronal cytomegaly), moderate (polymicrogyria, white matter heterotopia), or mild (absence of the findings that constitute severe or medium abnormalities). Analyzing 77 children with surgically-treated cortical dysplasia and epilepsy, they found that histologic grade correlated with seizure frequency (more severe grade meant higher seizure frequency) and age at surgical intervention (more severe grade meant earlier intervention). Palmini et al[111] reviewed a surgical series of 30 patients and proposed a three-tiered histological grading system: grade I consisted of abnormalities of lamination; grade II contained dysmorphic neurons; and grade III contained balloon cells. They found that seizure control after surgery correlated with histological grade (and the extent of resection) of the FCD. In this context, it is noteworthy that Urbach et al[154] recently reported favorable postsurgical outcomes (i.e., seizure control) for focal cortical dysplastic lesions containing balloon cells.

Stages of Development

A useful classification system for cortical dysplasia groups pathological entities according to the earliest stage of cortical development during which the pathological process became manifest.[13,14] It is based on abnormalities of three main stages of cortical development: group I includes abnormalities of neuronal and glial proliferation; group II consists of abnormalities of neuronal migration; and group III is comprised of abnormalities of cortical organization. Each group is subdivided into focal and diffuse processes, and in the case of proliferation abnormalities, whether proliferation is increased or decreased. This classification system has provided a point of discussion for possible etiologies for the different cortical dysplastic syndromes (see below).

Association of Cortical Dysplasia and Epilepsy

The close association of cortical dysplasia and epilepsy is evident from the frequent presence of dysplasia in people who require surgical treatment for intractable epilepsy. The incidence of cortical dysplasia in people that undergo surgery for intractable epilepsy has been reported as 39% in children[58] and 14% across all age groups.[120] However, these percentages are likely to vary among surgical epilepsy centers, referral patterns, and regional attitudes regarding epilepsy surgery in children and people with mental retardation. Since detection of cortical dysplasia often requires detailed MR imaging techniques or microscopic analysis of the tissue (from surgery or autopsy), estimates of asymptomatic cortical dysplasia (or cortical dysplasia that produces impaired brain function outside of epilepsy) remain uncertain. Incidence of cortical dysplasia in the general population is not known. Barkovich et al[9] reviewed 537 brain MR studies in children and found 13 (2%) cases of cortical malformations; 12 of these 13 children had seizures and 8 had developmental delay. Leventer et al[95] reviewed all brain MR scans in children from the Royal Children's Hospital over a 6-year period (total number not provided) and identified 109 cases of malformation of cortical development. Seizures were present in 75% of these children and 68% had developmental delay. Therefore, although the number of people with asymptomatic cortical dysplasia is difficult to estimate, the incidence of seizures in cortical dysplasia is quite high among patients who are referred for MR studies of the brain. It is also appears that, at least in some cases, areas of cortical dysplasia are sufficient to produce seizures—since surgical resection of these lesions can abolish the seizures.[79,112,154]

What Are the Processes that Lead to Dysplasia?

As indicated above, recent advances in clinical and experimental research have shown that brain malformations often represent disorders of cortical development during the embryonic, fetal, and perinatal periods.[105,146] These structural malformations are extremely varied, depending on the underlying processes and critical timing of the developmental aberration.[71] In each of these disorders, some defect in the proliferation, migration, and/or organization of neurons (from their place of birth to the cerebral cortex) is the apparent basis of the dysplasia. Depending on the severity and pattern of the lesion, the epilepsy-related clinical phenotype may manifest at almost any developmental stage, from newborn to adult.

Developmental studies on human neocortex[104] and on animal models[29] have defined three broad developmental processes as key to understanding the disorders of cortical dysplasia: 1) proliferation and differentiation (specification); 2) neuronal migration and further cell differentiation; and 3) cortical organization, including synaptogenesis and circuitry formation.[72,158] Each of these developmental stages is defined by specific patterns of gene expression and associated multiple signaling mechanisms. Disruption or modification of these signaling pathways can lead to developmental aberrations and formation of cortical dysplasia.[74,146] Detailed knowledge about the processes occurring at these critical time points in the developing brain is essential to our understanding of the pathogenesis of cortical dysplasia (see Table 1).

Key Steps of Normal Neuronal Proliferation/Migration and Cortical Organization

The development of the six-layered neocortex depends on a series of precisely timed proliferative, migratory and maturational events.[2,103,123] During earliest development, cortical cells arise from a proliferative neuroepithelium, the ventricular zone, which contains the precursors of neurons and glia.[123] These cells undergo many cycles of proliferation, followed by waves of postmitotic neuroblasts migrating to a final destination, to form the layered structure of the neocortex (for review see ref. 124). Characteristic features of the early cortical development include the following: 1) first-born neurons come to reside in the deepest layers of the cortical plate, while the last-born neurons settle in the most superficial layers (inside to outside order[4]); 2) the majority of neurons migrate along radial glia into the cortical plate,[97,122] although a small proportion follows a tangential migration route[46,159] (see also ref. 119); 3) after arrival in the appropriate cortical layer, neurons form dendrites and axons, and initiate synaptic connectivity; 4) a substantial proportion of neurons undergo apoptosis. These early developmental events occur during the 8th to 14th weeks of gestation in human development, and between embryonic days 11 to 18 in the mouse[34,67] and E14 to E20 in the rat.[17] A defect in any of these key events can be linked to the pathogenesis of cortical dysplasia.[136]

Early cellular events within the ventricular zone—during proliferation/differentiation—and/or failed apoptosis in the cortical plate may result in pathologic cortical patterns. For example, neuronal precursors may fail to complete differentiation, or exhibit disturbances of cytoskeletal elements essential for the initiation of migration.[39,72] Indeed, it has been hypothesized that cortical dysplastic cells might arise as a clonal population of daughter progeny stemming from a solitary (or small number of) abnormal progenitor cells. Similarly, the dysmorphic cell types in the tubers of tuberous sclerosis complex might arise from proliferation of a single precursor cell which sustains a "second hit" mutation in one of the TSC genes.[163] Alternatively,

Table 1. *A speculative attempt to classify animal models and human syndromes based on the stages of cortical development as outlined by Walsh (1995), Schwartzkroin and Walsh (2000), and Gleeson and Walsh (2001)*

Stage of Development	Human Disorder	Implicated Human Genes	Animal Models
Pattern Formation	Schizencephaly	*Emx2*	
Neuronal/ Glial Proliferation	Hemimegalencephaly Microcephaly	Unknown	Flathead rat/ *Citron K* knockout, Tish rat
Neuronal / Glial Differentiation (Cell fate specification)	Tuberous Sclerosis Taylors-type focal cortical dysplasia	*TSC1* & *TSC2*	Eker rat (*TSC2* mutation)
Neuronal Migration Onset	Periventricular and Subcortical nodular heterotopia	*FLN1*	
Neuronal Migration Early/Ongoing	Type I Lissencephaly X-linked Lissencephaly (Subcortical band heterotopia) X-linked periventricular heterotopia Early acquired cortical dysplasia	*LIS1, DCX*	*LIS1* mutation, MAM rat, Irradiation, Tish Rat (?)
Neuronal Migration Late/Completion	Type II Lissencephaly (Cobblestone Lissencephaly) Lissencephaly with cerebellar hypoplasia ? Temporal lobe epilepsy with dentate granule cell dispersion	*RELN*	Reeler, *p35* and *cdk5* knockouts
Cortical Organization (including synaptogenesis and circuitry formation)	Polymicrogyria Microdysgenesis Focal cortical dysplasia (mild)	Unknown	

normal neuronal proliferation may be followed by failure of appropriate migration (e.g., because of disturbances in signaling pathways or defects within the radial glia that normally provide guidance).[34,39,129] Recent studies on neuronal migration disorders, focusing on the molecular and genetic bases of these disorders, suggest that key steps of neuronal migration depend both upon proper actin and microtubule cytoskeleton interaction, as well as on proper transduction of extracellular signals by migrating neurons.[40,72,117] The growing list of genes that are known to regulate this process of neuronal migration[1,73,158] promises to reveal important information about underlying molecular and cellular control mechanisms.

Identified Genetic Abnormalities in Humans and Animal Models

The pathogenesis of most forms of cortical dysplasia remains to be determined. However, cortical malformations are now increasingly categorized with respect to recent advances in molecular genetics and description of genetic syndromes.[13] For example, genetic loci for Miller-Dieker lissencephaly syndrome (17p13.3[125]), bilateral periventricular heterotopia (Xq28[56]), X-linked lissencephaly/subcortical band heterotopia (Xq22[51]), and tuberous sclerosis (16p13.3[57]) etc. have been identified, and have provided information not only about the genetic underpinnings of some forms of cortical dysplasia, but also about the underlying molecular etiologies.[72] The identification of genes for many of these disorders now allows for integration of cortical defects in mice and in humans into a cellular framework, in which key steps of cortical development are linked to gene-associated disorders (Table 1).

Tuberous Sclerosis—A Disorder of Disturbed Proliferation/Differentiation

Tuberous sclerosis complex (TSC) is a multi-organ, autosomal dominant syndrome[40,128] that causes characteristic malformations within the brain. The TSC lesions include tubers in the cerebral cortex, and subependymal nodules and giant cell astrocytomas along the lateral ventricle. In cortical tubers, the six-layered cortex is disordered (loss of lamination), and there are abnormal-appearing (dysmorphic) neurons and astrocytes, including giant cells and balloon cells (a pathological hallmark of TSC). In addition, heterotopic neurons are often identified in the deep white matter. Subependymal nodules located along the ventricular surface contain a heterogenous population of cell types; some of these cells can apparently become "transformed" into glial tumors. Mutations in two genes, *TSC1* (chromosome 9) and *TSC2* (chromosome 16) account for virtually all cases of tuberous sclerosis. These genes encode distinct proteins (hamartin and tuberin, respectively), which are widely expressed in the brain, and are thought to interact in a molecular cascade that modulates cellular differentiation.[163] The observed defects in cell type specification characteristic of TSC lesions may result from *TSC1* or *TSC2* mutations that interfere with the normal regulation of cell proliferation or specification.[40,146]

Some types of focal cortical dysplasia[151] show pathologies that closely resemble TSC lesions (see above). Sporadic focal cortical dysplasia has been associated with one or more spontaneous mutations at the TSC and/or other loci.[158] However, there are no current genetic investigations of focal cortical dysplasia that provide clues about the genetic bases of this dysplastic condition.

The Eker rat serves as an animal model to study the pathogenesis of TSC.[164] Rats carrying a null mutation at either of the *TSC2* alleles (homozygous *TSC2* and *TSC1* null mutations are embryonic lethal) exhibit glial hamartomas,[165] subependymal hamartoma-like regions, and dysmorphic neuronal cells (which are immunocytochemically reactive to both neuronal and glial markers).

Periventricular Heterotopia—A Disorder of Neuronal Migration Onset

Periventricular heterotopia are clusters of neurons and glial cells that form gray matter nodules lining the ventricle (or sometimes found in the white matter).[153] These "heterotopic" nodules (i.e., cells found outside their appropriate sites in cortical gray matter) are frequently associated with refractory epilepsy. Neurons in these nodules may make synaptic connections among themselves, between heterotopic nodules, and between heterotopic nodules and adjacent cortex.[78] In the human disorder, a fraction of postmitotic neurons appear incapable of leaving the ventricular zone, and it is thought that these abnormal neurons give rise to the periventricular heterotopia—i.e., the heterotopia are due to a primary defect in neuronal migration.[56,72] The genetic basis for X-linked periventricular nodular heterotopia has been identified as *FLN1*.[62] The role of *FLN1* gene product—filamin1—in neuronal migration is mediated through modulation of the actin cytoskeleton. However, filamin1 also binds to other proteins involved in neuronal migration (e.g., integrins, presenilin), so it is not clear which of these protein interactions is disrupted in the periventricular nodular heterotopic dysplasia.[72]

Further, in a recent clinical study, periventricular heterotopia without *FLN1* mutations were found in five human fetuses (gestational ages 21-34 weeks). Analysis of these brains showed disruption and abnormal projection pattern of the radial glia.[139] This observation suggests that at least one other mechanism for the development of periventricular heterotopia is a disruptive process that directly affects radial glia, resulting in a failure of cells to migrate from the ventricular zone.[139]

Lissencephaly and Double Cortex—Disorders of Ongoing Processes of Neuronal Migration

Neurological disorders caused by disturbances during ongoing neuronal migration typically affect the neocortex severely, and are associated with mental retardation and epilepsy. Migration along the radial glia over long distances towards the cortical plate may be interrupted, giving rise to dysplastic patterns such as the lissencephalies.[72,80] One of these disorders, Type-I lissencephaly, is characterized by a smooth cortex (see above). In double cortex, there is a normal six-layered outer neocortex, and an additional collection of neurons located "halfway to the cortex", in the subcortical white matter. The development of the normal neocortical architecture and additional heterotopia in the white matter suggest that some neurons have a defect in ongoing neuronal migration.[11,73] Double cortex is often associated with Lennox-Gastaut syndrome.[127]

Mutations of at least two genes have been implicated in the pathogenesis of these neurological disorders: *LIS1* (or PAFAH1β1) and doublecortin (*DCX*).[49,69,126] *LIS1*, an autosomal gene with normally two inherited copies, is located on chromosome 17p13.3. Patients with chromosome 17 lissencephaly have a mutation in one of those two copies. In patients carrying large mutations (including deletions involving *LIS1* and the surrounding genes *14-33 epsilon* and *MNT*), there is lissencephaly and additional dysmorphic features, which together constitutes the Miller-Dieker syndrome.[32,73] Mutations at the X-chromosome-linked lissencephaly locus cause a phenotype nearly identical to *LIS1* mutations in males, but affected females display subcortical band heterotopia.[49,69,72,118] Several putative functions of *LIS1* and *DCX* genes have been hypothesized; for example, the proteins encoded by the *LIS1* and *DCX* genes are microtubule-associated proteins[63,70,140] that are highly developmentally regulated, and thought to be involved in regulation of the microtubule cytoskeleton. How mutations in these genes affect neuronal migration remains to be determined (reviewed by refs. 72 and 146).

There are few animal models [e.g., a *Lis1* mutation mouse "knockout",[80] and a spontaneous "telencephalic internal structural heterotopia" (tish) mutation in rat[94]], which resemble some aspects of human neuronal migration disorders. The mouse *Lis1* model displays a disorganized cortex (e.g., poor layer specificity, with superficial neurons placed in deeper layers).[61] In electrophysiological studies, hippocampal slices obtained from *Lis1* mutant mice exhibit an abnormally high sensitivity to high potassium levels.[61] The tish rat reveals morphological characteristics similar to those of humans with subcortical band heterotopia, although the genetic basis of the rat abnormality is clearly different from the *DCX* mutation in humans. A striking feature of this rat cortex is a layer of gray matter localized below the normal gray of frontal and parietal cortex.[94] This band of heterotopic neurons has connections with both the overlying neocortex and with subcortical regions (e.g., thalamus);[145] some animals develop behavioral and electrographic seizures.[25]

Other Lissencephalies (Lissencephaly with Cerebellar Hypoplasia, Cobblestone Lissencephaly)—Disorders Affecting Completion of Neuronal Migration

Some disorders both in humans and in mice show abnormalities of the architecture of the developing cerebral wall. These conditions reflect disruption of neuronal migration during its later stages. For example:

Lissencephaly with Cerebellar Hypoplasia—Which Involves Disturbed Penetration of Migrating Neurons into the Preplate

Some patients with lissencephaly display severe cerebellar hypoplasia; however, cerebral gyri and sulci are present, and most of the patients have acquired microcephaly and severe cognitive developmental delay, frequent seizures, and no ability to sit or stand without support.[73] The morphological pattern in the neocortex of these patients appears similar to that seen in the reeler mouse.[23] As in the reeler mouse, migrating cortical neurons are unable to enter the preplate,[107] resulting in an inverted cortical lamination. Recent observations in patients also suggest an underlying defect similar to that seen in the reeler mouse;[45] in particular, a mutation in the gene for reelin—the basis of the reeler mutation—has been identified in two families.[81]

Cobblestone Lissencephaly (Type-II Lissencephaly)—Disturbed Signals to Stop Migration

In cobblestone lissencephaly, the neocortex lacks gyri and sulci, but its surface has a "cobblestone-like" appearance due to a defect in the final step of neuronal migration. In this disorder, migrating neurons fail to arrest at the normal "stopping point", penetrating the pia and forming nodular heterotopia in the overlying subarachnoidal tissue.[64] The gene for the Fukuyama muscular dystrophy associated with recessive Type-II lissencephaly has been identified (named fukutin), but its function is not yet clear. Preliminary evidence shows that the protein is expressed in layer-I neurons of the developing neocortex, suggesting that it acts as a stop-signal for migrating neurons.[73,137]

Recently described gene mutations in animal models have revealed additional genes that regulate crucial steps in corticogenesis. These genes, which encode cdk5 (cyclin-dependent kinase 5)[68] and its regulator, p35,[24] appear to disrupt migration. *Cdk5* and *p35* mutations produce phenotypes in mice with an inverted neocortex, similar (but not identical) to that seen in the reeler mouse. In the *cdk5* mutant, the subplate is in the middle of the cortical plate;[68] early cortical-plate neurons can penetrate the subplate, but later cortical-plate neurons cannot.[72] In contrast, in the *p35* mutant, the subplate is beneath the cortical plate;[92] the underlying defect appears to be an incompletion of migration (i.e., neurons cannot migrate through the cortical plate). *p35* mutants develop behavioral and electrographic seizures and display abnormal neuronal features such as dispersion of hippocampal pyramidal neurons and dentate granule cells, abnormal dendritic and axonal arbors, and mossy fiber sprouting (often associated with temporal lobe epilepsy).[160]

Prenatal Brain Insults Induce Development of Cortical Dysplasia

Disorders of neuronal migration in the cerebral cortex in humans occur in the first half of gestation. Beside genetic factors, early prenatal events may play crucial pathogenetic roles leading to cortical dysplasia.[110] Earlier reports in the clinical literature support the etiologic role of specific prenatal events in the genesis of different neuronal migration disorders [e.g., cytomegalovirus infection associated with polymicrogyria;[106] vascular abnormalities or maternal respiratory or cardiac arrest associated with schizencephaly and/or polmicrogyria[15]]. Other case reports support the importance of environmental factors in the pathogenesis of neuronal migration disorders.[31,86,150] In some of these cases, severe respiratory infection (and/or viral infection) of the mothers was thought to cause fetal circulatory failure or hypoxia, leading to Type-I lissencephaly. In a recent study, 63 patients with Type-I lissencephaly were analyzed; 40% with nonchromosomal forms of lissencephaly were linked with environmental prenatal factors.[52] Palmini et al[110] have identified several specific environmental factors associated with development of neuronal migration disorders and epilepsy (e.g., maternal ingestion of medication, uterine abnormalities, significant maternal trauma, and exposure to irradiation). Choi et al[31] described 2 cases of full-term newborn infants exposed to methylmercury in utero (at 6 to 20 weeks of gestation) following maternal ingestion, leading to numerous heterotopia in the white matter of the cerebral and cerebellar cortices, and disorganized cortical lamination. The role of methylmercury as an etiologic factor for cortical dysplasia has been experimentally confirmed;[88,115] rat fetuses transplacentally exposed to methylmercury at E13 developed leptomeningeal glioneuronal heterotopia similar to severe malformations in humans.

Hemorrhagic lesions during early fetal development play an important role in developing cortical dysplasia. Blood vessels of the periventricular zone and the external glial limitans membrane are particularly vulnerable to prenatal hypoxia, circulatory disturbances, and/or trauma; damage to these vessels often causes focal hemorrhagic lesions.[98] The time point and size of the hemorrhagic lesion, the local disruption of the radial glia, and the degree of white matter involved are crucial factors in pathogenesis of the neurological sequelae. In periventricular hemorrhagic injury, the lesion disrupts migration above the lesion (resulting in cellular heterotopia—acquired heterotopia—at any level above the injured site), and will cause post-injury reorganization of the still-differentiating gray matter (leading to acquired neocortical dysplasia).[98,99]

Recent neuropathological studies on human infants who survived a variety of perinatally acquired encephalopathies (and later died of unrelated causes) have shown that there is direct damage to the still developing neocortex, as well as indirect or postinjury impact on its subsequent structural and functional differentiation. The post-injury differentiation of the primarily undamaged cortical region adjacent to the injured site has been shown to undergo progressive alterations (e.g., neuronal, synaptic, glial), resulting in a variety of congenital and/or acquired cortical dysplastic abnormalities.[101]

What Makes a Dysplastic Brain (or Brain Region) Epileptic?

The question about the bases of epileptogenicity—posed within the above context of cortical dysplasia—is really no different from the issue that has generated so much work in the general epilepsy research field. What is it, indeed, that makes epilepsy? As has been argued for many years, one can initially separate theoretical "causes" into two categories—one focusing primarily on the abnormal properties of individual cells, and the other emphasizing the abnormal connectivity of the cellular aggregate. Although many epilepsy researchers find this categorization rather artificial (and perhaps not very useful), it may have a more obvious role in helping us think about cortical dysplasia. In some forms of dysplasia, there are fairly obvious abnormalities in the features of individual neurons (e.g., balloon cells) (Fig. 1); are these cells responsible for seizure initiation and/or discharge maintenance? In other forms of dysplasia, the individual cells appear relatively normal, but their organization and/or connectivity are obviously pathologic (Fig. 2); is it this organizational feature of the dysplasia that supports the epileptic state?

Abnormal Cells? ("The Epileptic Neuron")

Processes that Give Rise to "Dysplastic" Brain Cells

Some (but not all) forms of cortical dysplasia present with abnormal cell types. How are those cells generated? And what are their properties, especially with respect to electrical activity?

One vulnerable step during brain development is cell proliferation. Cell division at the proliferative ventricular zone is normally highly regulated, with some daughter cells leaving the zone, and others reentering to make additional cells.[22] Interference with this process—be it from external insult, or from "internal" abnormalities (e.g., absence of appropriate gene products)—can lead to the generation of too many or too few cells. The timing of such interference will also have an impact on the nature of the developmental abnormality, since cells of different cortical layers leave the proliferative zone to migrate through the cortical mantle at different developmental stages. Even the location of such interference may have a specific impact; for example, since inhibitory interneuron populations are formed in, and migrate from, the medial ganglionic eminence[3,119] (not the ventricular zone proliferative location of the primary (excitatory) cell populations), they may be differentially affected by some insults. Such a

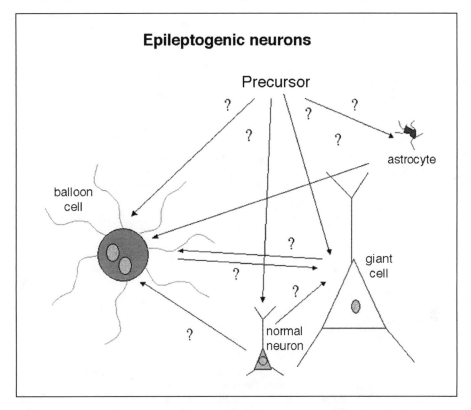

Figure 1. Epileptogenic cells. In some forms of cortical dysplasia, the abnormal cortex is characterized by dysmorphic cells that are poorly-specified (exhibit positive markers for both glia and neurons), and have multiple nuclei and abnormal processes (e.g., the "balloon cells" of tuberous sclerosis and Taylor-type focal cortical dysplasia). "Giant cells" are also a feature of these dysplastic lesions. The origin of such abnormal cells, and their functional properties—i.e., their roles in generating seizure activity—remain to be elucidated.

location-specific disruption may contribute to the loss of interneurons seen in the irradiated rat model of cortical dysplasia.[134]

It is important to note, however, that any inappropriate proliferative process that gives rise to a cortical dysplastic pattern seems to be restricted to the developing brain; it is not—as in a cancerous tumor—an ongoing process that gives rise to continuing cell generation in the mature brain.

Other features of proliferation may also determine cellular abnormalities. For example, incomplete cell division may yield a multi-nucleated cell. Or the DNA duplication process may be inappropriately regulated, yielding cells with multiple gene copies. The consequences of such pathologies are unclear, but may be key to understanding the balloon cell/giant cell cytopathologies.

Another key step in this developmental process is cell specification. Is a given cell destined to become a neuron or a glial cell? That specification is provided, presumably, by the precursor cell and by signals from the surrounding tissue. As indicated above, abnormalities in such signaling, even for a brief developmental window, could result in cells that have been inappropriately and/or incompletely specified. Specification may also become disturbed if/when daughter cells don't divide completely—perhaps resulting in a cell that contains multiple specification instructions.

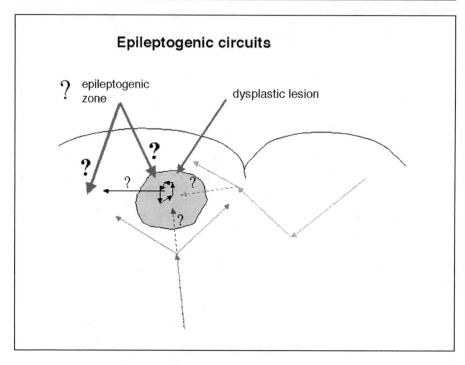

Figure 2. Epileptogenic circuits. In some forms of cortical dysplasia, the primary defect involves loss of cells and/or abnormal localization of poorly organized (but otherwise normal?) cell groups. As a result, abnormal connectivity develops between the affected brain region and surrounding cortical (and/or subcortical) brain regions. In these brains, it is not always clear where seizure activity is triggered (or generated) with respect to the dysplastic lesion.

Finally, where do these cells go? The migratory patterns of newly-born cells is determined by a complex set of signals, and interference with any of these signals may result in incomplete migration (premature stop signals), aberrant (heterotopic/ectopic) localization, or generation of superfluous structures in inappropriate locations (e.g., periventricular heterotopia). In such cases, the migrating cells (and thus, the cells that make up the dysplasia) could be relatively "normal," even though their migratory instructions are aberrant.

Abnormal Features of Neurons in Dysplastic Brain

Abnormal Cell Size and Shape

While normal neurons appear to come in a wide range of shapes and sizes, cytologically-abnormal cells in dysplastic brain fall significantly beyond the normal range. Perhaps the most dramatic example is that of the "balloon cell," characteristic of, e.g., the tubers of tuberous sclerosis.[151,155] These cells are generally spherical, and may measure >100 μm in diameter; they have multiple processes radiating from the cell body—but the nature of these processes (dendritic? axonal? neither?) remains unclear. Indeed, it is unclear whether to consider these cells as neurons or glia, since balloon cells stain positively for markers for both (e.g., synaptophysin and GFAP).[156] Preliminary electrophysiological studies indicate that these cells are capable of action potential generation,[102] but there are no studies to indicate whether these cells receive functional synaptic input or send electrical information to other cells. Another abnormal cell type—but one that seems to be more typically neuronal—is the "giant cell," often found together with balloon cells in some types of dysplasia (e.g., Taylor's focal dysplasia). The genetic characteristics and electrical properties of these unusually large cells remain to be determined.

Disoriented/Displaced Neurons

As described above, a large number of dysplastic syndromes are associated with heterotopic cells and/or cell clusters (e.g., periventricular heterotopia), and even with additional cortical layers (e.g., subcortical band heterotopia). Individual neurons in such misplaced structures often appear relatively normal; as a population, however, these cell groupings are typically rather disorganized.[33] In contrast, there are conditions in which individual cell structure is clearly aberrant. An example of this latter pathology can be seen in the dispersed granule cells of the *p35* knockout mouse.[160] The granule cells of these animals are not only misplaced, but many show gross dendritic and axonal abnormalities, including inverted and/or disoriented dendritic trees, basal dendrites, and recurrent axonal collaterals. Also sometimes noted in dysplastic tissue is an abnormality in numbers and/or localization of inhibitory synaptic terminals[149] (as well as changes in the colocalized peptides that are often expressed in GABA cells[148]).

Cells with Aberrant/Missing Channels/Receptors

Analysis of gene expression and protein products in dysplastic neurons has already begun to yield evidence of abnormalities (compared to cells from "normal" brain regions). Studies include immunocytochemical and in situ analyses, as well as single-cell PCR determinations of gene expression. These investigations have shown, for example, that cells in dysplastic brain regions express different patterns of mRNAs for AMPA, NMDA, and $GABA_A$ receptors[5,38]—and results from electrophysiological/functional studies reflect these receptor abnormalities.[96,102] In dysplastic tissue, long-duration (NMDA-mediated) excitatory synaptic events may become more prominent, up-regulation of AMPA receptor subunits may play a role in seizure spread, and GABA-mediated inhibition is sometimes much reduced.[48,77,96] Interpretation of these observations is complicated somewhat by the fact that the patients (or animals) from whom this material is obtained usually have had long histories of seizure activity—so that these changes may be due, at least in part, to the seizures themselves. However, in one animal model of dysplasia, evidence of potassium channel loss has been observed in dysplastic brain regions, at both the functional/electrophysiological and mRNA levels.[21] While these animals do not have spontaneous seizures, they are significantly more seizure-prone than normal rats (i.e., with no dysplastic abnormality).

Incomplete/Immature/Multiple Cell Type Specification

As suggested above, cells in focal cortical dysplasia and in tubers of tuberous sclerosis appear abnormal at several levels. For example, balloon cells are often multi-nucleated, suggesting abnormalities in cell division. This observation, in turn, suggests that these cells are polyploid, and express multiple gene copies. How those abnormalities might be functionally translated remains unclear. Problems associated with this proliferation stage of development also appear to extend to cell type determination. Studies of dysplastic tissue have found abnormalities in intracellular signaling elements (e.g., Dvl-1, Notch-1, Wnt) that are known to play a significant role in determining cell fate.[35,36] As noted above, some dysplastic cells stain for both glial and neuronal markers;[156] the neuronal markers tend to be characteristic of immature neurons.[39,42,43] The function of these cells, that are neither (or both) neuronal nor glial, remains a mystery. That they are characteristic of focal dysplasia associated with seizure initiation, and of epilepsies that are rather resistant to traditional pharmacological treatment, suggest that grossly dysmorphic cells may play an important role in determining the epileptogenicity of these developmental disorders.

Glial Abnormalities

Glia also must be affected by interference with cell proliferation, cell type determination, and migration during brain development. Thus, it is no surprise that studies of glial function in dysplastic brain have shown significant changes in glial voltage-gated ion channel expression,[18] as well as reductions in such key features as glutamate transporters.[41] Other key features of glia may also be abnormal in various dysplastic lesions, including: enhanced glial proliferation; disruption of radial glia (thought to provide the scaffolding for neuronal migration);[132] and disturbance of the glial-pial barrier.

Abnormal Circuits? ("The Epileptic Aggregate")

Dysplastic Patterns Associated with Abnormal Circuits

All cortical dysplastic abnormalities are associated with some reorganization of cortical circuitry. Such reorganization may be a consequence of too many or too few cells, cells with different structural properties (e.g., recurrent axon collaterals, disoriented dendrites), and/or displaced cells that must establish novel contacts. Although it is certainly possible to imagine epileptogenic processes that depend strictly on changes in cell properties (i.e., without altered circuitry), that situation does not apply to the dysplasia condition—which is defined by structural alterations. The question, then, is not whether structural reorganization occurs, but rather what aspects of the reorganization account for the epileptogenicity. Circuitry issues that have been studied in other epileptic conditions provide some guidance in this consideration. We have learned, for example, that the following types of changes can give rise to epileptogenicity: loss of inhibitory synaptic input to excitatory (principal) neurons; loss of excitatory input to inhibitory neurons (disinhibition); enhancement of recurrent excitatory collateral connectivity; sprouting of excitatory (and inhibitory?) axon collaterals; changes in the relationships between glial and neurons; and changes in spatial relationship involving the extracellular environment. How might these circuitry issues play out under dysplastic lesion conditions?

Cell Loss

As described above, a variety of dysplastic syndromes are defined by loss/thinning of specific cortical layers, the occurrence of micro-gyri, and the loss of specific cell types. With the reduction of cell number, afferent axons face a significant problem in finding postsynaptic targets. When they cannot find appropriate targets, they are likely to make aberrant synaptic contacts. This situation has been investigated in, for example, the freeze focus microgyrus model, which results in the loss of cells in cortical layers IV and V.[83,84] As a consequence, incoming (and local) axons make additional/abnormal synaptic contacts in the "normal" cortex surrounding the microgyrus, which begins to initiate hyperexcitable electrical activity. The fate of normal circuitry within the generally lissencephalic cortex has not yet been elucidated. This type of brain must cope not only with absence of appropriate cortical targets for sub-cortical afferents, but also a significant decrease of cortical efferents. The net functional effects appear to be rather variable, and include seizure activity in a significant percentage of cases.

Additional/Misplaced Cells and/or Cell Regions

Pachygyria is a condition of multiple (often fused) cortical gyri. The thickness of the cortical mantle is often increased in these brains, and the number of cells in cortex is significantly higher than in normal cortex. These anatomical changes are presumably due to a defect in proliferation, perhaps involving an inadequate "stop" signal so that neurogenesis continues beyond a normal termination time. Just as lissencephaly results in a "dilemma" for cellular connectivity, so in the pachygyria brain there is likely to be some confusion about how cells should interact with which targets. The net functional result at this point seems unpredictable.

An even more complex pathology is seen with subcortical band heterotopia ("double cortex") syndrome, an X-linked lissencephaly which results in a relatively normal cortex, below which (within the white matter) is a band of poorly organized neurons.[69] The rat "tish" model[94]

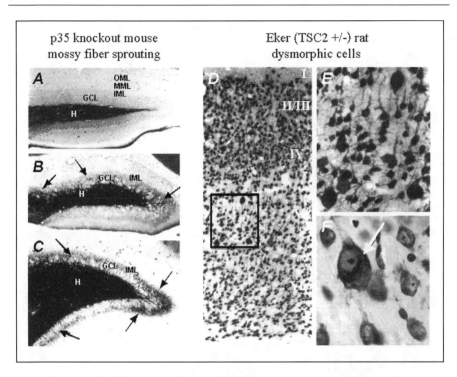

Figure 3. Circuitry and cellular abnormalities in models of dysplasia. A-C: Mossy fiber sprouting in dentate gyrus of *p35* knockout mouse. Deletion of the *p35* gene gives rise to loss of normal cortical lamination and to dispersed granule cells in the dentate gyrus. *p35* knockout animals also show mossy fiber sprouting, visualized with the Timm stain for heavy metals. A. Timm stain of normal dentate gyrus, showing mossy fiber terminals confined to the dentate hilus (dark area); B. Mossy fiber sprouting (arrows—darkly-stained processes reaching through the granule cell layer and into the inner molecular layer) in dorsal hippocampus of *p35* knockout mouse; C. Mossy fiber sprouting in ventral hippocampus (*p35* knockout), where exuberant sprouting forms a supragranular layer of mossy fibers (arrows). Abbr.: IML—inner molecular layer; MML—middle molecular layer; OML—outer molecular layer; GCL—granule cell layer; H—hilus. From Wenzel and Schwartzkroin, unpublished. D-F: Giant cells in irradiated Eker rat cortex. Eker rat, a model of tuberous sclerosis with a mutation in one allele of the *TSC2* gene, shows relatively normal cortical neuronal cell types. However, when Eker rats are subjected to a "second hit"—a single dose of irradiation during P1-3—the cortex shows dysmorphic "giant cells." D. Low magnification NeuN-stained section through Eker rat cortex (layers I-VI); E. Higher magnification of NeuN-positive neurons in cortex (boxed area in D), showing the large, dysmorphic neurons more clearly; F. Nissl-stained, high-magnification view of "giant cell" (arrow) in Eker rat cortex. From Wenzel and Schwartzkroin, unpublished.

shows a similar sub-cortical band phenotype, which has been studied with respect to regional connectivity and hyperexcitability. The "extra" band of neurons establishes connections with subcortical structures (e.g., thalamus) and with the overlying cortex. Seizure activity in these animals appears to be initiated within the overlying "normal" cortex.[25] However, the aberrant connectivity patterns within dysplastic gray matter—and its connectivity to "normal" cortex (perhaps including positive feedback loops)—may provide a basis for this seizure susceptibility.

Other forms of heterotopia also appear to be the result of aberrant/failed migration of neurons. Periventricular heterotopia, for example, involves a collection of disorganized but relatively normal appearing neurons, arrested near their site of origin. In periventricular heterotopia, connectivity of such cell collections remains unexplored. However, in the methylazoxymethanol (MAM) animal model of heterotopia, the cells in the heterotopic cluster (often at the ventricular wall) appear to be of an origin consistent with a future neocortical fate.[28] The cells in the heterotopia establish synaptic connectivity not only with neocortex but also to the underlying hippocampus (and to other structures?), and thus serve as a possible "disseminators" of electrical excitability—i.e., an aberrant nexus for the spread of seizure activity.

Circuitry-Related Changes in Tissue Excitability

As is virtually always the case in talking about basic mechanisms of the epilepsies, discussion always starts with an evaluation of the balance—the relative levels—of excitation vs. inhibition. Traditionally, the focus here has been on the relative strengths of excitatory and inhibitory postsynaptic events. However, investigators have been able to assess a variety of related processes (and structures) that may result in epileptogenicity. In dysplastic cortex, these observations include:

Changes in GABA and Glutamate Markers

Immunocytochemical and in situ hybridization studies have shown changes in the patterns of receptor subunit composition, although these results are somewhat variable from study to study. For example, in one immunocytochemical analysis, "dysmorphic" neurons showed increased immunoreactivity for GluR1 and GluR2/3, prominent $GABA_AR\beta$ immunoreactivity, and varying NR2 immunoreactivity.[89] Other immunocytochemical (ICC) studies have emphasized the increased NR1 and NR2(A&B) immunoreactivity.[5] Analysis of mRNA in dysmorphic cells, in contrast, showed up-regulation of GluR4 and NR2B&C, but down-regulation of NR2A and $GABA_AR\beta1$ sub-units; $GABA_AR\alpha1$&2, $GABA_AR\beta2$, and GluR1 mRNAs were reduced in all cells of the dysplastic brain.[38] Electrophysiological studies of dysplastic tissue have suggested alterations in the AMPA-mediated excitatory influences and in $GABA_A$ receptor-mediated inhibition,[96,167] and in NMDA (NR2B-mediated) excitation.[48] Analyses of transmitter uptake mechanisms in dysmorphic brain cells have shown an increase in neuronal glutamate transporter (EAAT3/EAAC1),[41] and a decrease in GABA transporter (GAT1).[149]

Modulation of Extracellular Environment

Even cursory analysis of a tissue section through dysplastic cortex reveals often dramatic changes in the relationship between neurons and glia, in the patterns/separations of fiber bundles and cell groups, as well as in the orientation of dendritic processes. These changes, in turn, significantly alter the extracellular environment of neuronal populations. How such changes may impact neuronal function remains an open question.

Sprouting and Aberrant Wiring

Sprouting—the development of abnormal connectivity due to the generation of new axon collaterals—has been identified in a number of models of cortical dysplasia. This phenomenon is most easily seen in the mossy fibers of the hippocampal dentate granule cells (e.g., in the *p35* model),[160] but also characterizes a number of systems in which "normal" axonal targets have been removed. Both excitatory and inhibitory axons may sprout, so that the net effect of these changes is not always obvious. However, the resulting abnormal synaptic arrangements provide potential mechanisms for recurrent excitation and/or disinhibition, and thus for seizure initiation. A different type of aberrant wiring pattern is seen in models of neuronal heterotopia (e.g., the MAM-exposed rat and the tish "double cortex" rat). Here, abnormal connections provide additional pathways for spread of excitability from/to the abnormal cell structures,[30] providing additional avenues for synchronization and spread of seizure activities. These aberrant patterns of connectivity (and perhaps the related changes in receptor/transporter molecules) may result from "normal" brain mechanisms that are "attempting to compensate" for the dysplastic lesion.

In all these conditions, it remains unclear whether the dysplastic structure is the origination of seizure activity, serves as a low-threshold trigger zone, and/or simply facilitates spread.

Seizure Variability in Dysplastic Brain—What Makes a Given Dysplastic Lesion Epileptogenic?

While it is possible to provide some experimentally-based suggestions with respect to the question "What makes a dysplastic brain epileptic?", there are really no data to guide a response to the question "Why is there so much seizure variability across different forms of dysplasia, and from case to case within a restricted category of dysplasia?" The comments below, therefore, are rather subjective proposals of which factors that might contribute to the epileptogenicity arising from a given dysplastic lesion.

Given that there is so much variability from case to case, it seems unlikely that there is a single factor that determines the epileptogenicity of a dysplastic lesion. Among the more salient variables that could contribute to epileptogenicity are:

- Functionality of the lesion itself—i.e., to what degree does the combination of abnormal organization and abnormal cell types coalesce in a structure that is intrinsically hyperexcitable?
- Location and connectivity of the lesion—i.e., how does the dysplastic lesion interact with the rest of the brain?
- Genetically determined predisposition of the "host" brain—i.e., what is the seizure-sensitivity of the "background" on which the lesion is imposed?
- Developmental issues—i.e., when does the lesion develop and how does it interact with the surrounding developing brain?

What Are the Qualitative and Quantitative Variables That Determine the Epileptogenicity of the Dysplastic Lesion Itself?

Does Size of the Lesion Make a Difference?

Dysplastic lesions may be small/discrete, or comprise an entire hemisphere. In theory, even if the lesion itself is not epileptogenic, a large size would give rise to many alterations involving surrounding "normal" brain. Yet, it is not clear whether lesion size, per se, is related to net epileptogenicity. In an analysis of the MAM rat model, there was only a very weak correlation between extent of dysplasia and alterations in seizure threshold.[7]

How Is Epileptogenicity Related to Presence (or Absence) of Abnormal Cell Types?

There appear to be two fairly distinct "types" of dysplasia—those that involve only disorganization of normal-appearing neurons, and those in which abnormal ("dysmorphic") cells are present. This distinction perhaps reflects the developmental timing of the dysplastic process. Does presence and/or number of abnormal cell types determine epileptic propensity? Given that cell-associated abnormalities in receptors/channels, etc. are most apparent in these dysmorphic cells, multiplication of such cell types would, intuitively, lead to increased functional abnormality. Whether such functional abnormality means epilepsy is unclear.

What Is the Nature of the Internal Connectivity Within the Aberrant Structure?

If dysplastic lesions are (or can be) the initiator zones for seizure genesis, then the properties of cells intrinsic to the aberrant structure, and the patterns of connectivity within the structure, must be critical to its epileptogenic nature. Enhanced inhibition among neurons within a dysplastic lesion would, for example, not provide obvious support for a "low-threshold" seizure initiation zone. A high level of feedback excitation, however, would be an obvious means of initiating hyperexcitable activities.

What Is the Nature of the Connectivity Between Lesion and Surrounding Tissue?

Where Relative to the Lesion Is the Seizure Initiation Zone?

Although the above discussion assumes the dysplastic lesion to be the seizure generation/initiation zone, that may not always be the case. Current studies present rather different views on that issue, and it may be that the site of seizure initiation varies with respect to the dysplastic lesion (perhaps depending on the type of lesion and/or its location). Instructively, in tuberous sclerosis, it appears that only some tubers are "active" (i.e, seizure generators)—although resection of active tubers may lead to "activation" of previously inactive tubers.[90] Many of the human MRI/EEG coregistration studies are consistent with the view that seizure activity arises from the dysplastic zone. However, many of the studies on animal models suggest that hyperexcitability occurs primarily in the surrounding "normal" brain tissue (e.g., ref. 83). This latter possibility makes intuitive sense if one considers the likelihood that the lesion area may be significantly "deafferented" (and/or be primarily inhibitory), and that the surrounding tissue experiences considerable reorganization (with increased afferent input) as a "response" to this deafferentation. As indicated above, this reorganization may consist not only of gross anatomical features, but also changes in cell properties (different receptors, channels, transporters, etc.) and/or changes in the number/position/nature of synaptic contacts.

What Brain Regions Are Most Likely to Support Lesion-Associated Epileptogenicity?

Even in the normal brain, some regions are more likely to be the site of seizure initiation than other regions (e.g., parieto-temporal vs. frontal). This propensity undoubtedly reflects the nature of the region's internal and external connectivity. How, then, does the presence of a dysplastic lesion "interact" with the epileptic propensity of its host brain region? Do dysplasias in temporal lobe more often give rise to seizures than similar lesions in frontal cortex? Also of interest are the connectivity patterns between the lesion (and of the host cortex) and sub-cortical structures. In nondysplastic brain, some seizure types (e.g., spike-and-wave associated with absence) appear to depend strongly on cortical-subcortical interactions. Does that type of interaction also influence the epileptogenicity of dysplastic lesions (e.g., ref. 87)?

Is There a Genetic Predisposition? What Is the Role of a 2nd Hit?

Does Background Seizure Predisposition Influence the Epileptogenicity of a Dysplastic Lesion?

Current data suggest that even epileptogenic treatment and/or single gene mutations are more or less likely to lead to a seizure phenotype, depending on the background strain of the animal that carries the mutation.[143] It seems likely that such genetically-based variability in background predisposition must contribute strongly to dysplasia-associated seizure activity in the human epileptic population. However, this possibility has not been systematically explored, even within genetically-defined experimental animal models. Of particular interest is the possibility that some type of "two-hit mechanism"[163]—e.g., genetic predisposition and developmental trauma—is much more likely to lead to dysplasia-associated epilepsy than a single dysplasia-inducing factor.

Does the "Double Hit" Hypothesis Explain the Appearance of Abnormal (Dysmorphic) Cell Types?

Only a couple of the current animal models of cortical dysplasia exhibit dysmorphic cell pathological characteristics. And only a few of these models are spontaneously epileptic (most exhibit reduced seizure thresholds). Imposing a second insult—e.g., irradiation or toxin exposure—on an already dysplastic tendency may give rise to aberrant cell types. Does such an insult also alter epileptogenicity?

When During Development Is the Lesion Generated?

How Does the Timing of Dysplasia Generation with Respect to the Maturational Sequence of Brain Development Influence Epileptogenicity?

Clearly, the timing of expression of the dysplastic influence determines (at least in part) the cell types, location, and regional preference of the lesion. For example, expression of a dysplastic influence during cell proliferation may result in incomplete cell division and the appearance of dysmorphic cells with multiple nuclei. In contrast, if the disruption occurs during cell migration, normal-appearing cells may end up in heterotopic positions. Developmental timing determines not only the "type" of dysplasia, but also the location (and, probably, connectivity) of the lesions. As discussed above, these factors are all likely to contribute to the epileptogenicity of any given dysplastic pattern.

Does Aberrant (e.g., Postnatal) Neurogenesis (or Gliogenesis) Contribute to the Dysplasia and to Its Epileptogenic Propensity?

Current studies have shown that neurogenesis continues in many parts of the brain, even after the major periods of neuron (and glial) formation during early development. Can such late-developing cells give rise to (or contribute to) dysplastic patterns?[113] If so, the nature of these cell types, and their connectivity patterns, undoubtedly influence the excitability of the dysplastic lesion, and the likelihood that such an aberrant structure would support epileptogenic function.

The list of factors that potentially contribute to epileptogenicity is perhaps discouragingly long and complex. The list, however, is "hypothetical." It may be that we can identify one or two factors that provide the predominant influence with respect to dysplastic epileptogenicity. The question simply awaits careful and systematic study.

What Do Animal Models of Cortical Dysplasia Tell Us about Human Epilepsies?

Investigators have turned to animal models of cortical dysplasia to answer questions that cannot be addressed in human studies. It is clear that no one study or model will answer all questions regarding cortical dysplasia and epilepsy. Each model, therefore, should be evaluated based on the pathological process that it is attempting to mimic and the specific question that it is trying to address. Comparisons of animal models to human disease can be made based on similarities in histological features, common gene alterations, and abnormalities in common stages of cortical development. Animal models can be classified as injury-based, spontaneous mutations, and transgenics.

Injury-Based Models

In utero irradiation produces diffuse cortical dysplasia in rats. If pregnant females are exposed to 200 cGy of external irradiation on E16 or 17, the offspring show microcephaly, diffuse cortical dysplasia, periventricular and subcortical heterotopia, hippocampal heterotopia, and agenesis/hypoplasia of the corpus callosum.[37,134] The dysplastic cortex is reduced in depth, with loss of lamination and of spatial orientation of the cortical neurons (Fig. 4A). Spontaneous electrographic seizures have been detected during long-term EEG recordings,[91] and in vitro slices (through regions of dysplasia) show enhanced epileptiform bursting compared to controls, in the presence of partial $GABA_A$ receptor blockade.[135] There is also a selective reduction in density of parvalbumin- and calbindin-immunoreactive neurons in dysplastic cortex.[133] Pyramidal cells in dysplastic cortex show reduced frequency of miniature IPSCs compared to controls,[167] but there is no difference in miniature EPSCs. These studies suggest that in utero irradiation results in a selective impairment of inhibition in dysplastic cortex due, at least in part, to a loss of (or failure to produce) cortical inhibitory neurons. Reduced inhibition could enhance the epileptogenic potential of dysplastic cortex in this model.

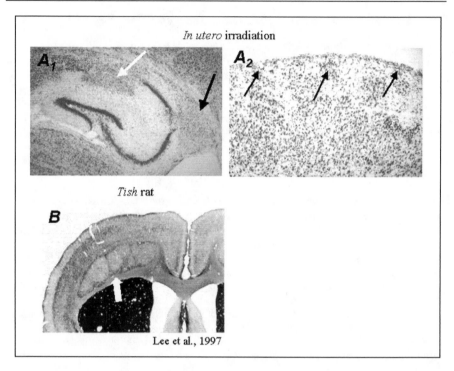

Figure 4. Rat models of dysplasia. A) In utero irradation in rat. In utero irradiation induces regions of dysplasia in neocortex, hippocampus, and around the ventricles. 1. Nissl-stained section, showing heterotopic neurons in the hippocampus (white arrow) and the periventricular region (dark arrow). 2. Higher magnification section through neocortex, illustrating the loss of normal lamination, including collections of neurons that extend through layer I to the pial surface (arrows). From Roper, unpublished. B) Tish rat. Coronal section through the brain of a tish rat (acetylcholinesterase (AChE) staining), showing the heterotopia (white arrow), a large structure located within the subcortical white matter. The heterotopic cell region resembles the "double cortex" of subcortical band heterotopia (but is due to a different gene abnormality and a different aberrant developmental process). Reprinted from Lee et al, 1997, with permission of the authors and publisher. Copyright 1997 by the Society for Neuroscience.

Methylazoxymethanol acetate (MAM) is a DNA methylating agent. When administered to pregnant rats on E15, the offspring show microcephaly, cortical dysplasia, periventricular heterotopia, and hippocampal heterotopia.[147] Although they do not have spontaneous seizures, MAM-treated rats have an increased propensity for seizures in a number of seizure models.[7,26,65,66] Increased numbers of bursting neurons have been reported in the dysplastic cortex, periventricular heterotopia, and hippocampal heterotopia.[6,138] In hippocampal heterotopia, Baraban and colleagues have found: increased epileptiform bursting in slices bathed in elevated extracellular K^+;[6] loss of Kv4.2 (A-type) potassium channels in heterotopic neurons[21] (Fig. 5A-C); and the ability of heterotopic cell regions to generate epileptiform bursting independent of surrounding hippocampal cells.[8] Heterotopic neurons in the hippocampus share many features with layer II/III neocortical neurons based on morphology, birthdating, and neurochemical profiles.[28] These cells also form an abnormal functional bridge between adjacent hippocampal neurons and overlying neocortical neurons.[8,27]

The perinatal freeze lesion model produces a focal area of cortical dysplasia that resembles four-layered polymicrogyria. Application of a cold probe to the skull of a rat pup (P0 or P1) results in focal necrosis of layers IV and V; however, later-generated neurons are able to migrate through this region to create the more superficial layers of the neocortex.[55] The injury creates

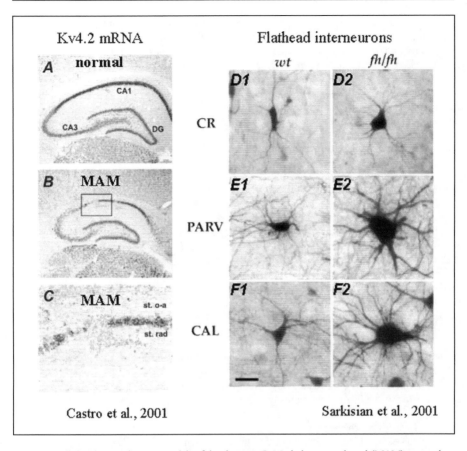

Figure 5. Cellular abnormalities in models of dysplasia. A-C: Methylazoxymethanol (MAM)-exposed rat hippocampus. MAM-treated rats have heterotopic neurons in/near the hippocampus, which exhibit electrophysiological properties different from their normotopic neighbors. A. Expression of mRNA for the Kv4.2 potassium channel subunit in control hippocampus. B. Kv4.2 mRNA expression in MAM-exposed rat. This subunit is absent from the heterotopic neurons (box) in CA1. C. Higher magnification of boxed area in B. Abbr.: CA1 and CA3 - hippocampal subfields; DG – dentate gyrus; st. o-a – stratum oriens-alveus; st. rad – stratum radiatum. Reprinted from Castro et al, 2001, with permission of the authors and publisher. Copyright 2001 by the Society for Neuroscience. D-F: Interneurons in flathead rat. The flathead rat shows a selective loss of inhibitory interneurons, and the surviving interneurons are grossly abnormal. Three types of inhibitory interneurons were imaged with antibodies against calretinin (CR), parvalbumin (PARV), and calbindin (CAL). In all three types, the somal size was significantly larger in flathead rats (fh/fh) compared to wild-type (wt) rats. Reprinted from Sarkisian et al, 2001, with permission of the authors and publisher. Copyright 2001 by Oxford University Press.

an infolding of the cortical surface that is called a microsulcus. Although the animals do not appear to have spontaneous seizures, the cortex adjacent to the microsulcus, the paramicrogyral zone (PMZ), is intrinsically hyperexcitable.[82] One hypothesis to explain this hyperexcitability points to redirection of thalamic and callosal afferents away from the microgyrus and into the PMZ, resulting in an imbalance of excitatory/inhibitory connections.[84] This hypothesis is supported by findings of enhanced excitatory drive onto inhibitory neurons in the PMZ,[121] an increase in the relative preponderance of spontaneous EPSCs over IPSCs,[84] and alterations in barrel receptor fields that suggest reorganization of thalamic afferents.[85] Other studies in freeze-lesioned rats have focused on alterations in post-synaptic receptors. DeFazio and Hablitz[48] showed increased sensitivity to ifenprodil, a selective NR2B agonist, in PMZ pyramidal neurons. This sensitivity is characteristic of immature cortical neurons, and therefore may represent preservation of an immature phenotype in dysplastic neurons in this model. DeFazio and Hablitz[47] also showed that neurons of the PMZ are not sensitive to zolpidem, a selective agonist of the type I benzodiazepine receptor—a feature also characteristic of an immature receptor profile. Finally, Redecker et al[125] showed specific alterations in GABA receptor subunits throughout the ipsilateral cortex and hippocampus (and even in homologous regions of the contralateral hemisphere); these changes, at least in some cases, again suggested an immature GABA receptor profile. These studies support the concept of cortical dysplasia as a state of arrested development, and show that a focal perinatal injury can have widespread effects on cortical development.

Spontaneous Mutants

The *telencephalic internal structure heterotopia* (tish) rat is a spontaneous mutant that is characterized by a large mass of gray matter that runs deep to the frontoparietal cortex (Fig. 4B); the affected animals exhibit spontaneous seizures.[94] Although this structural abnormality resembles subcortical band heterotopia in humans, the two conditions are genetically distinct (tish is an autosomal recessive trait, whereas subcortical band heterotopia is X-linked). During cortical development, tish rats have one preplate but two proliferative zones and two cortical plates. The cortical plate that develops within the preplate shows relatively normal lamination, but the "cortex" that develops deep to the preplate shows absence of lamination.[93] As with irradiated and MAM-treated rats, heterotopic cells in tish rats retain the capacity to form long-range reciprocal connections with cortical and subcortical targets.[144,145] Although the mechanisms of seizure generation in tish rat are not known, in vitro slice experiments have suggested that seizure activity may actually start in the "normal" cortex and then be synaptically transmitted to the heterotopic cortex.[25]

The *flathead* rat was discovered as a spontaneous mutant with microcephaly and spontaneous seizures.[131,142] It is now known to result from a mutation of the *citron kinase (K)* gene and, as such, is the rat counterpart to the *Citron K* knockout mouse.[50] *Citron K* is normally expressed in neuronal progenitor cells, and is involved in cytokinesis. Neuronal progenitors in *Citron K* knockout mice become stuck at the G2-M transition; this defect results in an increased number of tetraploid neurons, and appears to initiate apoptosis in a large number of these cells.[50,141] Knockout animals show a preferential loss of late-generated neurons, so that the cerebellum and dentate gyrus are most severely affected. In addition, progenitors of GABAergic interneurons in the ganglionic eminence show dramatic increases in apoptosis, resulting in a significant reduction in the density of cortical interneurons.[141] Interestingly, surviving interneurons have giant proportions (Fig. 5D-F) with somal areas averaging 400 μm^2 for parvalbumin-containing cells;[141] the *flathead/Citron K* knockout is among the few animal models with quantitatively documented "giant" (i.e., "dysmorphic") neurons.

Transgenic Animals

Lis1 +/- mice show major abnormalities of cortical development and seizures that correlate with the degree of loss of function of the gene.[80] Structural changes include diffuse cortical dysplasia and disorganization of the hippocampal cell layers. Birthdating studies in Lis1 transgenics have shown impairment of neuronal migration in the neocortex and the hippocampus.[20,61] The Lis1 protein likely regulates neuronal migration through its interactions with the microtubule system (for review, see ref. 59). Functional abnormalities in hippocampi from *Lis1* +/- mice include increased epileptiform bursting in the presence of elevated extracellular K+ and hyperexcitability of the Schaffer collateral-CA1 synapse.[61]

p35 -/- mice have disruption of cortical lamination due to an inability of later-generated neurons to bypass earlier-generated neurons in the cortical plate.[92] These animals also show hippocampal abnormalities, and have spontaneous seizures.[160] *p35* is a neuron-specific activator of cyclin-dependent kinase 5 (cdk5) that is expressed in migrating neurons but not neuronal progenitors;[152] this *p35*/cdk5 complex is important in late neuronal migration and lamination. Investigations of *p35* knockouts have focused on the hippocampal dysplasia, which include heterotopic pyramidal neurons in the strata oriens and radiatum in CA3, the formation of a second pyramidal cell layer in the stratum oriens in CA1, broadening of the dentate granule cell layer (with dispersion of granule cells into the molecular layer and hilus, and sprouting of mossy fibers into the inner molecular layer), and the development of abnormal dendritic orientations and prominent basal dendrites (in some granule cells). Since many of these abnormal features have been reported in human mesial temporal lobe epilepsy, it is tempting to hypothesize that disorders of neuronal migration—such as found in the *p35* knockout mouse— provide an initial substrate for epileptogenesis in temporal lobe epilepsy. This hypothesis is supported by the recent finding of decreased reelin mRNA expression in Cajal-Retzius cells from hippocampi of patients with temporal lobe epilepsy and granule cell dispersion in the dentate gyrus.[76]

Comparison with Human Cortical Dysplasia

In spite of all that has been learned from animal models of cortical dysplasia, direct comparison with human syndromes remains tenuous. For example, there is currently no good model of severe focal cortical dysplasia characterized by dysmorphic, giant neurons and balloon cells. Given this limitation, it remains important to study dysplastic cells taken directly from human tissue.

The inability to reproduce a human dysplastic disorder in a rat or mouse is perhaps not surprising, given that many of these disorders are a function of a developmental defect – and that rodent and human have significant differences (temporal and spatial) that determine the patterns of developmental change. Further, if the epileptogenicity of dysplastic lesions is determined, at least in part, by genetically-dictated predisposing factors, the different genetic make-ups (and gene expressions) of human vs. rodent neurons (and glia) will have important consequences for dysplasia-based epileptogenicity. It is nonetheless worthwhile to take advantage of specific aspects of animal model systems in order to gain a better understanding of the developmental mechanisms that lead to dysplasia, and of the "causal" relationships between various patterns of dysplasia and epileptogenicity.

Injury-based models may reflect human cortical dysplasia that result from in utero or perinatal injury. The perinatal freeze lesion model looks very similar to human four-layered polymicrogyria. In addition to those characteristics described above, reactive astrocytes in the microsulcus show reduced dye coupling and reduced inwardly-rectifying K$^+$ channels. This type of change may represent another important mechanism of epileptogenesis in injury-based developmental lesions.[18] Irradiated rat and MAM-treated rat represent cortical injuries earlier in development. The disorganized and heterotopic neurons that are seen in these models mimic many of the dysplastic features that were described in children with "acquired" cortical dysplasia.[101] The findings from the irradiated rat suggest that inhibitory interneurons may have an increased susceptibility to some types of in utero injury—which would in turn have major implications for epilepsy associated with injuries of this type.

The connection between *Lis1* +/- mice and human lissencephaly is obvious, since the model was based on the human gene defect. However, even here there are differences in dysplastic phenotype (and, presumably, epileptogenicity) based on the specific nature of the gene mutation.[20,80] The more severely affected mice resemble some aspects of human lissencephaly type I, although the classic four-layered arrangement of the cortex has not been described in the mouse. In this lissencephalic condition, it remains difficult to isolate the specific abnormalities that result in epilepsy. Modulators of late neuronal migration such as reelin and *p35* can produce cortical dysplasia if they are disrupted experimentally. Mutations in the *RELN* gene have been implicated in the rare syndrome of human lissencephaly with cerebellar hypoplasia, but mutations of *p35* and the *citron K* gene have yet to be implicated in human disease. It is noteworthy, however, that these latter mouse models do exhibit spontaneous seizure activity.

Concluding Comments

Our current sophisticated characterization of developmental/structural brain lesions—made possible in large measure by increased sensitivity of our tools for detecting these abnormalities—has led to an awareness of cortical dysplasia as a major structural "contributor" to many epileptic syndromes. Working with human clinical material, investigators have generated a detailed characterization of cortical dysplasias, along with an analysis of the genetic mutations underlying many of these abnormalities. Our current challenge is to elucidate the epileptic— and epileptogenic—processes associated with cortical dysplasias, so that we can develop more effective therapies (including "prevention") for these epileptic conditions. Toward that end, development of animal models provides an important window for investigation. What constitutes a "useful" model for studying these dysplastic syndromes? Different animal models provide rather different perspectives on the dysplasias seen in human epileptic patients. Some models are based on known genetic abnormalities (e.g., lissencephaly or tuberous sclerosis). Other models are designed in an effort to "reproduce" the structural abnormalities (e.g., heterotopia, double cortex) seen in human brain. And still other models emphasize the generation of chronic, spontaneous seizures. Few rat or mouse systems meet all the phenotypic requirements of the "ideal" model. It is important to remain aware of the strengths—and limitations—of each new model, as we work toward elucidating the causal connections between cortical dysplasia and epilepsy.

References

1. Allen KM, Walsh CA. Genes that regulate neuronal migration in the cerebral cortex. Epilepsy Res 1999; 36:143-154.
2. Allendoerfer KL, Shatz CJ. The subplate, a transient neocortical structure: its role in the development of connections between thalamus and cortex. Ann Rev Neurosci 1994; 17:185-218.
3. Anderson SA, Eisenstat DD, Shi L et al. Interneuron migration from basal forebrain to neocortex: dependence on *DLX* genes. Science 1997; 278:474-476.
4. Angevine JB, Sidman RL. Autoradiographic study of the cell migration during histogenesis of the cerebral cortex in the mouse. Nature 1961; 192:766-768.
5. Babb TL, Ying Z, Mikuni N et al. Brain plasticity and cellular mechanisms of epileptogenesis in human and experimental cortical dysplasia. Epilepsia 2000; 41(Suppl 6):S76-S81.
6. Baraban SC, Schwartzkroin PA. Electrophysiology of CA1 pyramidal neurons in an animal model of neuronal migration disorders: prenatal methylazoxymethanol treatment. Epilepsy Res 1995; 22:145-156.
7. Baraban SC, Schwartzkroin PA. Flurothyl seizure susceptibility in rats following prenatal methylazoxymethanol treatment. Epilepsy Res 1996; 23:189-194.
8. Baraban SC, Wenzel HJ, Hochman DW et al. Characterization of heterotopic cell clusters in the hippocampus of rats exposed to methylazoxymethanol in utero. Epilepsy Res 2000; 39:87-102.
9. Barkovich AJ, Chuang SH, Norman D. MR of neuronal migration anomalies. AJNR 1987; 8:1009-1017.
10. Barkovich AJ, Gressens P, Evrard P. Formation, maturation and disorders of brain neocortex. AJNR 1992; 13:423-446.
11. Barkovich AJ, Jackson Jr DE, Boyer RS. Band heterotopias: a newly recognized neuronal migration anomaly. Radiology 1989; 171:455-458.
12. Barkovich AJ, Kjos BO. Schizencephaly: correlation of clinical findings with MR characteristics. AJNR 1992; 13:85-94.
13. Barkovich AJ, Kuzniecky RI, Dobyns WB et al. A classification scheme for malformations of cortical development. Neuropediatrics 1996; 27:59-63.
14. Barkovich AJ, Kuzniecky RI, Jackson GD et al. Classification system for malformations of cortical development: Update 2001. Neurology 2001; 57:2168-2178.
15. Barkovich AJ, Rowley H, Bollen A. Correlation of prenatal events with the development of polymicrogyria. AJNR Am J Neuroradiol 1995; 16(suppl 4):822-827.
16. Barth PG. Disorders of neuronal migration. Can J Neurol Sci 1987; 14:1-16.
17. Bayer SA, Altman J, Russo RJ. Time tables of neurogenesis in human brain based on experimentally determined patterns in the rat. Neurotoxicology 1993; 14:83-144.
18. Bordey A, Lyons SA, Hablitz JJ et al. Electrophysiological characteristics of reactive astrocytes in experimental cortical dysplasia. J Neurophysiol 2001; 85:1719-1731.

19. Brunelli S, Faiella A, Capra V et al. Germline mutations in the homeobox gene *EMX2* in patients with severe schizencephaly. Nat Genet 1996; 12:94-96.
20. Cahana A, Escamez T, Nowakowski RS et al. Targeted mutagenesis of *LIS1* disrupts cortical development and *LIS1* homodimerization. Proc Natl Acad Sci USA 2001; 98:6429-6434.
21. Castro PA, Cooper EC, Lowenstein DH et al. Hippocampal heterotopia lack functional Kv4.2 potassium channels in the methylazoxymethanol model of cortical malformations and epilepsy. J Neurosci 2001; 21:6626-6634.
22. Caviness Jr VS, Hatten ME, McConnell et al. Developmental neuropathology and childhood epilepsies. In: Schwartzkroin PA, Moshe SL, Noebels JL, Swann JW, eds, Brain development and epilepsy. New York: Oxford University Press, 1995:94-121.
23. Caviness Jr VS, Sidman RL. Time of origin of corresponding cell classes in the cerebral cortex of normal and reeler mutant mice: an autoradiographic analysis. J Comp Neurol 1973; 148:141-151.
24. Chae T, Kwon YT, Bronson R et al. Mice lacking *p35*, a neuronal specific activator of Cdk5, display cortical lamination defects, seizures, and adult lethality. Neuron 1997; 18:29-42.
25. Chen Z-F, Schottler F, Bertam E et al. Distribution and initiation of seizure activity in a rat brain with subcortical band heterotopia. Epilepsia 2000; 41:493-501.
26. Chevassus-au-Louis N, Ben-Ari Y, Vergnes M. Decreased seizure threshold and more rapid rate of kindling in rats with cortical malformation induced by prenatal treatment with methylazoxymethanol. Brain Res 1998; 812:252-255.
27. Chevassus-au-Louis N, Congar P, Represa A et al. Neuronal migration disorders: heterotopic neocortical neurons in CA1 provide a bridge between the hippocampus and the neocortex. Proc Natl Acad Sci USA 1998; 95:10263-10268.
28. Chevassus-au-Louis N, Rafiki A, Jorquera I et al. Neocortex in the hippocampus: an anatomical and functional study of CA1 heterotopias after prenatal treatment with methylazoxymethanol in rats. J Comp Neurol 1998; 394:520-536.
29. Chevassus-au-Louis N, Baraban SC, Gaiarsa J-L et al. Cortical malformations and epilepsy: new insights from animal models. Epilepsia 1999; 40:811-821.
30. Chevassus-au-Louis N, Jorquera I, Ben-Ari Y et al. Abnormal connections in the malformed cortex of rats with prenatal treatment with methylazoxymethanol may support hyperexcitability. Dev Neurosci 1999; 21:385-392.
31. Choi BH, Lapham LW, Amin-Zaki L et al. Abnormal neuronal migration, deranged cerbral cortical organization, and diffuse white matter astrocytosis of human fetal brain: a major effect of methylmercury poisoining in utero. J Neuropath Exp Neurol 1978; 37:719-733.
32. Chong SS, Pack SD, Roschke AV et al. A revision of the lissencephaly and Miller-Dieker syndrome critical regions in chromosome 17p13.3. Hum Mol Genet 1997; 6:147-155.
33. Colacitti C, Sancini G, Franceschetti S et al. Altered connections between neocortical and heterotopic areas in methylazoxymethanol-treated rat. Epilepsy Res 1998; 32:49-62.
34. Copp AJ, Harding BN. Neuronal migration disorders in humans and in mouse models—an overview. Epilepsy Res 1999; 36:133-141.
35. Cotter DR, Honavar M, Everall I. Focal cortical dysplasia: a neuropathological and developmental perspective. Epilepsy Res 1999; 36:155-164.
36. Cotter D, Honavar M, Lovestone S et al. Disturbance of Notch-1 and Wnt signaling proteins in neuroglial balloon cells and abnormal large neurons in focal cortical dysplasia in human cortex. Acta Neuropathol (Berl) 1999; 98:465-472.
37. Cowan D, Geller LM. Long-term pathological effects of prenatal X-irradiation on the central nervous system of the rat. J Neuropathol Exp Neurol 1960; 19:488-527.
38. Crino PB, Duhaime AC, Baltuch G et al. Differential expression of glutamate and $GABA_A$ receptor subunit mRNA in cortical dysplasia. Neurology 2001; 56:906-913.
39. Crino PB, Eberwine J. Cellular and molecular basis of cerebral dysgenesis. J Neurosci Res 1997; 50:907-916.
40. Crino PB, Henske EP. New developments in the neurobiology of the tuberous sclerosis complex. Neurology 1999; 53:1384-1390.
41. Crino PB, Jin H, Shumate MD et al. Increased expression of the neuronal glutamate transporter (EAAT3/EAAC1) in hippocampal and neocortical epilepsy. Epilepsia 2002; 43:211-218.
42. Crino PB, Trojanowski JQ, Dichter MA et al. Embryonic neuronal markers in tuberous sclerosis: Single-cell moeluclar pathology. Proc Natl Acad Sci USA 1996; 93:14152-14157.
43. Crino PB, Trojanowski JQ, Eberwine J. Internexin, MAP1B, and nestin in cortical dysplasia as markers of developmental maturity. Acta Neuropathol (Berl) 1997; 93:619-627.
44. Cusmai R, Wheless JW, Berkovic S et al. Genetic and neuroradiological heterogeneity of double cortex syndrome. Ann Neurol 2000; 47:265-9.

45. D'Arcangelo G, Miao GG, Chen S-C et al. A protein related to extracellular matrix proteins deleted in the mouse mutant reeler. Nature 1995; 374:719-723.
46. DeCarlos JA, Lopez-Mascaraque L, Valverde F. Dynamics of cell migration from the lateral ganglionic eminence in the rat. J Neurosci 1996; 16:8313-8323.
47. DeFazio RA, Hablitz JJ. Reduction of zolpidem sensitivity in a freeze lesion model of neocortical dysgenesis. J Neurophysiol 1999; 81:404-407.
48. DeFazio RA, Hablitz JJ. Alterations in NMDA receptors in a rat model of cortical dysplasia. J Neurophysiol 2000; 83:315-321.
49. des Portes V, Pinard JM, Billuart P et al. A novel CNS gene required for neuronal migration and involved in X-linked subcortical laminar heterotopia and lissencephaly syndrome. Cell 1998; 92:51-61.
50. DiCunto F, Imarisio S, Hirsch E et al. Defective neurogenesis in citron kinase knockout mice by altered cytokinesis and massive apoptosis. Neuron 2000; 28:115-127.
51. Dobyns WB, Andermann E, Andermann F et al. X-linked malformations of neuronal migrations. Neurology 1996; 47:331-339.
52. Dobyns WD, Ledbetter DH. Clinical and molecular studies in 62 patients with type I lissencephaly. Ann Neurol 1990; 28:240.
53. Dobyns WB, Reiner O, Carrozzo R et al. Lissencephaly. A human brain malformation associated with deletion of the *LIS1* gene located at chromosome 17p3. JAMA 1993; 270:2838-2842.
54. Duong T, De-Rosa MJ, Poukens V et al. Neuronal cytoskeletal abnormalities in human cerebral cortical dysplasia. Acta Neuropathol (Berl) 1994; 87:493-503.
55. Dvorak K, Feit J. Migration of neuroblasts through partial necrosis of the cerebral cortex in newborn rats: contribution to the problems of morphological development and developmental period of cerebral microgyria. Acta Neuropathol 1977; 38:203-212.
56. Eksioglu YZ, Sheffer IE, Cardenas P et al. Periventricular heterotopia: an X-linked dominant epilepsy locus causing aberrant cerebral cortical development. Neuron 1996; 16:77-87.
57. European Chromosome-16 Tuberous Sclerosis Consortium. Identification and characterization of the tuberous sclerosis gene on chromosome 16. Cell 1993; 75:1305-1315.
58. Farrell MA, DeRosa MJ, Curran JG et al. Neuropathologic findings in cortical resections (including hemispherectomies) performed for the treatment of intractable childhood epilepsy. Acta Neuropathol 1992; 83:246-259.
59. Feng Y, Walsh CA. Protein-protein interactions, cytoskeletal regulation and neuronal migration. Nature Rev Neurosci 2001; 2:408-416.
60. Ferrer I, Catala I. Unlayered polymicrogyria: structural and developmental aspects. Anat Embryol 1991; 184:517-528.
61. Fleck MW, Hirotsune S, Gambello MJ et al. Hippocampal abnormalities and enhanced excitability in a murine model of human lissencephaly. J Neurosci 2000; 20:2439-2450.
62. Fox JW. Mutations in filamin 1 prevents migration of cerebral cortical neurons in human periventricular heteropia. Neuron 1998; 21:1315-1325.
63. Francis F, Koulakoff A, Boucher D et al. Doublecortin is a developmentally regulated, microtubule-associated protein expressed in migrating and differentiating neurons. Neuron 1999; 23:247-256.
64. Gelot A, Billette de Villemeur T, Bordarier C et al. Developmental aspects of type II lissencephaly. Comparative study of dysplastic lesions in fetal and post-natal brains. Acta Neuropathol (Berl) 1995; 89:72-84.
65. Germano IM, Sperber EF. Increased seizure susceptibility in adult rats with neuronal migration disorders. Brain Res 1997; 777:219-222.
66. Germano IM, Zhang YF, Sperber EF et al. Neuronal migration disorders increase susceptibility to hyperthermia-induced seizures in developing rats. Epilepsia 1996; 37:902-210.
67. Gillies K, Price DJ. The fates of cells in the developing cerebral cortex of normal and methylazoxymethanol acetate-lesioned mice. Eur J Neurosci 1993; 5:73-84.
68. Gilmore EC, Ohshima T, Goffinet AM et al. Cyclin-dependent kinase 5-deficient mice demonstrate novel developmental arrest in cerebral cortex. J Neurosci 1998; 18:6370-7.
69. Gleeson JG, Allen KM, Fox JW et al. Doublecortin, a brain-specific gene mutated in human X-linked lissencephaly and double cortex syndrome, encodes a putative signaling protein. Cell 1998; 92:63-72.
70. Gleeson JG, Lin PT, Flanagan LA et al. Doublecortin is a microtubule-associated protein and is expressed widely by migrating neurons. Neuron 1999; 23:257-271.
71. Gleeson JG, Luo RF, Grant PE et al. Genetic and neuroradiological heterogeneity of double cortex syndrom. Ann Neurol 2000; 47:265-269.

72. Gleeson JG, Walsh CA. Neuronal migration disorders: from genetic diseases to developmental mechanisms. Trends Neurosci 2000; 23:352-359.
73. Gleeson JG. Neuronal migration disorders. Ment Retard Dev Disabil Res Rev 2001; 7:167-171.
74. Golden JA. Cell migration and cerebral cortical development. Neuropathol Appl Neurobiol 2001; 27: 22-28.
75. Granata T, Farina L, Faiella A et al. Familial schizencephaly associated with *EMX2* mutation. Neurology 1997; 48:1403-1406.
76. Haas CA, Dudeck O, Kirsch M et al. Role for reelin in the development of granule cell dispersion in temporal lobe epilepsy. J Neurosci 2002; 22:5797-5802.
77. Hablitz JJ, DeFazio RA. Altered receptor subunit expression in rat neocortical malformations. Epilepsia 2000; 41(Suppl 6):S82-S85.
78. Hannan AJ, Servotte S, Katsnelson A et al. Characterization of nodular neuronal heterotopia in children. Brain 1999; 122:219-238.
79. Hirabayashi S, Binnie CD, Janota I et al. Surgical treatment of epilepsy due to cortical dysplasia: clinical and EEG findings. J Neurol Neurosurg Psych 1992; 56:765-770.
80. Hirotsune S, Fleck MW, Gambello MJ et al. Graded reduction of *Pafahb1* (*LIS1*) activity results in neuronal migration defects and early embryonic lethality. Nat Genet 1998; 19:333-339.
81. Hong SE, Shugart YY, Huang DT et al. Autosomal recessive lissencephaly with cerebellar hypoplasia is associated with human *RELN* mutations. Nat Genet 2000; 26:93-96.
82. Jacobs KM, Gutnick MJ, Prince DA. Hyperexcitability in a model of cortical maldevelopment. Cereb Cortex 1996; 6:514-523.
83. Jacobs KM, Hwang BJ, Prince DA. Focal epileptogenesis in a rat model of polymicrogyria. J Neurophysiol 1999; 81:159-173.
84. Jacobs KM, Kharazia VN, Prince DA. Mechanisms underlying epileptogenesis in cortical malformations. Epilepsy Res 1999; 36:165-188.
85. Jacobs KM, Mogensen M, Warren Eet al. Experimental microgyri disrupt the barrel field pattern in rat somatosensory cortex. Cereb Cortex 1999; 9:733-744.
86. Jellinger K, Rett A. Agyria-pachygyria (lissencephaly syndrome). Neuropediatrie 1976; 7:66-91.
87. Juhasz C, Chugani HT, Muzik O et al. Hypotheses from functional neuroimaging studies. In Schwartzkroin PA, Rho JM eds, Epilepsy, Infantile spasms, and Developmental Encephalopathy. San Diego: Academic Press, 2002:37-55.
88. Kakita A, Wakabayashi K, Su M et al. Experimentally induced leptomeningeal glioneuronal heterotopia and underlying cortical dysplasia of the lateral limbic area in rats treated transplacentally with methylmercury. J Neuropathol Exp Neurol 2001; 60:767-777.
89. Kerfoot C, Vinters HV, Mathern GW. Cerebral cortical dysplasia: giant neurons show potential for increased excitation and axonal plasticity. Dev Neurosci 1999; 21:260-270.
90. Koh S, Jayakar P, Dunoyer C et al. Epilepsy surgery in children with tuberous sclerosis complex: presurgical evaluation and outcome. Epilepsia 2000; 41:1206-1213.
91. Kondo S, Najm I, Kunieda T et al. Electroencephalographic characterization of an adult rat model of radiation-induced cortical dysplasia. Epilepsia 2001; 42:1221-1227.
92. Kwon YT, Tsai L-H. A novel disruption of cortical development in *p35* -/- mice distinct from reeler. J Comp Neurol 1998; 395:510-522.
93. Lee KS, Collins JL, Anzivino MJ et al. Heterotopic neurogenesis in a rat with cortical heterotopia. J Neurosci 1998; 18:9365-9375.
94. Lee KS, Schottler F, Collins JL et al. A genetic animal model of human neocortical heterotopia associated with seizures. J Neurosci 1997; 17:6236-6242.
95. Leventer RJ, Phelan EM, Coleman LT et al. Clinical and imaging features of cortical malformations in childhood. Neurology 1999; 53:715-722.
96. Luhmann HJ, Raabe K, Qu M et al. Charactrization of neuronal migration disorders in neocrotical structures: extracellular in vitro recordings. Eur J Neurosci 1998; 10:3085-3094.
97. Luskin MB, Shatz CJ. Neurogenesis of the cat's primary visual cortex. J Comp Neurol 1985; 242: 611-631.
98. Marin-Padilla M. Developmental neuropathology and impact of perinatal brain damage. I: Hemorrhagic lesions of neocortex. J Neuropathol Exp Neurol 1996; 55:758-773.
99. Marin-Padilla M. Developmental neuropathology and impact of perinatal brain damage. II: White matter lesions of the neocortex. J Neuropathol Exp Neurol 1997; 56:219-235.
100. Marin-Padilla M. Developmental neuropathology and impact of perinatal brain damage. III: Gray matter lesions of the neocortex. J Neuropathol Exp Neurol 1999; 58:407-429.
101. Marin-Padilla M, Tsai R J-C, King MA et al. Abnormalities of neuronal morphology in an animal model of cortical dysgenesis with a comparison to human acquired cortical dysplasia. Epilepsia 2002; 43(Suppl 7):24.

102. Mathern GW, Cepdea C, Hurst RS et al. Neurons recorded from pediatric epilepsy surgery patients with cortical dysplasia. Epilepsia 2000; 41(Suppl 6):S162-S167.
103. McConnell S. Constructing the cerebral cortex: neurogenesis and fate determination. Neuron 1995; 15: 761-768.
104. Meyer G, Schaaps JP, Moreau L et al. Embryonic and early fetal development of the human neocortex. J Neurosci 2000; 20:1858-1868.
105. Mischel PS, Nguyen LP, Vinters HV. Cerebral cortical dysplasia associated with pediatric epilepsy. Review of neuropathologic features and proposal for a grading system. J Neuropathol Exp Neurol 1995; 54:137-153.
106. Norman MG, Roberts M, Sirois J et al. Lissencephaly. Can J Neurol Sci 1976; 3:39-46.
107. Ogawa M, Miyata T, Nakajima K et al. The reeler gene-associated antigen on Cajal-Retzius neurons is a crucial molecule for laminar organization of cortical neurons. Neuron 1995; 14:899-912.
108. Palmini A. Disorders of cortical development. Curr Opin Neurol 2000; 13:183-192.
109. Palmini A, Andermann F, Olivier A et al. Focal neuronal migration disorders and intractable epilepsy: a study of 30 patients. Ann Neurol 1991; 30:741-749.
110. Palmini A, Andermann E, Andermann F. Prenatal events and genetic factors in epileptic patients with neuronal migration disorders. Epilepsia 1994; 35:965-973.
111. Palmini A, Gambardella A, Andermann F et al. Operative strategies for patients with cortical dysplastic lesions and intractable epilepsy. Epilepsia 199b; 35(Suppl 6):S57-S71.
112. Palmini A, Gambardella A, Andermann F et al. Intrinsic epileptogenicity of human dysplastic cortex as suggested by corticography and surgical results. Ann Neurol 1995; 37:476-487.
113. Parent JM, Lowenstein DY. Seizure-induced neurogenesis: are more new neurons good for an adult brain? Prog Brain Res 2002; 135:121-131.
114. Pavone L, Curatolo P, Rizzo R et al. Epidermal nevus syndrome: a neurologic variant with hemimegalencephaly, gyral malformation, mental retardation, seizures, and facial hemihypertrophy. Neurology 1991; 41:266-271.
115. Peckham NH, Choi BH. Abnormal neuronal distribution within the cerebral cortex after prenatal methylmercury intoxication. Acta Neuropathol (Berl) 1988; 76:222-226.
116. Pilz DT, Macha ME, Precht KS et al. Fluorescence in situ hybridization analysis with *LIS1* specific probes reveals a high deletion mutation rate in isolated lissencephaly sequence. Genet Med 1998; 1:29-33.
117. Pilz D, Stoodley N, Golden JA. Neuronal migration, cerebral cortical development, and cerebral cortical anomalies. J Neuropathol Exp Neurol 2002; 61:1-11.
118. Pinard JM, Motte J, Chiron C et al. Subcortical laminar heterotopia and lissencephaly in two families: a single X-linked dominant gene. J Neurol Neurosurg Psychiat 1994; 57:914-920.
119. Pleasure SJ, Anderson S, Hevner R et al. Cell migration from the ganglionic eminences is required for the development of hippocampal GABAergic interneurons. Neuron 2000; 28:727-740.
120. Prayson RA, Estes ML. Cortical dysplasia: a histopathologic study of 52 cases of partial lobectomy in patients with epilepsy. Hum Pathol 1995; 26:493-500.
121. Prince DA, Jacobs KM, Salin PA et al. Chronic focal neocortical epileptogenesis: does disinhibition play a role? Can J Physiol Pharmacol 1997; 75:500-507.
122. Rakic P. Mode of cell migration to the superficial layers of fetal monkey neocortex. J Comp Neurol 1972; 145:61-83.
123. Rakic P. Timing of major ontogenetic events in the visual cortex of the rhesus monkey. In: Buchwald NA, Brazier MA, eds. Brain Mechanisms in Mental Retardation. New York: Academic Press, 1975:3-40.
124. Rakic P. Specification of cerebral cortical areas. Science 1988; 241:170-176.
125. Redecker C, Luhmann HJ, Hagemann G et al. Differential downregulation of $GABA_A$ receptor subunits in widespread brain regions in the freeze-lesion model of focal cortical malformations. J Neurosci 2000; 20:5045-5053.
126. Reiner O, Carrozzo R, Shen Y et al. Isolation of a Miller-Dieker lissencephaly gene containing G protein β-subunit-like repeats. Nature 1993; 364:717-721.
127. Ricci S, Cusmai R, Fariello G et al. Double cortex. A neuronal migration anomaly as a possible cause of Lennox-Gastaut syndrome. Arch Neurol 1992; 49:61-4.
128. Richardson EP. Pathology of tuberous sclerosis. Ann NY Acad Sci 1991; 615:128-139.
129. Rio C, Rieff HI, Qi P et al. Neuregulin and erbB receptors play a critical role in neuronal migration. Neuron 1997; 19:39-50.
130. Robain O, Floquet C, Heldt N et al. Hemimegalencephaly: a clinicopathological study of four cases. Neuropathol Appl Neurobiol 1988:125-135.
131. Roberts MR, Bittman K, Li W-W et al. The flathead mutation causes CNS-specific developmental abnormalities and apoptosis. J Neurosci 2000; 20:2295-2306.

132. Roper SN, Abraham LA, Streit WJ. Exposure to in utero irradiation produces disruption of radial glia in rats. Dev Neurosci 1997; 19:521-528.

133. Roper SN, Eisenschenk S, King MA. Reduced density of parvalbumin- and calbindin D28k-immunoreactive neurons in experimental cortical dysplasia. Epilepsy Res 1999; 37:63-71.

134. Roper SN, Gilmore RL, Houser CR. Experimentally induced disorders of neuronal migration produce an increased propensity for electrographic seizures in rats. Epilepsy Res 1995; 21:205-219.

135. Roper SN, King MA, Abraham LA et al. Disinhibited in vitro neocortical slices containing experimentally induced cortical dysplasia demonstrate hyperexcitability. Epilepsy Res 1997; 26:443-449.

136. Rorke LB. A perspective: the role of disordered genetic control of neurogenesis in the pathogenesis of migration disorders. J Neuropathol Exp Neurol 1994; 53:105-117.

137. Saito Y, Mizuguchi M, Oka A et al. Fukutin protein is expressed in neurons of the normal developing human brain but is reduced in Fukuyama-type congenital muscular dystrophy brain. Ann Neurol 2000; 47:756-764.

138. Sancini G, Franceschetti S, Battaglia G et al. Dysplastic neocortex and subcortical heterotopias in methylazoxymethanol-treated rats: an intracellular study of identified pyramidal neurons. Neurosci Lett 1998; 246:181-185.

139. Santi MR, Golden JA. Periventricular heterotopia may result from radial glial fiber disruption. J Neuropathol Exp Neurol 2001; 60:856-862.

140. Sapir T, Elbaum M, Reiner O. Reduction of microtubule catastrophe events by *LIS1*, platelet-activating factor acetylhydrolase subunit. EMBO J 1997; 16:6977-6984.

141. Sarkisian MR, Frenkel M, Li W et al. Altered interneuron development in the cerebral cortex of the flathead mutant. Cereb Cortex 2001; 11:734-743.

142. Sarkisian MR, Rattan S, D'Mello SR et al. Characterization of seizures in the flathead rat: a new genetic model of epilepsy in early postnatal development. Epilepsia 1999; 40:394-400.

143. Schauwecker PE. Complications associated with genetic background effects in models of experimental epilepsy. Prog Brain Res 2002; 135:139-148.

144. Schottler F, Fabiato H, Leland JM et al. Normotopic and heterotopic cortical representations of mystacial vibrissae in rats with subcortical band heterotopia. Neuroscience 2001; 108:217-235.

145. Schottler F, Couture D, Rao A et al. Subcortical connections of normotopic and heterotopic neurons in sensory and motor cortices of the tish mutant rat. J Comp Neurol 1998; 395:29-42.

146. Schwartzkroin PA, Walsh CA. Cortical malformations and epilepsy. Ment Retard Dev Disabil Res Rev 2000; 6:268-280.

147. Singh SC. Ectopic neurons in the hippocampus of the postnatal rat exposed to methylazoxymethanol during foetal development. Acta Neuropathol (Berl) 1977; 44:197-202.

148. Spreafico R, Pasquier B, Minotti L et al. Immunocytochemical investigation on dysplastic human tissue from epileptic patients. Epilepsy Res 1998; 32:34-48.

149. Spreafico R, Tassi L, Colombo N et al. Inhibitory circuits in human dysplastic tissue. Epilepsia 2000; 41(Suppl 6):S168-S173.

150. Stewart RM, Richman DP, Caviness Jr VS. Lissencephaly and pachygyria: an architectonic and topographical analysis. Acta Neuropathol (Berl) 1975; 31:1-12.

151. Taylor DC, Falconer MA, Bruton CJ et al. Focal dysplasia of the cerebral cortex in epilepsy. J Neurol Neurosurg Psychiat 1971; 34:369-387.

152. Tsai L-H, Delalle I, Caviness Jr VS et al. *p35* is a neural-specific regulatory subunit of cyclin-dependent kinase 5. Nature 1994; 371:419-423.

153. Uher BF, Golden JA. Neuronal migration defects of the cerebral cortex: a destination debacle. Clin Genet 2000; 58:16-24.

154. Urbach H, Scheffler B, Heinrichsmeier T et al. Focal cortical dysplasia of Taylor's balloon cell type: A clinicopathological entity with characteristic neuroimaging and histopathological features and favorable postsurgical outcome. Epilepsia 2002; 43:33-40.

155. Vinters HV. Histopathology of brain tissue from patients with infantile spasms. In: Schwartzkroin PA, Rho JM, eds. Epilepsy, infantile spasms and developmental encephalopathy. San Diego: Academic Press, 2002:63-76.

156. Vinters HV, Fisher RS, Cornford FE et al. Morphological substrates of infantile spasms based on surgically resected cerebral tissue. Childs Nerv Syst 1992; 8:8-17.

157. Walsh CA. Neuronal identity, neuronal migration and epileptic disorders of the cerebral cortex. In: Schwartzkroin PA, Noebels JL, Moshe SL, Swann JW, eds. Brain Development and Epilepsy. New York: Oxford University Press, 1995:122-143.

158. Walsh CA. Genetic malformations of the human cerebral cortex. Neuron 1999; 23:19-29.

159. Walsh CA, Cepko CL. Widespread disperson of neuronal clones across functional regions of the cerebral cortex. Science 1992; 255:434-440.

160. Wenzel HJ, Robbins CA, Tsai L-H et al. Abnormal morphological and functional organization of the hippocampus in a *p35* mutant model of cortical dysplasia associated with spontaneous seizures. J Neurosci 2001; 21:983-998.
161. Yakovlev PI, Wadsworth RC. Schizencephalies. A study of the congenital clefts in the cerebral mantle. I. Clefts with fused lips. J Neuropathol Exp Neurol 1946; 5:116-130.
162. Yakovlev PI, Wadsworth RC. Schizencephalies. A study of the congenital clefts in the cerebral mantle. II. Clefts with hydrocephalus and lips separated. J Neuropathol Exp Neurol 1946; 5:169-206.
163. Yeung RS. Tuberous sclerosis as an underlying basis for infantile spasm. In Schwartzkroin PA, Rho JM, eds. Epilepsy, infantile spasms and developmental encephalopathy. San Diego : Academic Press, 2002:315-332.
164. Yeung RS, Xiao GH, Jin F et al. Predisposition to renal carcinoma in the Eker rat is determined by germ-line mutation of the tuberous sclerosis 2 (*TSC2*) gene. Proc Natl Acad USA 1994; 91:11413-11416.
165. Yeung RS, Katsetos CD, Kleinszanto A. Subependymal astrocytic hamartomas in the Eker rat model of tuberous sclerosis. Am J Pathol 1997; 151(5):1477-1486.
166. Ying Z, Babb TL, Mikuni N et al. Selective coexpression of NMDAR2A/B and NMDAR1 subunits proteins in dysplastic neurons of human epileptic cortex. Exp Neurol 1999; 159:409-418.
167. Zhu WJ, Roper SN. Reduced inhibition in an animal model of cortical dysplasia. J Neurosci 2000; 20:8925-8931.

Malformations of Cortical Development:
Molecular Pathogenesis and Experimental Strategies

Peter B. Crino

Abstract

Malformations of cortical development (MCD) are developmental brain lesions characterized by abnormal formation of the cerebral cortex and a high clinical association with epilepsy in infants, children, and adults. Despite multiple anti-epileptic drugs (AEDs), treatment of epilepsy associated with MCD may require cortical resection performed to remove the cytoarchitecturally abnormal region of cortex. Single genes responsible for distinct MCD including lissencephaly, subcortical band heterotopia, and tuberous sclerosis, have been identified and permit important mechanistic insights into how gene mutations result in abnormal cortical cytoarchitecture. The pathogenesis of MCD such as focal cortical dysplasia, hemimegalencephaly, and polymicrogyria, remains unknown. A variety of new techniques including cDNA array analysis now allow for analysis of gene expression within MCD.

Introduction

Malformations of cortical development (MCD) include a heterogeneous group of disorders also referred to as cortical dysplasias or neuronal migration disorders. MCD are characterized by disruption of the normal hexalaminar structure of the cerebral cortex (Fig. 1) and aberrant individual cellular morphologies (for reviews, see refs. 2, 23, 27, 115). MCD can uniformly affect broad regions of the cerebral cortex as in classical lissencepaly and hemimegalencephaly, or may be restricted to focal areas such as tubers in the tuberous sclerosis complex (TSC) or Taylor-type focal cortical dysplasia (FCD). In some of MCD, the normal six-layered organization of the cerebral cortex is replaced by a more primitive 4-layered arrangement, as in lissencephaly and polymicrogyria, whereas in FCD or tubers of TSC, there is a virtual loss of all lamination. MCD may also exhibit large collections of heterotopic neurons, as in syndromes of subcortical band heterotopia and periventricular nodular heterotopia. The morphology of individual neurons in many MCD subtypes is abnormal, suggesting a pervasive disruption of many steps important in neuronal development. MCD have been previously categorized on the basis of structural pathological features such as gyral patterning (agyria, pachygyria, polymicrogyria) or extent of cortical involvement (diffuse or focal). The developmental pathogenesis of many MCDs remains to be defined; however, recent advances in molecular biology and positional cloning strategies have permitted the identification of gene mutations responsible for select MCDs that have led to more precise molecular diagnostic classification of several disorders. As a consequence of these findings and understanding the functions of mutated genes in normal development, a recent scheme has been proposed that classifies MCD according to distinct neurodevelopmental stages including proliferation, migration, cortical organization, as well as disorders that have not yet been specified (Table 1).[6]

Recent Advances in Epilepsy Research, edited by Devin K. Binder and Helen E. Scharfman.
©2004 Eurekah.com and Kluwer Academic / Plenum Publishers.

> Disorganized cytoarchitecture
>
> Disruption of cortical lamination
>
> Disruption of radial organization
>
> Abnormal morphologies of neurons
> - dendritic arborizations
> - axon projections
>
> Aberrant synaptic connectivity

Figure 1. Malformations of cortical development (MCD).

Important Clinical Issues in Patients with MCD

The single unifying feature of all MCD is a high clinical association with epilepsy.[2,19] Malformations of cortical development are a recognized cause of often intractable epilepsy in infants, especially infantile spasms (IS), children and even adults.[44] Studies from most major pediatric epilepsy surgery centers report that MCD are the most common neuropathologic abnormality encountered when cortical resection is performed to treat IS and related intractable seizure disorders of infancy or childhood.[36,92,110,112] For example, it is estimated that MCD may account for 20% of all epilepsies[11,55] and in select extensive MCD such as lissencephaly, hemimegalencephaly, and TSC, seizures may occur in 70-90% of affected patients. More anatomically restricted malformations such as focal heterotopia or focal dysplasias are also associated with medically intractable seizures.[105] Estimates are that nearly 30% of cortical specimens resected as treatment for neocortical epilepsy contain some type of MCD. In fact, recent advances in neuroimaging have demonstrated that many cases of "cryptogenic" epilepsy are actually associated with small regions of cortex containing subtle cytoarchitectural abnormalities (microdysgenesis). Finally, a subgroup of adult patients with temporal lobe epilepsy exhibit radiographic and histopathologic evidence of MCD either alone or in combination with hippocampal sclerosis ("dual pathology" patients) suggesting that MCD is clinically relevant in both pediatric and adult epilepsy patients.[57]

From a clinical perspective, virtually all seizure subtypes (e.g., generalized tonic-clonic, complex partial, atonic, myoclonic, atypical absence seizures and infantile spasms) have been described within the broad family of MCD. Anti-epileptic drug therapy in patents with MCD often fails and surgical therapy provides the only hope for seizure remission.[35,76,88] Unfortunately, even with extensive attempts at localization of the seizure focus and skilled neurosurgical technique, seizure cure following surgical resection of focal CDs is successful in less than 50% of patients. Even worse, a subgroup of patients may not be surgical candidates at all. Unfortunately, the mechanisms of seizure initiation and epileptogenesis is unknown in most MCD and the existing animal models for many MCD do not fully recapitulate the human condition. Thus, challenging issues are to explain the origin of the cytoarchitectural abnormalities in MCD and to understand how developmental malformations of the cerebral cortex produce intractable epilepsy.

Developmental Contextual Background for MCD

Development of the human cerebral cortex is initiated at gestational week 7 and continues through week 24 (for reviews, see refs. 51,79,96). The cortex is derived from progenitors that

Table 1. Classification scheme for MCD (after Barkovich et al, 2001)[6]

I. Proliferative Disorders	II. Migration Disorders
Tuberous Sclerosis Complex	Lissencephaly Type I
	-Miller-Dieker
Focal Cortical Dysplasia	-XLIS
Hemimegalencephaly	Band Heterotopia
Microcephaly	Nodular Heterotopia
DNET	Cobblestone Lissencephaly
	Type II
Ganglioglioma	-Walker-Warburg syndrome
Gangliocytoma	-Muscle-eye-brain disease
	-Fukuyama Muscular Dystrophy
III. Organization Disorders	
Polymicrogyria	
Schizencephaly	
Microdysgenesis	
IV. Malformations, not otherwise classified	
Mitochondrial	
Peroxisomal	

DNET = dysembryoplastic neuroectodermal tumor; XLIS = X-linked lissencephaly

reside in the embryonic ventricular zone (VZ) or from the ganglionic eminence (GE) in four broad stages: 1) mitosis and proliferation of neural progenitor cells in the VZ and GE; 2) commitment to a neural lineage and exiting mitotic phases of the cell cycle; 3) dynamic migration of postmitotic neurons out of the VZ and GE; and 4) the establishment of cortical laminae. The process of cortical lamination involves radial migration of nascent neurons from the VZ along radial glial fibers into the evolving cortical plate through an "inside-out gradient". Neurons destined to reside in deeper cortical laminae (e.g., layer VI), arrive first in the nascent cortical plate, and subsequent waves of neurons destined for more superficial layers (e.g., V, IV, III), must migrate through each preceding and established layer. A small proportion of cells from the VZ migrate via nonradial pathways. In addition, cells move in a tangential pathway from the GE into the cortical plate. As we study the pathogenesis of each MCD subtype, it is important to place the putative molecular events responsible for the malformation into an appropriate developmental context, since the genesis of each malformation is inexorably linked to cortical developmental processes. The genes responsible for MCD have important functions in one or more stages of normal cortical development. Thus, the histopathologic features of each MCD render a fascinating view of the role that responsible gene plays in cortical development by demonstrating the sequelae of loss of encoded protein function. In addition, the effect of each gene mutation also highlights the developmental epoch in which that gene contributes to corticogenesis; these critical time-points provide a framework to understand the interface between gene mutations and abnormal neural development. For example, gene mutations that alter mitosis will result in an effect confined to the proliferative stages of cortical development but may have little effect on post-mitotic neurons. Similarly, a gene mutation that alters cytoskeletal assembly during dynamic phases of neuronal migration will have distinct effects in an actively migrating neuron versus a neuron that has already achieved its laminar destination. In sum, a defined molecular event occurring within a specific developmental context results in a characteristic malformation (for reviews, see refs. 23, 27).

Table 2. Genetics of MCD syndromes

MCD Subtype	Locus	Gene
•Miller-Dieker Lissencephaly	17p13	LIS-1
•Subcortical Band Heterotopia	Xq22	DCX
•XLIS	Xq22	DCX
•Periventricular Heterotopia	Xq28	FLN-1
•Tuberous Sclerosis Complex	16p13	tuberin
	9q34	hamartin
•Fukuyama Muscular Dystrophy	9q31	fukutin
•Muscle-eye-brain disease	1p32	POMGnT1

DCX, doublecortin; FLN-1, filamin-1;
POMGnT1, protein O-mannose beta-1,2-N-acetylglucosaminyltransferase
  XLIS = X-linked lissencephaly

Molecular Neurobiology of MCD

Subtle developmental malformations may be observed in a variety of neurological and psychiatric disorders without a definitive molecular correlate, such as dyslexia, autism, and schizophrenia. Additionally, MCD may be observed in the setting of large chromosomal rearrangements such as trisomy syndromes. The molecular pathogenesis of these MCD subtypes remains to be defined. The greatest progress has come from analysis of human pedigrees with inherited (autosomal and sex-linked) forms of MCD in which positional cloning strategies in have led to the identification of at least 7 genes (Table 2) directly responsible for human MCD that are associated with epilepsy. To better understand the function of the encoded gene products, the role of these genes in normal development is being studied experimentally in transgenic or knockout mouse strains. This chapter will focus on representative MCD from each category (proliferation, migration, organization, and not yet specified) in an attempt to highlight how the molecular pathogenesis of many MCD is becoming more clearly understood.

Disorders of Cellular Proliferation

The tuberous sclerosis complex is an autosomal dominant disorder characterized by cerebral cortical tubers that are highly associated with epilepsy. Tubers are the likely source of seizure initiation in TSC patients[52,68] since resection of these lesions is often associated with seizure control. Histologically, the normal hexalaminar structure of cortex is lost within a tuber and neurons exhibiting aberrant somatodendritic morphologies and cytomegaly (dysplastic neurons) are abundant within tubers. Giant cells are a unique cell type and a defining feature of tubers.[114] These bizarre cell types exhibit extreme cytomegaly, shortened processes of unknown identity (axons versus dendrites), and often are bi- or multinucleate. Extensive astrocytosis is a variable feature of tubers (for review, see ref. 26).

TSC results from mutations in one of two nonhomologous genes *TSC1*[109] and *TSC2*.[108] The *TSC1* gene encodes a protein, hamartin, and has virtually no homology to known vertebrate genes. The *TSC2* gene encodes a 200kD protein, tuberin, that is structurally distinct from hamartin. Both hamartin and tuberin mRNA and protein are widely expressed in normal tissues including brain, liver, adrenal cortex, cardiac muscle, skin, and kidney.[63,66,80] Identification of an encoded coiled-coil domain in the carboxy region of hamartin[109] raised the possibility of a functional protein-protein interaction with tuberin (and other proteins) that has recently been demonstrated in *Drosophila*.[90] Hamartin also interacts with the ezrin-radixin-moesin (ERM) family of actin-binding proteins and may contribute to cell-cell interactions, cell adhesion, and cell migration.[71] Tuberin contains a hydrophobic N-terminal domain and a conserved 163-amino-acid carboxy terminal region that exhibits sequence homology to the catalytic domain of a GTPase activating protein (GAP) for Rap1. As a member of the superfamily of Ras-related proteins, Rap1 likely functions in regulation of DNA synthesis and cell cycle transition. Tuberin displays GAP activity for Rap1 and colocalizes with Rap1 in the Golgi apparatus in several cell lines.[118,119] The GAP activity of functional tuberin may modulate the effects of Rap1 on G- to S-phase transition during cell division. Mutations in *TSC2* might result in constitutive activation of Rap1 leading to enhanced cell proliferation or incomplete cellular differentiation. Recent studies in *Drosophila* suggest that hamartin and tuberin form a functional heteromeric complex that is an important component of a pathway that modulates insulin receptor or insulin-like growth factor mediated signaling.[46,90] This pathway functions downstream of the cell signaling molecule Akt to regulate cell growth and potentially cell size. Thus, loss of hamartin or tuberin function following *TSC1* or *TSC2* mutations may result in enhanced proliferation of neural and astrocytic precursor cells and increased cell size characteristic of dysplastic neurons and giant cells commonly found in tubers of TSC. Of course, enhanced cell size may compromise neuronal migration and account for the loss of lamination within tubers. Alternatively, loss of hamartin or tuberin function may independently compromise neural migration via an interaction with ERM or actin binding proteins. Recently it has been shown that hamartin and tuberin interact with the G2/M cyclin-dependent kinase CDK1 and its regulatory cyclins A and B.[17] Abnormalities of radial glial cells have also been implicated in the pathogenesis of TSC.[87]

An interesting issue in TSC is whether the cellular constituents of tubers collectively reflect the cellular manifestations of TSC gene haploinsufficiency (heterozygosity) versus complete loss of gene function in tubers (loss of heterozygosity). For example, all somatic cells in TSC patients (including neurons), contain a single mutated copy of either TSC gene and are thus haploinsufficient or heterozygous. With the interesting exception of tubers, it has been shown that a "second hit" somatic mutation occurs in the second unaffected TSC allele (loss of heterozygosity) in a variety of TSC lesions resulting in loss of function of either hamartin or tuberin. A "second-hit" mutation has not been conclusively identified in tubers and recent studies suggest that haploinsufficiency effects alone may lead to tuber formation. Indeed, several studies have demonstrated tuberin expression in tubers despite a known *TSC2* mutation genotype.[62] Alternatively, only a select population of cell types in tubers (e.g., giant cells) may actually have sustained two mutational hits and thus these cells represent the effects of loss of function in maturing developing neurons. This important concept implies that only a select population of cell types in tubers (e.g., giant cells) may actually have sustained two mutational hits and reflect loss of gene function during brain development. Other adjacent cell types in tubers may be mere innocent (and haploinsufficient) bystanders whose migration into cortex has been interrupted.

Focal cortical dysplasias (FCD) are identified frequently in resected surgical specimens from patients with neocortical epilepsy.[106] The precise developmental epoch in which FCD are generated is unknown (they are classified as disorders of proliferation) and there is little data on whether FCD reflects an abnormality in cell proliferation, migration or laminar destination.[41,25] The search for candidate genes responsible for FCD is an area of intense research. There are two competing hypotheses regarding the formation of FCD. The first states these lesions result from an abnormality affecting a single neural precursor cell which in turn undergoes successive rounds of division to yield a clonal progeny that comprise the cellular constituents of the FCD. This scenario might develop if a somatic mutation were to occur in a single progenitor cell. One intriguing possibility is that a mutation in one of the known MCD genes occurs as a somatic event in a neural precursor, that then gives rise to a clonal population of neurons within the FCD. The alternative hypothesis is that an external event affects the development of multiple precursor cells that yield multiple nonclonal cell types as progeny. Histologically, multiple cell types are present within FCD, including neurons with dysmorphic features (dysplastic neurons), neurons within the subcortical white matter (heterotopic neurons), and neurons with excessively large cell bodies (neuronal cytomegaly). Thus, a pivotal question is whether these cell types all reflect a central pathogenetic process affecting their laminar distribution and morphology or there is a select cell type that is actually the "dysplastic" cell type. Indeed, we are not sure whether cortical dysplasia reflects a regional or cellular abnormality or in fact, if FCD reflects an abnormality in radial glial cells. These considerations are relevant since the precise phenotype of cells within FCD has not been clearly defined. While FCDs rarely, if ever, have a familial inheritance pattern, the histologic features of FCD suggest a consistent or uniform etiology.[83,25] It has been speculated that FCD results from early somatic mutations in one of the known MCD genes, or in a yet to be defined, or even novel gene. Alternatively, some suggest that FCD is a late-occurring event, possibly even a postnatal event (see ref. 74) resulting from external injury such as trauma or hypoxia-ischemia. An interesting recent study demonstrates discordant incidence of select FCD in monozygotic twins, suggesting that these lesions result from acquired factors, such as prenatal insults or postfertilization genetic abnormalities.[10] Furthermore, it is not clear at what point in cortical development FCD occurs.

An interesting feature of many dysplastic, heterotopic, and "balloon" or enlarged neurons is the expression of a variety of cytoskeletal genes and proteins, many of which are expressed under normal circumstances only in neural precursor cells. These proteins are normally expressed on a regulated developmental schedule during corticogenesis and are necessary for appropriate neuronal differentiation, migration, and process outgrowth. For example, expression of select intermediate filament (IF) proteins such as nestin, α-internexin, and vimentin, proteins typically found in immature neurons, has been reported in subpopulations of dysplastic and heterotopic neurons within FCD.[29,120] However, dysplastic neurons contain abnormal accumulations of highly- and nonphosphorylated neurofilament (NF) protein isoforms which are normally expressed in more differentiated neurons.[34] Thus, detection of select IFs in FCD that are normally expressed at either early or late stages of cortical development may yield clues to the maturational phenotype of dysplastic neurons and supports the hypothesis that these cells may retain components of an immature developmental phenotype.

Hemimegalencephaly (HMEG) is characterized by massive enlargement of one cerebral hemisphere with abnormal genesis of both cortical and subcortical structures. The opposite hemisphere may be histologically normal but subtle malformations or even focal dysplasias have been reported radiographically or in rare autopsy specimens. The cortex shows a complete loss of laminar organization and the presence of dysplastic, heterotopic, and large "balloon" neurons. The cellular phenotype of these cells has not been defined although these cells express many distinct intermediate filament proteins.[29,32,61] Rarely, HMEG may be associated with TSC as well as other syndromes including Proteus, Cowden, and linear sebaceous nevus syndromes. Early studies suggested that HMEG was a proliferative disorder and that DNA polyploidy could be demonstrated. However, more recent studies have not supported this hypothesis and the mechanism of the malformation remains to be defined. The most curious aspect of HMEG is the clear asymmetry of the malformation which suggests a very early and pervasive abnormality in cortical development.

Disorders of Neuronal Migration

Lissencephaly is a severe developmental brain malformation characterized by loss of the normal gyral patterns in the cerebral hemispheres, marked disorganization of the cerebral cortical cytoarchitecture, and a high association with profound neurologic deficits and epilepsy. The lissencephalies comprise a group of disorders sharing similar pathologic features that result from mutations in distinct genes. Recent evidence has shown that the identified genes in each syndrome play important roles in normal neural migration and thus, these MCD result from defects in neural migration. There are two pathological subtypes of lissencephaly, Type I (classical) and type II (cobblestone). In type I lissencephaly, the cortex is thickened and without gyri. The normal hexalaminar structure is reduced to a 4-layered pattern including a marginal, superficial cellular, sparsely cellular, and deep cellular laminae. Cells in the deeper layers are dysmorphic and exhibit features of pyramidal, fusiform, or rounded neurons without clear radial orientation. One of the first *MCD* genes discovered was *LIS-1* on chromosome 17p13.3

in patients with the autosomal recessive Miller-Dieker lissencephaly syndrome.[98] This syndrome is an autosomal disorder characterized by classical lissencephaly, profound mental retardation, epilepsy, and craniofacial dysmorphism. Hemizygous deletions within the *LIS-1* gene are associated with Miller-Dieker lissencephaly in over 90% of patients. The encoded LIS-1 protein is the β subunit of platelet activating factor acetylhydrolase (PAFAH1β). Platelet activating factor (PAF) is a potent phospholipid messenger molecule whose intra- and extracellular levels are controlled by PAFAH, which functions as a degradative enzyme for PAF. LIS-1 is a β subunit component of a trimeric α1/α1/β complex. PAFAH1β contains 7 repetitive stereotyped tryptophan and aspartate repeats (WD40 repeats) that likely function in protein-protein interactions. Stimulation of the PAF receptor disrupts neuronal migration in vitro[8] and mutations in the *LIS-1* gene may lead to defective nucleokinesis (movement of the neuronal nucleus during dynamic phases of neural migration). LIS-1 is expressed in Cajal-Retzius cells, the ventricular neuroepithelium, a subset of thalamic neurons, and in the subplate in fetal brain specimens.[20,97] The LIS1 protein interacts with microtubules[101] and may function in destabilization of these cytoskeletal elements during nuclear migration (for review, see ref. 39). A fungal homolog of LIS-1, *NudF*, has been identified in *Aspergillus nidulans* and interacts with a downstream protein NUDE.[58] These proteins interact with cytoplasmic dynein/dynactin subunits and may contribute to nuclear transport. Two LIS-interacting proteins, Nudel and a mammalian homolog NudE, identified by a yeast two-hybrid screen, are components of the dynein motor complex and microtubule-organizing centers[37] that are critical for neuronal migration. In addition, LIS-1-dynein interactions may also regulate cell division. In mutant *LIS-1* mouse strains, there is a range of disorganization of cortical cytoarchitecture including abnormal hippocampal and cortical lamination[56] and electrophysiologic studies have demonstrated hyperexcitability in this tissue.[42]

X-linked lissencephaly (XLIS) is a classical lissencephaly although a few abnormally large gyri (pachygyri) may be noted. While the neuropathologic features of XLIS are virtually indistinguishable from Miller-Dieker lissencephaly and mental retardation and epilepsy are invariably present in affected patients, there are no associated craniofacial abnormalities in XLIS. Mutations in the doublecortin gene (*DCX*) on chromosome Xq22 in hemizygous males results in lissencephaly[48,33] whereas *DCX* gene mutations in females result in the subcortical band heterotopia syndrome (see below). The mutational spectrum includes deletion, nonsense, missense, and splice donor site mutations,[77] many of which are clustered in two regions of the open reading frame.[49] The 40kDa DCX protein is normally expressed in post-mitotic neurons during the limited time window surrounding neuronal migration (human gestational weeks 12-20; ref. 94) and thus mutational effects will be exerted only during this dynamic phase of cortical development.[50] A small proportion of mature neurons also express DCX.[84] DCX is a microtubule-associated protein that specifically interacts with, stabilizes, and stimulates tubulin polymerization via a binding domain ("beta-grasp" superfold motif) that has been shown to be disrupted in patients with DCX mutations.[107] DCX coprecipitates with LIS-1 protein[15] suggesting that these two molecules may reflect a pivotal common pathway in assembly of the neuronal cytoskeleton during dynamic phases of neuronal migration.[49] A recent study has also shown that DCX interacts directly with the AP-1 and AP-2 adaptor complexes involved in clathrin-dependent protein sorting and potentially vesicular trafficking.[45] Thus, in addition to effects on neural migration, mutations in DCX might compromise movement of select proteins or vesicle-bound molecules such as neurotransmitters that may enhance excitability and foster seizures (see below). A related molecule, Doublecortin-like kinase (DCLK), shares sequence similarity to DCX in its N-terminal region and is also colocalized with microtubules.[14] DCLK is also a substrate for the cysteine protease calpain.[13] DCX contains a consensus substrate site for c-Abl, a nonreceptor tyrosine kinase that also modulates cytoskeletal assembly.[48] The association with c-Abl may provide an important mechanistic link to specific cell pathways since a mutation in the mouse *disabled1* gene, a c-Abl binding protein, also results in abnormal neuronal migration.

Although *LIS-1* and *doublecortin* mutations account for the majority of classical lissencephaly syndromes, other lissencephalies exist that do not result from mutations at either of these loci. A recent study reported two consanguineous pedigrees in which an autosomal recessive lissencephaly syndrome associated with cerebellar hypoplasia was mapped to chromosome 7q22 by linkage analysis.[59] The investigators postulated that the mutational locus for this syndrome would be at or near 7q22 since this chromosomal region contains the *reelin* (*RELN*) gene and mutations in this gene in mice result in neocortical migration abnormalities and cerebellar hypoplasia, similar to those seen in the patient cohort.[30] Reelin is a secreted protein that modulates neuronal migration by binding to several cell-surface molecules including the very low density lipoprotein receptor, the apoprotein E receptor 2, $\alpha 3\beta 1$ integrin, and protocadherins. *RELN* is encoded by 65 exons and spans more than 400 kilobase pairs of genomic DNA. A precise 85 base pair deletion corresponding to exon 36 was identified in one pedigree that resulted in abnormal splicing of exon 35 to exon 37. In the second pedigree, a second distinct mutation was identified in which 148 base pairs corresponding to exon 42 were deleted. Both mutations produced a translational frameshift followed by a premature termination codon and resembled naturally occurring mouse *reelin* alleles. Western blot analysis of serum from affected patients demonstrated reduced or absent reelin protein expression. Thus, other lissencephaly candidate genes may exist.

Periventricular nodular heterotopia (PH) and subcortical band heterotopia (SBH) are X-linked disorders characterized by differential phenotypes in males and females (for review, see ref. 72). Nodules of abnormal neurons and astrocytes separated by layers of myelinated fibers are identified along the lateral ventricles beneath the cortex in female PH patients.[54,91] PH in females results from mutations in the *filamin1* (*FLN1*) gene, which is located on chromosome Xq28.[43] Hemizygous males die in utero although rare male cases occur (see below). The encoded protein filamin1 is an actin-cross-linking phosphoprotein that modulates actin reorganization necessary for cellular locomotion. In *Drosophila*, filamin is required for ring canal assembly and actin organization during oogenesis.[73] The precise mechanisms by which loss of filamin1 function in PH leads to nodular accumulations of neurons remain to be fully defined. A likely scenario is that neurons within the nodules are unable to migrate successfully out of the embryonic ventricular zone as a consequence of actin cytoskeletal dysfunction and remain trapped within the nodules. While *DCX* gene mutations in males cause lissencephaly, *DCX* gene mutations in females are associated with the SBH ("double cortex") syndrome in which there is a bilaterally symmetric band of cortical neurons extending through the underlying white matter of the centrum semiovale (see ref. 5). The subcortical bands contain heterotopic neurons which are of small pyramidal shape without clear radial orientation. Neurons may be arrayed into clusters, sheets, or wide bands. Interestingly, however, the overlying cortex exhibits normal cytoarchitecture. The SBH is separated from the overlying cortex and underlying ventricles by normal white matter. A recent study has suggested that PH may result from radial glial fiber disruption.[100]

The sexual dimorphism of the PH and SBH/XLIS syndromes likely reflects differential expression of the mutant or normal *(FLN1)* or *DCX* gene alleles within select populations of neurons during brain development. Females with PH or SBH carry one normal and one mutant allele for either the *filamin 1* or *DCX* genes. As a consequence of X-chromosome inactivation (Lyonization), one of these alleles is no longer used for gene transcription. It has been speculated that neuronal migration will be compromised in neurons in which the normal allele is inactivated and the mutant gene is expressed. Neurons expressing the mutant allele will become the cellular constituents of either the nodules in PH or the band heterotopia in SBH. In contrast, those cells that inactivate the mutant allele and express the normal allele will migrate to the appropriate destination and come to comprise the overlying "normal" cortex. In males, only a single X-chromosome is present and thus if the mutant gene is inherited the effect is either lissencephaly (XLIS) or a lethal state in PH. Sporadic male cases of PH have been reported that result from truncating mutations in *FLN1*.[102]

The Fukuyama muscular dystrophy syndrome (FCMD) and muscle-eye-brain disease (MEB) are rare, autosomal recessive disorders (see refs. 21,22) that exhibit "cobblestone" (Type II) lissencephaly in which there is a complete loss of regional and laminar organization that is distinct from classical (type I) lissencephalies such as the Miller-Dieker syndrome. FCMD is seen primarily in Japan and is associated with a debilitating muscular dystrophy as well as seizures. The FCMD gene encodes the protein fukutin, which maps to chromosome 9q31[67] and may function as a secreted protein. MEB is associated with retinal dysplasia, congenital myopathy, and lissencephaly. The *MEB* gene, POMGnT1,[122] which encodes an acetylglucosaminyltransferase (POMGNT1) that participates in O-mannosyl glycan synthesis, is located on chromosome 1p32-34.[21]

Disorders of Cortical Organization

Polymicrogyria (PMG) and schizencephaly (SCHZ) are fascinating disorders that remain poorly understood. Both malformations have a high association with epilepsy. In PMG, the cortex exhibits multiple small microgyri which reveal a four-layered lamination pattern. It is not clear that PMG is a malformation of development per se and may instead reflect a destructive process occurring during later stages of corticogenesis. PMG has been reported in associated with a variety of neurological syndromes but a specific molecular mechanism has not been identified. Recent studies have identified both X-linked and autosomal forms of PMG[89] suggesting that some PMG subtypes are in fact inherited.

Schizencephaly is a rare MCD characterized by a full-thickness cleft within the cerebral hemispheres. The cleft may be unilateral or bilateral and the missing portions of the cerebral hemispheres may be replaced by cerebrospinal fluid. The walls of the clefts are lined by dysplastic cortex. The clefts can be small or quite large. Radiographically, SCHZ is characterized as either open- or closed-lip although as of yet these distinctions have no clear molecular correlate. While a proportion of SCHZ may be a result of intrauterine hypoxic-ischemic injury there has been evidence that mutations in the homeobox gene *Emx2* may be associated with some forms of SCHZ.[12]

Epileptogenesis and MCD: How Does Cortical Maldevelopment Lead to Epilepsy?

A pivotal question directly relevant to both clinical and laboratory investigations of MCD is why cortical malformations are so highly associated with epilepsy. An overriding hypothesis that has been supported by clinical studies is that in patients with focal types of dysplasias, the seizures emanate from the malformation rather than surrounding cortex. For example, in Taylor-type FCD, tubers, and polymicrogyria, intracranial grid recordings have clearly demonstrated that the ictal onset zone is within the malformation.[86,93] An important recent study using human depth electrode recording showed that seizures in patients with PH may emanate directly from the heterotopia[69] but these individuals may have multifocal epileptiform abnormalities. Interestingly, in a rat model for SBH, seizures were shown to emanate from the overlying cortex[18] and not the heterotopia, although this finding has not been confirmed in humans. Surgical resection in SBH patients does not seem to be as effective as more focal resections in other MCD.[7]

Several theories have been proposed based on studies in human MCD and in animal models of MCD that include altered synaptic connectivity, aberrant expression of molecules that mediate synaptic transmission, and an imbalance between excitatory and inhibitory impulses associated with the dysplasia. Solid evidence has been provided in favor of each of these hypotheses and it is likely that, broadly speaking, epilepsy in MCD reflects contributions from multiple contigent and convoluted factors. However, it is unclear at this time whether the mechanisms of epileptogenesis in MCD will be similar or distinct across multiple MCD subtypes. In other words, epilepsy in MCD may be more akin to a final common electrophysiologic or clinical pathway for a variety of malformations. There may be important differences in epileptogenesis between individual MCD syndromes. As a corollary to this question, within each MCD why are their distinct seizure phenotypes? For example, how can we mechanistically explain or investigate the occurrence of infantile spasms and complex partial epilepsy in tuberous sclerosis or in two distinct MCD subtypes such as tuberous sclerosis and lissencephaly? Implicit in this question is whether clinical epilepsy syndromes or semiologic subtypes are specific for each MCD subtype or whether they reflect maturational contextual differences in cortical development. This point is critical since the clear failures in antiepileptic drug treatment for many patients with MCD may imply an essential flaw in drug utility as a direct consequence of the pathoanatomic differences among these different brain lesions. Indeed, an inherent question in this idea is whether future anti-epileptic drug design may be predicated on individual cellular and molecular differences in epileptogenesis among different MCD subtypes. A final point is that it is often a tacit assumption that aberrant cortical cytoarchitecture alone is responsible for seizures in MCD. This is an especially important consideration since MCD have been well-documented in patients with no history of epilepsy. Certainly, disorganized synaptic connectivity within MCD and between MCD and the surrounding cortex may enhance the likelihood of recurrent and propagated synchronous, paroxysmal discharges. However, an alternative hypothesis is that some of the genes responsible for individual MCD may in fact also serve as epilepsy susceptibility loci that confer hyperexcitability to neurons regardless of their laminar positioning.

The mechanisms of epileptogenesis in MCD have been investigated in human tissue and animal models of MCD. Data addressing the physiological properties of neurons with MCD remain sparse largely as a consequence of technical limitations of studying resected tissue in vitro, including limited viability of the tissue slice, the paucity of connections between the slice and surrounding cortex, and lack of true control tissue. Surprisingly little is known about mechanisms controlling excitatory and inhibitory tone in several important MCD syndromes including lissencephaly, band heterotopia, and nodular heterotopia. However, one MCD subtype that has been studied using cellular, electrophysiologic and recently, molecular strategies, is Taylor-type FCD since this MCD subtype is frequently identified in neocortical epilepsy surgical specimens as well as in temporal lobe resections. The cellular constituents of FCD may include a variable population of dysplastic neurons, heterotopic neurons, and large "balloon" neurons intermingled with neuronal subtypes of indeterminate laminar destiny. It is important to remember that it has yet to be defined whether all neurons within foci of FCD are in fact abnormal or within an incorrect laminar position, and thus how we view the physiologic responses of these cells may depend on how we weight the contributions of each cell type. One hypothesis that has received experimental support is that there is decreased inhibitory and enhanced excitatory tone in FCD.[3,103] The few studies that have addressed this hypothesis directly with field potential and intracellular recording within human FCD specimens have shown that dysplastic cortex is hyperexcitable[75,78] and that a subset of dysplastic neurons generate repetitive ictal discharges in response to the K+ channel blocker 4AP that can be blocked by NMDA receptor antagonists.[3] These electrophysiological responses may result from several possible causes. First, several reports have demonstrated a reduction in the number of parvalbumin, somatostatin, and GAD65 immunolabeled inhibitory interneurons in FCD.[40,103,104] In cortical tubers, there are few GABAergic neurons identified by GAD65 immunolabeling and the expression of the vesicular GABA transporter (VGAT) mRNA, a marker for GABAergic neurons, is also reduced.[116] Interestingly, using proton magnetic resonance spectroscopy to assay tuber samples, an increase in GABA was defined[1] suggesting a possible compensatory response. Neurons within nodular subcortical heterotopias are largely calbindin-immunoreactive, suggesting a GABAergic phenotype that has failed to migrate into cortex.[54] Taken together, these studies suggest that there is a paucity of GABAergic interneurons in MCD that represents reduction in the genesis of GABAergic cells, a selective failure of GABAergic cell migration from the median eminence during development, or enhanced death of GABAergic neurons. A second contributory factor to hyperexcitability in MCD is a selective reduction in GABA$_A$ receptor subunits in FCD which argues that the synaptic machinery to modulate neural inhibition within FCD may be diminished. Reduced expression of GABA$_A$ α1, α2, β1, and β2 receptor subunit mRNAs is observed in human FCD[24] and GABAergic terminals identified in close proximity to balloon neurons in FCD do not appear to make functional synapses.[47] The reduction in GABAergic neurons and loss of GABAergic receptor subunits would alter critical inhibitory synaptic control and render neurons in FCD more susceptible to sudden and prolonged excitability. A third factor related to hyperexcitability is that recent reports have demonstrated the expression of several glutamate receptor subunits is enhanced in FCD.[24,65,121] For example, expression of the NR2B site has been shown to be increased in FCD.[24,81] This site has been shown to modulate sustained calcium-mediated depolarization and is an ideal candidate protein to account for recurrent hyperexcitability in FCD. The expression of the NR2B site was correlated with more widespread epileptiform abnormalities detected by surface grid electrodes.[85] Increased NR2B mRNA and protein expression has been defined in tubers of the tuberous sclerosis complex by cDNA array and receptor ligand pharmacology.[116] Increased expression of several other glutamate receptor subunits including GluR1 and GluR2 have been defined in FCD.[24,65] Diminished coupling of NR1 with calmodulin has been reported in 3 FCD specimens, an interaction that is essential for inactivation of the NR1 complex.[81] Two recent studies suggest that the number of excitatory neurons, as evidenced by expression of the neuronal glutamate transporter EAAT3/EAAC1, is increased in tubers[116] and in FCD[23,27] suggesting that regional hyperexcitability in select MCD may result from enhanced numbers of excitatory neurons as well.

Corroborative studies in several animal models of MCD support the idea of an imbalance between excitatory and inhibitory tone in MCD.[60] These models include the freeze-induced microgyrus, methylazoxymethanol (MAM), in utero radiation, and several spontaneous and engineered rodent strains such as the flathead, OTX-1, p35, tish, and NZB strains (see ref. 4). The histologic features of these individual strains are quite distinct and only very few of them e.g., the flathead strain, exhibit spontaneous seizures. Yet differential expression of select glutamate and GABA receptor subunits have been demonstrated nonetheless. For example, in the freeze microgyrus lesion model, selective upregulation of the NR2B site has been shown using patch clamp recordings[31] and there is a reduction in parvalbumin-immunolabeled neurons within the microgyrus.[60] Interestingly, in a mouse model of cortical dysplasia generated by in utero irradiation, there is a reduction in the numbers of cortical parvalbumin- and calbindin-immunoreactive neurons.[99] In rats treated with methylazoxymethanol (MAM), altered expression of GluR2, NR2A, and NR2B were reported within the heterotopic cell islands in the hippocampus and neocortex.[95] In the mouse Lis1 mutant strain, hyperexcitability in the CA3 hippocampal sector reflected the cytoarchitectural disruption observed, including displaced somatostatin- and parvalbumin-immunoreactive neurons.[42] Disruption of ion channel function may also contribute to epileptogenesis of MCD, although few data regarding expression of these channels in human brain tissue are available. In the freeze lesion model, loss of an inwardly-rectifying potassium current as well as a reduction in gap junction coupling have been demonstrated[9] although similar findings in humans are unknown. In the MAM model, selective reduction in Kv4.2 potassium channel was observed in heterotopia.[16] These findings suggest that animal models of cortical malformations can provide important insights into epileptogenesis associated with MCD.

Experimental Strategies for Studying Epilepsy in MCD

Future studies addressed at the molecular pathogenesis of MCD will require implementation of novel strategies. For example, the generation of animal models that closely recapitulate human pathologic abnormalities, direct electrophysiological recording from human specimens obtained intraoperatively, identification of candidate gene loci in family MCD pedigrees, and the application of gene and protein array technologies to both animal and human specimens are several attractive approaches. Gene array analysis provides a rapid strategy to screen numerous candidate gene families in MCD. Alterations in mRNA expression can provide clues to the developmental mechanisms leading to abnormal cortical cytoarchitecture in MCD. Analysis of gene expression can be performed in whole tissue samples obtained intraoperatively or at post-mortem examination and in fact the first application of gene expression analysis in human epilepsy was implemented in cortical tubers. An additional advance was made when it was shown that mRNA expression could be determined in single microdissected cells in fixed, immunohistochemically labeled sections.[28] The implementation of single cell mRNA expression analysis in human MCD permits assay of differential gene expression in distinct cell types so that cellular heterogeneity can be addressed. Indeed, this strategy was used for the first time to assay MCD associated with epilepsy including FCD,[24] TSC[28,116,70] and hemimegalencephaly[29] that have yielded clues as to the cellular and molecular heterogeneity of neurons in these malformations.

Experimental Methodology

The experimental protocol to amplify cellular poly (A) mRNA from fixed tissue using an oligo-dT primer was first described using resected surgical epilepsy specimens[28] and has subsequently been outlined in detail.[64] The approach uses an oligo-dT primer coupled to a T7 RNA polymerase promoter to convert poly (A) mRNA into cDNA directly on the fixed tissue section by in situ (reverse) transcription (IST). In general, tissue sections are immunolabeled prior to this so that cells can be characterized experimentally by phenotypic characteristics. Following immunolabeling, tissue sections are treated with proteinase K to reverse excessive fixation and then washed in diethyl pyrocarbonate (DEPC)-treated water. To initiate IST, an oligo-dT(24) primer-T7 RNA polymerase promoter is annealed to cellular poly (A) mRNA directly on the tissue section at room temperature. cDNA is synthesized on the section with avian myeloblastosis reverse transcriptase (AMVRT). At this point, cDNA can be extracted from cellular poly (A) mRNA and then processed for mRNA amplification and synthesis of a radiolabeled RNA probe. This pool of aRNA will reflect the total population of genes present in the whole section. Alternatively, single immunolabeled cells may be microdissected so that the mRNA can be amplified and used as a probe of cDNA arrays to define gene expression in that specific cell. Single cells can be microdissected from sections under light microscopy using a glass electrode or a stainless steel microscalpel guided by a joystick micromanipulator. Recent technologies also include the use of laser capture although the utility of this approach in single cells has yet to be demonstrated. Fixed cells are then aspirated into a second glass microelectrode and transferred to a microfuge tube reaction buffer and incubated at 40°C for 90 minutes to ensure cDNA synthesis in the single dissected cell.

cDNA extracted from whole sections or generated from single cells serves as a template for synthesis of double-stranded template cDNA with T4 DNA polymerase I. mRNA is amplified (aRNA) from the double-stranded cDNA template with T7 RNA polymerase. aRNA serves as a template for a second round of cDNA synthesis with AMVRT, dNTPs, and N(6) random hexamers. cDNA generated from aRNA is made double-stranded and serves as template for a second aRNA amplification incorporating ^{32}PCTP. Radiolabeled aRNA is then used to probe candidate cDNA arrays.

Gene Expression Analysis in Human MCD

Recent work has demonstrated that the expression of select gene families in MCD is distinct from normal cortex. Furthermore, the transcriptional profiles of individual genes in MCD differs in phenotypically defined cells. For example, the expression of select glutamate and GABA receptor subunit mRNAs is altered in tubers in TSC.[116] These changes in receptor subunits are distinct from those identified in human FCD.[24,27,23] Large scale cDNA screening efforts using cDNA microchips have been undertaken and preliminary results reveal alterations in select gene families. For example, increased expression of two genes, collapsin response mediator protein 4 and doublecortin has been detected in tubers and focal cortical dysplasias (Lee et al, submitted). These genes encode proteins that are associated with cellular proliferation and suggest that a subpopulation of cells in these MCD may be newly generated. Recent identification of select proinflammatory cytokine genes in tubers such as ICAM-1 and TNF-α suggests that inflammation may contribute to tubers and potentially epileptogenesis (Maldonado et al, submitted).

Conclusions

In summary, the ability to assay gene expression provides a unique strategy to assay the coordinate expression of multiple mRNAs in phenotypically defined cell types since interpretation of gene expression assays of whole brain regions is complicated by contributions from multiple distinct cell types within a tissue sample. Indeed, analysis of gene transcription in single disease-affected cells versus adjacent unaffected cells provides a unique opportunity to understand specifically which genes may be responsible for the pathologic features of the disease in question within select populations of cells. Furthermore, alterations in mRNA levels have been correlated with and thus can be used to predict, changes in expression of proteins in single cells.

Summary and New Directions: Targeted Therapy for Epilepsy in MCD

Prior to the 1990's, the molecular pathogenesis of MCD was largely the source of speculation. However, with the discovery of several genes responsible for MCD, including *LIS1*, *doublecortin*, *FLN1*, *TSC1* and *TSC2*, it is clear that single gene mutations may account for numerous subtypes of MCD. The identification of MCD genes now permits in-depth analysis of the proteins encoded by these genes and may aid in designing new therapies targeted at these molecules. Thus, in the future, we may hope to modulate pathways in select MCD syndromes so that specific agents can be used to treat seizures in, for example, PH or TSC. Perhaps even more exciting is the potential to design therapeutic strategies to actually abolish or prevent the development of these malfomations in utero.

Acknowledgements

Work supported by MH01658, NS39938, the Esther A. and Joseph Klingenstein Fund, the Tuberous Sclerosis Alliance/Center Without Walls, and Parents Against Childhood Epilepsy (PBC).

References

1. Aasly J, Silfvenius H, Aas TC et al. Proton magnetic resonance spectroscopy of brain biopsies from patients with intractable epilepsy. Epilepsy Res 1999; 35(3):211-217.
2. Andermann F. Cortical dysplasias and epilepsy: a review of the architectonic, clinical, and seizure patterns. Adv Neurol 2000; 84:479-496.
3. Avoli M, Bernasconi A, Mattia D et al. Epileptiform discharges in the human dysplastic neocortex: in vitro physiology and pharmacology. Ann Neurol 1999; 46:816-826.
4. Baraban S. Epileptogenesis in the dysplastic brain: a revival of familiar themes. Epilepsy Currents 2001; 1:22-29.
5. Barkovich AJ, Kuzniecky RI. Gray matter heterotopia. Neurology 2000; 55:1603-1608.
6. Barkovich AJ, Kuzniecky RI, Jackson GD et al. Classification system for malformations of cortical development: update 2001. Neurology 2001; 57(12):2168-78.
7. Bernasconi A, Martinez V, Rosa-Neto P et al. Surgical resection for intractable epilepsy in "double cortex" syndrome yields inadequate results. Epilepsia 2001; 42:1124-1129.
8. Bix GJ, Clark GD. Platelet-activating factor receptor stimulation disrupts neuronal migration in vitro. J Neurosci 1998; 18:307-318.
9. Bordey A, Lyons SA, Hablitz JJ et al. Electrophysiological characteristics of reactive astrocytes in experimental cortical dysplasia. J Neurophysiol 2001; 85:1719-1731.
10. Briellmann RS, Jackson GD, Torn-Broers Y et al. Causes of epilepsies: insights from discordant monozygous twins. Ann Neurol 2001; 49:45-52.
11. Brodtkorb E, Nilsen G, Smevik O et al. Epilepsy and anomalies of neuronal migration: MRI and clinical aspects. Acta Neurol Scand 1992; 86:24-32.

12. Brunelli S, Faiella A, Capra V et al. Germline mutations in the homeobox gene EMX2 in patients with severe schizencephaly. Nat Genet 1996; 12:94-6.
13. Burgess HA, Reiner O. Cleavage of doublecortin-like kinase by calpain releases an active kinase fragment from a microtubule anchorage domain. J Biol Chem 2001; 276:36397-36403.
14. Burgess HA, Reiner O. Doublecortin-like kinase is associated with microtubules in neuronal growth cones. Mol Cell Neurosci 2000;16:529-41
15. Caspi M, Atlas R, Kantor A et al. Interaction between LIS1 and doublecortin, two lissencephaly gene products. Hum Mol Genet 2000; 9:2205-2213.
16. Castro PA, Cooper EC, Lowenstein DH et al. Hippocampal heterotopia lack functional Kv4.2 potassium channels in the methylazoxymethanol model of cortical malformations and epilepsy. J Neurosci 2001; 21:6626-6634.
17. Catania MG, Mischel PS, Vinters HV. Hamartin and tuberin interaction with the G2/M cyclin-dependent kinase CDK1 and its regulatory cyclins A and B. J Neuropathol Exp Neurol 2001; 60:711-723.
18. Chen ZF, Schottler F, Bertram E et al. Distribution and initiation of seizure activity in a rat brain with subcortical band heterotopia. Epilepsia 2000; 41:493-501.
19. Chou K, Crino PB. Epilepsy and cortical dysplasias. Curr Treat Options Neurol 2000; 2:543-552.
20. Clark GD, Mizuguchi M, Antalffy B et al. Predominant localization of the LIS family of gene products to Cajal-Retzius cells and ventricular neuroepithelium in the developing human cortex. J Neuropathol Exp Neurol 1997; 56:1044-1052.
21. Cormand B, Avela K, Pihko H et al. Assignment of the muscle-eye-brain disease gene to 1p32-p34 by linkage analysis and homozygosity mapping. Am J Hum Genet 1999; 64:126-135.
22. Cormand B, Pihko H, Bayes M et al. Clinical and genetic distinction between Walker-Warburg syndrome and muscle-eye-brain disease. Neurology 2001; 56:1059-1069.
23. Crino, PB, Miyata H, Vinters HVV. Neurodevelopmental disorders as a cause of seizures: neuropathologic, genetic, and mechanistic considerations. Brain Pathol 2002; 12:212-33.
24. Crino PB, Duhaime AC, Baltuch G et al. Differential expression of glutamate and GABA-A receptor subunit mRNA in cortical dysplasia. Neurology 2001; 56:906-913.
25. Crino PB, Eberwine J. Cellular and molecular basis of cerebral dysgenesis. J Neurosci Res 1997; 50:907-916.
26. Crino PB, Henske EP. New developments in the neurobiology of the tuberous sclerosis complex. Neurology 1999; 53:1384-1390.
27. Crino PB, Jin H, Robinson M et al. Increased expression of the neuronal glutamate transporter (EAAT3/EAAC1) in hippocampal and neocortical epilepsy. Epilepsia 2002; 43:211-8.
28. Crino PB, Trojanowski JQ, Dichter MA et al. Embryonic neuronal markers in tuberous sclerosis: single-cell molecular pathology. Proc Natl Acad Sci USA 1996; 93:14152-14157.
29. Crino PB, Trojanowski JQ, Eberwine J. Internexin, MAP1B, and nestin in cortical dysplasia as markers of developmental maturity. Acta Neuropathol 1997; 93:619-627.
30. D'Arcangelo G, Miao GG, Chen SC et al. A protein related to extracellular matrix proteins deleted in the mouse mutant reeler. Nature 1995; 374:719-723.
31. DeFazio RA, Hablitz JJ. Alterations in NMDA receptors in a rat model of cortical dysplasia. J Neurophysiol 2000; 83:315-321.
32. De Rosa MJ, Secor DL, Barsom M et al. Neuropathologic findings in surgically treated hemimegalencephaly: immunohistochemical, morphometric, and ultrastructural study. Acta Neuropathol 1992; 84:250-260.
33. des Portes V, Pinard JM, Billuart P et al. A novel CNS gene required for neuronal migration and involved in X-linked subcortical laminar heterotopia and lissencephaly syndrome. Cell 1998; 92:51-61.
34. Duong T, De Rosa MJ, Poukens V et al. Neuronal cytoskeletal abnormalities in human cerebral cortical dysplasia. Acta Neuropathol 1994; 87:493-503.
35. Engel Jr J. Surgery for seizures. N Engl J Med 1996; 334:647-652.
36. Farrell MA, DeRosa MJ, Curran JG et al. Neuropathologic findings in cortical resections (including hemispherectomies) performed for the treatment of intractable childhood epilepsy. Acta Neuropathol 1992; 83:246-259.
37. Faulkner NE, Dujardin DL, Tai CY et al. A role for the lissencephaly gene LIS1 in mitosis and cytoplasmic dynein function. Nat Cell Biol 2000; 2:784-791.
38. Feng Y, Olson EC, Stukenberg PT et al. LIS1 regulates CNS lamination by interacting with mNudE, a central component of the centrosome. Neuron 2000; 28:665-679.
39. Feng Y, Walsh CA. Protein-protein interactions, cytoskeletal regulation and neuronal migration. Nat Rev Neurosci 2001; 2:408-16.

40. Ferrer I, Oliver B, Russi A et al. Parvalbumin and calbindin-D28k immunocytochemistry in human neocortical epileptic foci. J Neurol Sci 1994; 123:18-25.
41. Ferrer I, Pineda M, Tallada M et al. Abnormal local-circuit neurons in epilepsia partialis continua associated with focal cortical dysplasia. Acta Neuropathol 1992; 83:647-652.
42. Fleck MW, Hirotsune S, Gambello MJ et al. Hippocampal abnormalities and enhanced excitability in a murine model of human lissencephaly. J Neurosci 2000; 20:2439-2450.
43. Fox JW, Lamperti ED, Eksioglu YZ et al. Mutations in filamin 1 prevent migration of cerebral cortical neurons in human periventricular heterotopia. Neuron 1998; 21:1315-1325.
44. Frater JL, Prayson RA, Morris III HH et al. Surgical pathologic findings of extratemporal-based intractable epilepsy: a study of 133 consecutive resections. Arch Pathol Lab Med 2000; 124:545-549.
45. Friocourt G, Chafey P, Billuart P et al. Doublecortin interacts with μ subunits of clathrin adaptor complexes in the developing nervous system. Mol Cell Neurosci 2001; 18:307-319.
46. Gao X, Pan D *TSC1* and *TSC2* tumor suppressors antagonize insulin signaling in cell growth. Genes Dev 2001; 15:1383-1392.
47. Garbelli R, Munari C, De Biasi S et al. Taylor's cortical dysplasia: a confocal and ultrastructural immunohistochemical study. Brain Pathol 1999; 9:445-461.
48. Gleeson JG, Allen KM, Fox JW et al. Doublecortin, a brain-specific gene mutated in human X-linked lissencephaly and double cortex syndrome, encodes a putative signaling protein. Cell 1998; 92:63-72.
49. Gleeson JG, Lin PT, Flanagan LA et al. Doublecortin is a microtubule-associated protein and is expressed widely by migrating neurons. Neuron 1999a; 23:257-271.
50. Gleeson JG, Minnerath SR, Fox JW et al. Characterization of mutations in the gene *doublecortin* in patients with double cortex syndrome. Ann Neurol 1999b; 45:146-153.
51. Golden JA. Cell migration and cerebral cortical development. Neuropathol Appl Neurobiol 2001; 27:22-28.
52. Guerreiro MM, Andermann F, Andermann E et al. Surgical treatment of epilepsy in tuberous sclerosis: strategies and results in 18 patients. Neurology 1998; 51:1263-1269.
53. Gutmann DH, Zhang Y, Hasbani MJ et al. Expression of the tuberous sclerosis complex gene products, hamartin and tuberin, in central nervous system tissues. Acta Neuropathol 2000; 99:223-230.
54. Hannan AJ, Servotte S, Katsnelson A et al. Characterization of nodular neuronal heterotopia in children. Brain 1999; 122:219-238.
55. Hauser WA, Annegers JF, Kurland LT. Incidence of epilepsy and unprovoked seizures in Rochester, Minnesota: 1935-1984. Epilepsia 1993; 34:453-468.
56. Hirotsune S, Fleck MW, Gambello MJ et al. Graded reduction of Pafah1b1 (Lis1) activity results in neuronal migration defects and early embryonic lethality. Nat Genet 1998; 19:333-339.
57. Ho SS, Kuzniecky RI, Gilliam F et al. Temporal lobe developmental malformations and epilepsy: dual pathology and bilateral hippocampal abnormalities. Neurology 1998; 50:748-54
58. Hoffmann B, Zuo W, Liu A et al. The LIS1-related protein NUDF of *Aspergillus nidulans* and its interaction partner NUDE bind directly to specific subunits of dynein and dynactin and to α- and γ-tubulin. J Biol Chem 2001; 276:38877-38884.
59. Hong SE, Shugart YY, Huang DT et al. Autosomal recessive lissencephaly with cerebellar hypoplasia is associated with human *RELN* mutations. Nat Genet 2000; 26:93-96.
60. Jacobs KM, Kharazia VN, Prince DA. Mechanisms underlying epileptogenesis in cortical malformations. Epilepsy Res 1999; 36:165-188.
61. Jahan R, Mischel PS, Curran JG et al. Bilateral neuropathologic changes in a child with hemimegalencephaly. Pediatr Neurol 1997; 17:344-349.
62. Johnson MW, Emelin JK, Park SH et al. Colocalization of *TSC1* and *TSC2* gene products in tubers of patients with tuberous sclerosis. Brain Pathol 1999; 9:45-54.
63. Johnson MW, Kerfoot C, Bushnell T et al. Hamartin and tuberin expression in human tissues. Mod Pathol 2001; 14:202-210.
64. Kacharmina JE, Crino PB, Eberwine J. Preparation of cDNA from single cells and subcellular regions. Methods Enzymol 1999; 303:3-18.
65. Kerfoot C, Vinters HV, Mathern GW. Cerebral cortical dysplasia: giant neurons show potential for increased excitation and axonal plasticity. Dev Neurosci 1999; 21:260-270.
66. Kerfoot C, Wienecke R, Menchine M et al. Localization of tuberous sclerosis 2 mRNA and its protein product tuberin in normal human brain and in cerebral lesions of patients with tuberous sclerosis. Brain Pathol 1996; 6:367-375.
67. Kobayashi K, Nakahori Y, Miyake M et al. An ancient retrotransposal insertion causes Fukuyama-type congenital muscular dystrophy. Nature 1998; 394:388-392.

68. Koh S, Jayakar P, Dunoyer C et al. Epilepsy surgery in children with tuberous sclerosis complex: presurgical evaluation and outcome. Epilepsia 2000; 41:1206-1213.
69. Kothare SV, VanLandingham K, Armon C et al. Seizure onset from periventricular nodular heterotopias: depth-electrode study. Neurology 1998; 51:1723-1727.
70. Kyin R, Hua Y, Baybis M et al. Differential cellular expression of neurotrophins in cortical tubers of the tuberous sclerosis complex. Am J Pathol 2001; 159:1541-1554.
71. Lamb RF, Roy C, Diefenbach TJ et al. The TSC1 tumour suppressor hamartin regulates cell adhesion through ERM proteins and the GTPase Rho. Nat Cell Biol 2000; 2:281-287.
72. Leventer RJ, Mills PL, Dobyns WB. X-linked malformations of cortical development. Am J Med Genet 2000; 97:213-220.
73. Li MG, Serr M, Edwards K et al. Filamin is required for ring canal assembly and actin organization during Drosophila oogenesis. J Cell Biol 1999; 146:1061-1074.
74. Lombroso CT. Can early postnatal closed head injury induce cortical dysplasia. Epilepsia 2000; 41:245-253.
75. Mathern GW, Cepeda C, Hurst RS et al. neurons recorded from pediatric epilepsy surgery patients with cortical dysplasia. Epilepsia 2001; 41(Suppl 6):S162-7.
76. Mathern GW, Giza CC, Yudovin S et al. Postoperative seizure control and antiepileptic drug use in pediatric epilepsy surgery patients: The UCLA experience, 1986-1997. Epilepsia 1999; 40:1740-1749.
77. Matsumoto N, Leventer RJ, Kuc JA et al. Mutation analysis of the DCX gene and genotype/phenotype correlation in subcortical band heterotopia. Eur J Hum Genet 2001; 9:5-12.
78. Mattia D, Olivier A, Avoli M. Seizure-like discharges recorded in human dysplastic neocortex maintained in vitro. Neurology 1995; 45:1391-1395.
79. McConnell SK. Constructing the cerebral cortex: Neurogenesis and fate determination. Neuron 1995; 15:761-768.
80. Menchine M, Emelin JK, Mischel PS et al. Tissue and cell-type specific expression of the tuberous sclerosis gene, TSC2, in human tissues. Mod Pathol 1996; 9:1071-1080.
81. Mikuni N, Babb TL, Ying Z et al. NMDA-receptors 1 and 2A/B coassembly increased in human epileptic focal cortical dysplasia. Epilepsia 1999; 40:1683-1687.
82. Mikuni N, Nishiyama K, Babb TL et al. Decreased calmodulin-NR1 coassembly as a mechanism for focal epilepsy in cortical dysplasia. Neuroreport 1999; 10:1609-1612.
83. Mischel PS, Nguyen LP, Vinters HV Cerebral cortical dysplasia associated with pediatric epilepsy: review of neuropathologic features and proposal for a grading system. J Neuropathol Exp Neurol 1995; 54:137-153.
84. Nacher J, Crespo C, McEwen BS. Doublecortin expression in the adult rat telencephalon. Eur J Neurosci 2001; 14:629-644.
85. Najm IM, Ying Z, Babb T et al. Epileptogenicity correlated with increased N-methyl-D-aspartate receptor subunit NR2A/B in human focal cortical dysplasia. Epilepsia 2000; 41:971-976.
86. Palmini A, Gambardella A, Andermann F et al. Intrinsic epileptogenicity of human dysplastic cortex as suggested by corticography and surgical results. Ann Neurol 1995; 37:476-487.
87. Park SH, Pepkowitz SH, Kerfoot C et al. Tuberous sclerosis in a 20-week gestation fetus: immunohistochemical study. Acta Neuropathol 1997; 94:180-186.
88. Peacock WJ, Wehby-Grant MC, Shields WD et al. Hemispherectomy for intractable seizures in children: a report of 58 cases. Childs Nerv Syst 1996; 12:376-384.
89. Piao X, Basel-Vanagaite L, Straussberg R et al. An autosomal recessive form of bilateral frontoparietal polymicrogyria maps to chromosome 16q12.2-21. Am J Hum Genet 2002; 70:1028-33.
90. Potter CJ, Huang H, Xu T. Drosophila Tsc1 functions with Tsc2 to antagonize insulin signaling in regulating cell growth, cell proliferation, and organ size. Cell 2001; 105:357-368.
91. Poussaint TY, Fox JW, Dobyns WB et al. Periventricular nodular heterotopia in patients with filamin-1 gene mutations: neuroimaging findings. Pediatr Radiol 2000; 30:748-755.
92. Prayson RA. Clinicopathological findings in patients who have undergone epilepsy surgery in the first year of life. Pathol Int 2000; 50:620-625.
93. Preul MC, Leblanc R, Cendes F et al. Function and organization in dysgenic cortex. Case report. J Neurosurg 1997; 87:113-121.
94. Qin J, Mizuguchi M, Itoh M et al. Immunohistochemical expression of doublecortin in the human cerebrum: comparison of normal development and neuronal migration disorders. Brain Res 2000; 863:225-232.
95. Rafiki A, Chevassus-au-Louis N, Ben-Ari Y et al. Glutamate receptors in dysplastic cortex: an in situ hybridization and immunohistochemistry study in rats with prenatal treatment with methylazoxymethanol. Brain Res 1998; 782:142-152.
96. Rakic P. Principles of neural cell migration. Experientia 1990; 46:882-891.

97. Reiner O, Albrecht U, Gordon M et al. Lissencephaly gene (LIS1) expression in the CNS suggests a role in neuronal migration. J Neurosci 1995; 15:3730-3738.
98. Reiner O, Carrozzo R, Shen Y et al. Isolation of a Miller-Dieker lissencephaly gene containing G protein beta-subunit-like repeats. Nature 1993; 364:717-721.
99. Roper SN, Eisenschenk S, King MA. Reduced density of parvalbumin- and calbindin D28-immunoreactive neurons in experimental cortical dysplasia. Epilepsy Res 1999; 37:63-71.
100. Santi MR, Golden JA. Periventricular heterotopia may result from radial glial fiber disruption. J Neuropathol Exp Neurol 2001; 60:856-862.
101. Sapir T, Elbaum M, Reiner O. Reduction of microtubule catastrophe events by LIS1, platelet-activating factor acetylhydrolase subunit. EMBO J 1997; 16:6977-6984.
102. Sheen VL, Dixon PH, Fox JW et al. Mutations in the X-linked filamin 1 gene cause periventricular nodular heterotopia in males as well as in females. Hum Mol Genet 2001; 10:1775-1783.
103. Spreafico R, Battaglia G, Arcelli P et al. Cortical dysplasia: an immunocytochemical study of three patients. Neurology 1998; 50:27-36.
104. Spreafico R, Tassi L, Colombo N et al. Inhibitory circuits in human dysplastic tissue. Epilepsia 2000; 41(Suppl 6):S168-73.
105. Tassi L, Pasquier B, Minotti L et al. Cortical dysplasia: electroclinical, imaging, and neuropathologic study of 13 patients. Epilepsia 2001; 42:1112-1123.
106. Taylor DC, Falconer MA, Bruton CJ et al. Focal dysplasia of the cerebral cortex in epilepsy. J Neurol Neurosurg Psychiatry 1971; 34:369-387.
107. Taylor KR, Holzer AK, Bazan JF et al. Patient mutations in doublecortin define a repeated tubulin-binding domain. J Biol Chem 2000; 275:34442-34450.
108. The European Chromosome 16 Tuberous Sclerosis Consortium. Identification and characterization of the tuberous sclerosis gene on chromosome 16. Cell 1993; 75:1305-1315.
109. van Slegtenhorst M, de Hoogt R, Hermans C et al. Identification of the tuberous sclerosis gene TSC1 on chromosome 9q34. Science 1997; 277:805-808.
110. Vinters HV. Surgical pathologic findings of extratemporal-based intractable epilepsy. A study of 133 consecutive cases. Arch Pathol Lab Med 2000; 124:1111-1112.
111. Vinters HV. Histopathology of brain tissue from patients with infantile spasms. Int Rev Neurobiol 2002; 49 (in press).
112. Vinters HV, De Rosa MJ, Farrell MA. Neuropathologic study of resected cerebral tissue from patients with infantile spasms. Epilepsia 1993; 34:772-779.
113. Vinters HV, Fisher RS, Cornford ME et al. Morphological substrates of infantile spasms: studies based on surgically resected cerebral tissue. Child's Nerv Syst 1992; 8:8-17.
114. Vinters HV, Park SH, Johnson MW et al. Cortical dysplasia, genetic abnormalities and neurocutaneous syndromes. Dev Neurosci 1999; 21:248-259.
115. Walsh CA. Genetic malformations of the human cerebral cortex. Neuron 1999; 23:19-29.
116. White R, Hua Y, Scheithauer B et al. Selective alterations in glutamate and GABA receptor subunit mRNA expression in dysplastic neurons and giant cells of cortical tubers. Ann Neurol 2001; 49:67-78.
117. Wick W, Grimmel C, Wild-Bode C et al. Ezrin-dependent promotion of glioma cell clonogenicity, motility, and invasion mediated by BCL-2 and transforming growth factor-β2. J Neurosci 2001; 21:3360-3368.
118. Wienecke R, Konig A, DeClue JE. Identification of tuberin, the tuberous sclerosis-2 product. Tuberin possesses specific Rap1GAP activity. J Biol Chem 1995; 270:16409-16414.
119. Wienecke R, Maize Jr JC, Shoarinejad F et al. Colocalization of the TSC2 product tuberin with its target Rap1 in the Golgi apparatus. Oncogene 1996; 13:913-923.
120. Yamanouchi H, Jay V, Otsubo H et al. Early forms of microtubule-associated protein are strongly expressed in cortical dysplasia. Acta Neuropathol 1998; 95:466-470.
121. Ying Z, Babb TL, Comair YG et al. Induced expression of NMDAR2 proteins and differential expression of NMDAR1 splice variants in dysplastic neurons of human epileptic neocortex. J Neuropathol Exp Neurol 1998; 57:47-62.
122. Yoshida A, Kobayashi K, Manya H et al. Muscular dystrophy and neuronal migration disorder caused by mutations in a glycosyltransferase, POMGnT1. Dev Cell 2001; 1:717-24.

CHAPTER 14

Functional Implications of Seizure-Induced Neurogenesis

Helen E. Scharfman

Abstract

The neurobiological doctrine governing the concept of neurogenesis has undergone a revolution in the past few years. What was once considered dubious is now well accepted: new neurons are born in the adult brain. Science fiction is quickly becoming a reality as scientists discover ways to convert skin, bone, or blood cells into neurons.

In the epilepsy arena, widespread interest has developed because of the evidence that neurogenesis increases after seizures, trauma, and other insults or injuries that alter seizure susceptibility.

This review discusses some of the initial studies in this field, and their often surprising functional implications. The emphasis will be on the granule cells of hippocampus, because they are perhaps more relevant to epilepsy than other areas in which neurogenesis occurs throughout life, the olfactory bulb and subventricular zone. In particular, the following questions will be addressed:

1. Do granule cells that are born in the adult brain become functional, and what are the limits of their function? Do they behave homogeneously? Results from our own laboratory have focused on cells that become established outside the normal boundaries of the granule cell layer, forming a group of "ectopic" granule cells in the hilar region.
2. Is increased neurogenesis beneficial, or might it actually exacerbate seizures? Evidence is presented that supports the hypothesis that new granule cells may not necessarily act to ameliorate seizures, and might even contribute to them. Furthermore, cognitive deficits following seizures might in part be due to new circuits that develop between new cells and the host brain.
3. How do the new cells interact with the host brain? Several changes occur in the dentate gyrus after seizures, and increased neurogenesis is only one of many. What is the interdependence of this multitude of changes, if any?
4. Is neurogenesis increased after seizures in man? Research suggests that the data from human epileptics are actually inconsistent with the studies in animal models of epilepsy, because there is little evidence of increased neurogenesis in epileptic tissue resected from intractable epileptics. Yet neurogenesis has been shown to occur in humans throughout adult life. What might be the reasons for these seemingly disparate results?

Introduction

Neurogenesis is the creation of nerve cells. New neurons normally develop from progenitor cells located in many areas of the immature nervous system during development. In the adult, it occurs primarily in three locations: the subgranular zone of the dentate gyrus, the olfactory bulb, and the subventricular zone lining the ventricular walls.

Recent Advances in Epilepsy Research, edited by Devin K. Binder and Helen E. Scharfman. ©2004 Eurekah.com and Kluwer Academic / Plenum Publishers.

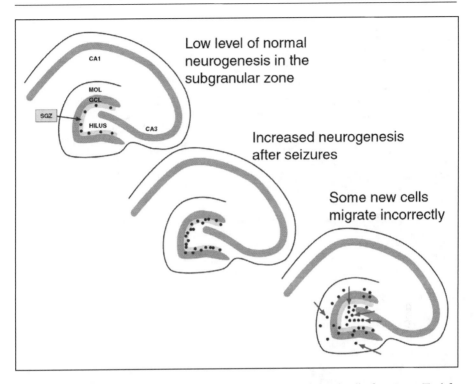

Figure 1. Schematic diagram of changes in neurogenesis of hippocampal granule cells after seizures. Top left: A low level of neurogenesis is normal in the adult dentate gyrus of the rat. A diagram of a transverse section of the rat hippocampus is shown. Neurogenesis normally occurs in a layer positioned immediately below the granule cell layer which is called the subgranular zone (SGZ). MOL = molecular layer, GCL = granule cell layer. Center: Increased neurogenesis occurs after seizures. This is thought to arise from the SGZ primarily. Lower right: After seizures induced by pilocarpine, some of the newly born cells appear to migrate from the SGZ incorrectly, entering the molecular layer and hilus (arrows).

In the dentate gyrus, the subgranular zone is a thin layer that lies just beneath the granule cell layer (Fig. 1). Ordinarily, cells are born there at a modest rate throughout life, although this rate gradually decreases with age.[56,76,90,116] Remarkably, seizures appear to increase the rate of neurogenesis. This appears to occur after brief seizures,[9] kindling,[60,71,92] and after status epilepticus induced by chemoconvulsants such as pilocarpine[26,73] or kainic acid[45,69] or status epilepticus induced by electrical stimulation.[82] Seizures following electroconvulsive shock also increase neurogenesis.[64,93] Other experimental insults that injure the brain cause increased neurogenesis also, such as experimental trauma or stroke.[61,62,94] Importantly, a number of experiments have shown that various facets of the biological responses to such events (e.g., seizures, trauma) may contribute to the increase in neurogenesis, indicating that seizures per se may not be the only factor. For example, manipulations that increase neuronal activity, without necessarily leading to seizures, induce increased neurogenesis.[11,20,21,29,42] Changes in stress also increase neurogenesis.[42-44] Many of the neurons that are born in the dentate gyrus after seizures migrate correctly, that is, into the granule cell layer[73] (Fig. 1). However, others apparently do not. Thus, new cells appear in the molecular layer and hilus after seizures[73,87] (Fig. 1). Presumably, the new cells in the molecular layer were born in the subgranular zone and migrated into and past the cell layer. They may fail to stop because in the adult brain the signals for migration are weaker than in the developing animal, when virtually all cells know when to stop and the granule cell layer is a very well defined space. Regarding the granule cells that migrate into the

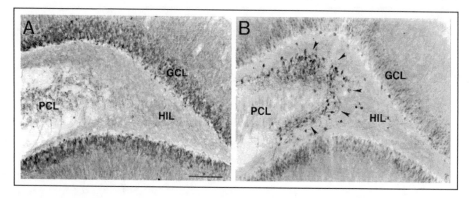

Figure 2. Ectopic hilar granule cells in the dentate gyrus after kainic acid-induced status epilepticus. A) A section from a saline-injected rat demonstrates calbindin-immunoreactivity predominantly in the cell bodies, dendrites and axons of dentate gyrus granule cells. PCL = pyramidal cell layer; HIL = hilus; GCL= granule cell layer. Calibration = 200 μm. B) A section from a kainic acid-injected rat, over 1 month after status epilepticus. Numerous calbindin-immunoreactive neurons surround the border of the hilus and pyramidal cell layer (arrows). Same calibration as A. Used with permission, ref. 87.

hilus, the sites of origin may also be the subgranular zone, and in this case the cells may simply miss the signal that tells them the correct direction to travel. Instead, they may migrate in the opposite direction than they normally would move. What is extraordinary is that they apparently continue to migrate in this direction, about 180° from normal, and do not stop for hundreds of microns. Apparently they do stop eventually, because clusters of cells appear to accumulate at the border between the hilus and area CA3 (Fig. 2). Therefore, it is likely that the adult brain has a potent signal at that location that eliminates any chance that a dentate neuron would cross into area CA3.

It is conceivable that the hilar granule cells actually don't arise from the subgranular zone. Instead, the site of origin may be the hilus itself, because this is an area that contributes, as late as postnatal day 10, to granule cells that are normally situated in the granule cell layer during adulthood;[2,3] vestiges of this tertiary matrix might still exist thereafter. It is not entirely clear that this matrix converts completely to the subgranular zone at maturation. However, regardless of where in the hilus they are born, these cells still seem to migrate until the CA3/hilar border, as if they have migrated there and then must stop.

Recent evidence indicates that cells from the periphery can be transformed into cells with the characteristics of neurons.[5,47,68,105] This raises an even more heretical possibility. Some of the new neurons that appear in the dentate gyrus could originate from the blood cells that enter the dentate gyrus through breaks in the blood-brain barrier during seizures. This would be consistent with the empirical observation that prolonged seizures (e.g., status epilepticus) lead to more neurogenesis than weak seizures, because severe seizures would be associated with greater breaks in the blood-brain barrier. In fact there does appear to be an increase in the endothelial cell staining of the hilus after pilocarpine-induced seizures (see chapter by Croll et al, this volume). But it is too soon to say whether these cells could become neurons.

Proving That Granule-Like Cells in the Hilus after Seizures Were Newly-Born Granule Cells

Our studies of granule-like cells in the hilus that appeared after seizures at first focused on whether in fact they were granule cells, and the evidence that they were born after seizures. The data were at first anatomical: multiple cells that were immunoreactive for a neuronal marker of granule cells, the calcium-binding protein calbindin D28K, were present in animals that had status epilepticus, but not saline-treated, age-matched controls.[87] Our hypothesis, that these were newly-born granule cells that had somehow ended up in the hilus rather than the granule cell layer, clearly required proof. Our hypothesis stemmed from the work of Parent et al,[73] who had already shown that some of the new granule cells that arose after seizures appeared to be in the hilus. Still, when we found very much larger numbers of calbindin-immunoreactive neurons that were quite far from the granule cell layer, actually on the border with area CA3 and the hilus (Fig. 2), it was not clear that these in fact were the same cells as shown in the study by Parent et al,[73] which were much closer to the granule cell layer.

One argument against the hypothesis was based on several papers which actually showed that some granule cells in the normal rat do exist in the hilus.[38,67] However, the number of these "ectopic" granule cells is extremely small, and in fact in our own saline-control tissue, there were rarely any hilar neurons that were calbindin-immunoreactive (Fig. 2). After ischemia, there appear to be some ectopic hilar cells that have the general morphology of granule cells, but again only a few.[52] This contrasted with our observations in rats that had status epilepticus and recurrent seizures, where in the average 50 μm section there were often as many as 100 cells (Fig. 2).

It is important to consider the validity of calbindin immunoreactivity to identify granule cells in this context. The immunoreactivity for calbindin appeared suitable to identify granule cells because, in saline-treated controls, granule cells were consistently immunoreactive for calbindin, as indeed many others have shown.[6,48,77,97] Yet calbindin is not a consistent marker of granule cells after seizures. Thus, seizures lead to a decrease in calbindin immunoreactivity in some granule cells,[7,106] and this was also evident in our epileptic tissue (Fig. 1 of ref. 89). Developing granule cells also have weak labeling for calbindin,[32,41,77] so the cells that were recently born might not be immunoreactive. Therefore, calbindin may not be a faithful marker of granule cells after seizures. Thus, the number of granule neurons in the hilus after seizures is difficult to estimate based entirely on calbindin immunoreactivity, and it could be underestimated if many were recently born.

Based on the considerations above, it was highly unlikely that the neurons we found in pilocarpine-treated rats were simply normal granule cells that were in the wrong place. However, another possibility was that the hilar cells we surmised were granule cells because of their calbindin-immunoreactivity and general morphology were actually a type of hilar interneuron. The primary argument against this was the data from saline-treated controls, which did not show many neurons in the hilus that were immunoreactive for calbindin. However, calbindin-immunoreactive hilar neurons have been reported by others.[95] Furthermore, some interneurons that normally have low levels of calbindin might increase expression after seizures. Indeed, this occurs for hilar neurons that express neuropeptide Y.[91,108]

Therefore, a series of anatomical experiments were conducted to address the hypothesis that the hilar neurons we suspected were granule cells were actually a type of interneuron. First, we found that markers of interneurons (parvalbumin, somatostatin, GABA, GAD, calretinin) did not coincide with the clusters of calbindin-immunoreactive, granule-like neurons in the hilus. Thus, in serial sections through the dentate gyrus, clusters of cells were present at the CA3/ hilar border that were immunoreactive for calbindin, but adjacent sections failed to demonstrate hilar neurons in that location that stained for markers of GABAergic neurons. Interestingly, neuropeptide Y (NPY) did not label the hilar calbindin-immunoreactive plexus, although it would not have been surprising if it had, because NPY labels some granule cell axons after seizures, and sometimes there is immunoreactivity in their somata.[91,108] Glial fibrillary acidic protein (GFAP), which labels adult glia, also did not label these clusters of calbindin-immunoreactive hilar neurons. Therefore, it was unlikely that the neurons in question were glia.

Although these results were useful, there still remained a possibility that the neurons we thought were newly-born granule cells were actually not. For example, perhaps they were VIP- or CCK-immunoreactive neurons; indeed there are so many possible types of interneuron it is difficult to completely test the spectrum of possibilities. Another possibility was that our antibodies performed differently in epileptic tissue than in control sections, or identified an odd cell type that had not previously been described. Knowing that the hilus contains a remarkable number of cell types,[4] and many undergo seizure-induced changes in gene expression, questions still remained.

Perhaps the most convincing evidence came from the results of intracellular recordings from slices of pilocarpine-treated rats with chronic seizures. Normally in an adult rat, or in an aged saline control, it is quite rare to encounter neurons in the hilus with electrophysiological characteristics of granule cells. Most neurons are either similar to mossy cells or interneurons.[84,85] Thus, granule cells have high resting membrane potential relative to other cell types (mossy cells, interneurons, and pyramidal cells). Strong spike frequency adaptation is another distinguishing feature. Granule cells also have an action potential duration that is consistent with a regular spiking neuron, the action potential has a high ratio of rate of rise to rate of decay (dv/dt ratio), and there is no significant inward rectification or "sag" in response to hyperpolarization from resting potential.[84,85,87,103,104,113] Other cell types may have one or another of these characteristics, but not all of them. Therefore, it was striking when many healthy hilar neurons were encountered that had electrophysiological characteristics of granule cells.

Furthermore, after intracellular injection of Neurobiotin into the hilar cells with the electrophysiological "signature" of granule cells, the labeled neurons had the morphology of a granule cell. Thus, there was a small, ovoid soma, dendrites with many spines, and a mossy fiber axon. There were no thorny excrescences, consistent with the fact that these cells did not have electrophysiological characteristics of mossy cells or pyramidal neurons. Some of the hilar cells with electrophysiology like granule cells had a polarized dendritic tree, like a normal granule cell (Fig. 3). However, others had "basal" dendrites and their dendritic tree was therefore bipolar. This is actually consistent with the description of granule cells in the normal location (i.e., in the granule cell layer) in epileptic tissue, which may have basal dendrites.[17,79,102,110] However, a very high percentage of hilar granule cells had basal dendrites relative to granule cells in the granule cell layer of epileptic tissue. This difference raises the possibility that there were fewer morphological constraints on the developing dendritic tree of the hilar "granule" cells as they differentiated, so they developed more basal dendrites. In summary, these correlative electrophysiological-anatomical data provided additional evidence that the calbindin-immunoreactive hilar neurons observed in fixed tissue sections were granule cells.

Particularly important was the mossy fiber axon, given that some interneurons have small ovoid somata and can have spiny bipolar dendrites, somewhat similar to the bipolar hilar cells we suspected were granule cells. But interneurons do not have a mossy fiber axon. The classical meaning of the term "mossy fiber" is a granule cell axon that innervates stratum lucidum, where its major branch projects to CA2 in a fiber that runs approximately 100-200 μm and parallel to the pyramidal cell layer. Along its length there are giant boutons at regular intervals, and these innervate the proximal dendrites of CA3 neurons.[1,13,24,36,37] There are also many collaterals in the hilus, where giant boutons innervate the excrescences of mossy cells, and in addition there are smaller, more conventional terminals that innervate dendrites of several types of hilar neurons. In epileptic tissue, mossy fiber axons develop additional collaterals that project to the inner molecular layer, referred to as "mossy fiber sprouting." Sprouted axons contact the dendrites of granule cells and interneurons in the inner molecular layer.[55,70,111] Notably, the cells impaled in the hilus of epileptic rats that had electrophysiological features of granule cells had an axon with the characteristics of mossy fibers. Because no other hippocampal neurons have such an axon, it was a strong argument that the hilar cells which were impaled indeed were a type of granule cell.

Thus, the combination of electrophysiology and morphology allowed us to easily distinguish granule cells from other cells in the hilus/CA3 region (see also refs. 84 85,86).

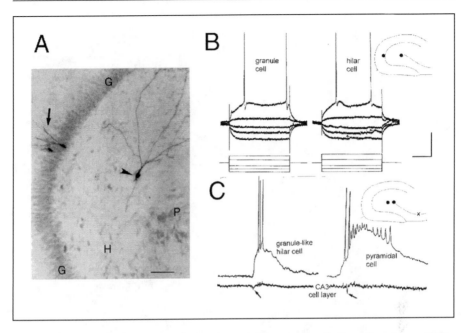

Figure 3. Morphology and electrophysiology of ectopic hilar granule cells born after seizures. A) The morphology of a hilar granule cell from a pilocarpine-treated rat (arrowhead) is shown after intracellular injection of Neurobiotin. For comparison, two granule cells in the granule cell layer (arrow) were also injected. G = granule cell layer; H = hilus; P = pyramidal cell layer. Calibration = 100 μm. B) Physiology of hilar granule cells. Responses to direct current injection (rectangular current pulses, as shown at the bottom) are superimposed for a granule cell located in the granule cell layer (left, "granule cell") and a granule-like neuron located in the hilus (right, "hilar cell") of the same slice. This slice was from a pilocarpine-treated rat which had status epilepticus and recurrent seizures. A diagram of the location of these cells is shown at top right. G = granule cell layer, H = hilus, P = pyramidal cell layer. Arrowheads mark spontaneous synaptic potentials. Calibration = 20 mV, 50 msec. C) Comparison of pyramidal cell and ectopic granule cell activity. Simultaneous extracellular recordings from the pyramidal cell layer of CA3b (bottom) and intracellular recordings are shown for a hilar granule cell (left) and a pyramidal cell (right). This slice was from a pilocarpine-treated rat that had status epilepticus and recurrent seizures. A diagram of the location of the cells (dots) and extracellular recording (x) is shown at top right. The arrows point to the population spike of the spontaneous burst discharge recorded extracellularly. Calibration same as B. Used with permission, ref. 87.

Taken together, the morphology and electrophysiology made a compelling case that the granule-like cells in the hilus actually were granule cells (Fig. 3A,B; ref. 87).

What was the evidence that these hilar granule cells were in fact born after seizures? To address this question, a marker of dividing cells, bromodeoxyuridine (BrdU), was used. BrdU was injected after status epilepticus using two different schedules (1 injection per day, 50 mg/kg, for 5 days, either days 3-8 after status or 4-11). In all cases, there were cells in the hilus that were immunoreactive for both BrdU, which stains the nucleus, and calbindin, which stains the cytoplasm (Fig. 4). Brdu/NeuN double-labeled cells were also present (Fig. 4). However, no BrdU profiles were double-labeled with markers of GABAergic neurons (parvalbumin, neuropeptide Y; Fig. 4). In addition, BrdU injections in the hours before status epilepticus, or in saline controls, have not provided evidence of new neurons in the hilar region. Furthermore, injection long after status (over one month) for 5 days (as described above, 1 injection per day, 50 mg/kg i.p.) did not result in labeled hilar neurons.

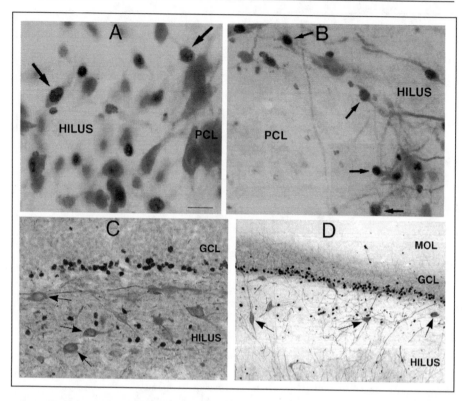

Figure 4. Ectopic hilar granule cells are born after status epilepticus. A-B: Following injection of BrdU after status epilepticus, tissue sections were double-labeled for BrdU and either NeuN (A) or calbindin (B). The results showed numerous double-labeled neurons (arrows) at the border of the hilus and pyramidal cell layer. Calibration = 30 μm (A) and 60 μm (B). PCL = end of pyramidal cell layer. From ref. 87 with permission: C-D. In a different animal, double-labeling with BrdU and neuropeptide Y (C) or BrdU and parvalbumin (D) demonstrated no double-labeled cells, but numerous hilar neurons that were immunoreactive for neuropeptide Y or parvalbumin (arrows). Calibration (in A) = 60 μm (C) and 120 μm (D).

Do All the New Granule Neurons Become Functional? Do They Behave the Same?

Our experiments with the pilocarpine and kainic acid models suggested that hilar granule cells that are born after seizures survive for long periods of time and become functional. Evidence for their long term survival came from the studies of BrdU-labeled neurons in animals that were not killed until over 1 year had passed since BrdU injection. In these animals, numerous BrdU-labeled neurons were found in the dentate gyrus granule cell layer as well as the hilus, and they were double-labeled with NeuN or calbindin. Thus, neurons that divided after status appeared to survive as long as one year.

Regarding the development of neuronal function, intracellular recordings from hilar granule cells in slices demonstrated that they could mature into normal, functioning neurons with action potentials and other physiological characteristics of adult neurons. A recent study in normal adult rats has also identified that neurons which are born in the dentate gyrus in the adult animal can develop action potentials and membrane properties like normal adult granule cells.[107] Therefore, in both the normal and pathological situation, granule cells that are born in the adult rat dentate gyrus develop functional characteristics of neurons, surviving at the very least to maturity.

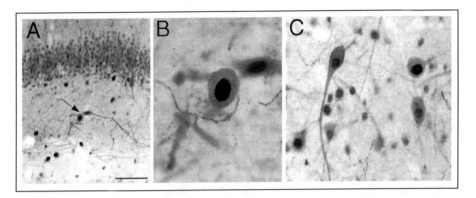

Figure 5. Ectopic granule cells in the hilus are active during a spontaneous seizure. Immunoreactivity for c-fos and calbindin in a pilocarpine-treated animal that had spontaneous seizures and was sacrificed 3 hours after a spontaneous seizure was observed. A-B) There are two double-labeled neurons in the hilus (arrow), enlarged in B. Calibration = 100 μm (A); 25 μm (B). C) Double-labeled hilar neurons in another animal are shown. Calibration (in A) = 50 μm. Used with permission, ref. 87.

Additional evidence from both in vitro and in vivo studies indicates that not only are these granule cells able to mature into functional neurons, but they also integrate into the host environment, so that they discharge in similar patterns as their neighbors.

In vivo studies used pilocarpine-treated rats at the stage when they displayed recurrent, spontaneous seizures. In these experiments, animals that had pilocarpine-induced status epilepticus were observed approximately 1-5 months after status, during the period of time when recurrent spontaneous seizures occur. Three hours after a spontaneous stage 5 limbic seizure (convulsion) was observed, each animal was perfused with fixative and compared to an animal that did not have a motor seizure in the last 8 hours, or a saline-treated, age-matched control, in which stage 5 seizures do not occur (and it is highly unlikely that subclinical seizure activity occurs). The immediate early gene c-fos was used as a marker of neural activity: we hypothesized that if the newly-born granule cells were active during a seizure, they would stain for c-fos, like other hippocampal neurons stain after a seizure. We thus used c-fos to mark a neuron that was active, regardless of the degree of activity, which c-fos immunoreactivity cannot describe. Indeed, in pilocarpine-treated rats with a recent seizure, there were numerous c-fos stained neurons in the hilus, and they could be double-labeled with markers of granule cells, such as calbindin and NeuN (see ref. 89; Fig. 5). There also were other cells in the hilus that were c-fos immunoreactive, and they could be demonstrated, for example, by double-labeling c-fos and parvalbumin or c-fos and NPY.[89]

Interestingly, at the 3-hour delay following a seizure, granule cells located in the granule cell layer were either not immunoreactive or less immunoreactive than the hilar granule cells, presumably because they become c-fos immunoreactive sooner and their immunoreactivity wanes by 3 hours.[30,40,59,115] Thus, hilar granule cells seemed to adopt a pattern of activity similar to their immediate neighbors, the hilar neurons, rather than the normal granule cells in the granule cell layer. These data suggest that the development of seizure-related activity of newly-born granule cells is highly influenced by their neighbors and local environment, although the development of their intrinsic properties seems independent of their immediate surroundings. Thus, there is a dissociation in the development of the interaction with host neurons and the maturation of neuronal intrinsic properties. The data suggest that there is little that can interfere with the membrane properties of these cells, the axon distribution, and the general shape of the neuron (somata size and shape, dendritic patterns), but a great deal of plasticity is possible for other variables, such as their final location and circuit-related behavior.

Do all newly born granule cells become functional? It is certainly not known at the present time that they all become mature, functional neurons. Indeed it is most likely that some do not survive, because there are reports that more BrdU-labeled neurons are detected after seizures if the animal is treated with a caspase inhibitor.[31] And even if they survive, whether all become functional or not is not clear. Some may develop to maturity, but others could have a different fate. Although there is little evidence at the present time, it is possible that some become glia, interneurons, or become dormant, perhaps retaining a capacity to differentiate at a future time.

Is Increased Neurogenesis Beneficial, or Might It Actually Increase Seizure Susceptibility?

Another line of evidence that the hilar granule cells function differently with respect to the host brain than normal granule cells came from in vitro studies. In these experiments, pilocarpine-treated rats that had status epilepticus and recurrent seizures were used to prepare hippocampal slices, and intracellular recordings were made from hilar granule cells as well as other neurons in the slice.

Although the hilar granule cells that were recorded in slices were remarkably similar to normal adult granule cells in their intrinsic properties (as described above), there was one aspect of their physiology that was quite different than the activity of a normal granule cell. Many of the hilar granule cells displayed spontaneous bursts of action potentials (Fig. 3C; see ref. 87). These bursts occurred at variable frequency from slice to slice (usually ~1/10 sec), but within any given slice the burst frequency was consistent. Each burst was composed of a variable number of action potentials and arose on a large depolarization (Fig. 3C). The depolarization appeared to be a giant EPSP because it increased in amplitude with hyperpolarization and action potentials were triggered at its peak. We have also found subsequently that bursts are blocked by the AMPA receptor antagonists CNQX (10 μM) but are prolonged if exposed to the GABA$_A$ receptor antagonist bicuculline (25 μM).

Insight into the mechanism underlying these burst discharges was obtained when simultaneous recordings were made in the area CA3 pyramidal cell layer, either intracellularly or extracellularly (Fig. 3C). The bursts of ectopic hilar granule cells were synchronous with bursts of area CA3 neurons. Many of the recordings of extracellular CA3 bursts were small in amplitude, in all likelihood because there was some damage and loss of cells in the CA3 region due to status and chronic seizures. However, all CA3 pyramidal cells that were recorded intracellularly demonstrated epileptiform burst discharges that were synchronized with the small field bursts recorded from the CA3 cell layer, indicating it was a robust phenomenon. Interestingly, 54% of slices from animals that were examined less than 6 months after status epilepticus did not exhibit bursts in area CA3, but 90% did if slices were from animals sacrificed >6 months after status. The delay may be due to the time required for the mechanisms underlying these types of burst discharges to fully develop, perhaps because additional recurrent connections that are not already present must sprout and form functional synapses. Some of this new circuitry may require synapses between hilar granule cells and CA3 pyramidal cells. Indeed, it may be that the new synapses that are required are the new mossy fibers of the ectopic granule cells and the hilar collaterals from CA3 pyramidal neurons onto ectopic granule cells. The first would require time to form synapses on CA3 pyramidal cells, and the latter would also require time to "find" the new hilar granule cells and form functional synapses. Since CA3 axons normally innervate the hilar region and some hilar neurons are lost after seizures, the CA3 axons may be in a state that facilitates innervation of new hilar neurons.

This of course assumes that the circuitry required for burst discharges to develop are chemical synapses, but one cannot rule out the potential role of gap junctions. The main argument for a role of chemical synapses lies in the relative timing of the bursts of CA3 neurons and the ectopic hilar granule cells. The peak of the first action potential of a given CA3 burst began at least 1 millisecond before the onset of the depolarization of a simultaneously-recorded hilar granule cell, and often there was a delay far greater than 1 millisecond; if there was no delay, gap junctions would be the more likely candidate. It is possible that ephaptic interactions are a factor because the clusters of ectopic granule cells can be quite dense, and there may be decreased efficacy in clearing extracellular potassium in epileptic tissue.[27] However, there seems to be quite a bit of damage in the CA3 cell layer in pilocarpine-treated rats, because it is very difficult to obtain a healthy intracellular impalement outside of CA3b, and the field potentials are quite small relative to control slices. This argues that ephaptic interactions would be less likely, at least in the CA3 cell layer.

These data address the long-standing controversy that seizures beget more seizures, a question that has clear implications for treating individuals who have just had their first seizure. Especially when the individual is a child, it is not clear that anticonvulsant drugs are needed to prevent another seizure, or desirable, given their side effects. It appears, at least in the pilocarpine model in adult rats, that the pathophysiology in the epileptic brain can indeed get "worse" as new circuits develop between excitatory neurons, in this case pyramidal cells and newly-born granule cells. Therefore, intervention to block this recurrent excitatory circuitry could be prudent. A major argument against this interpretation is that irradiation to block neurogenesis after pilocarpine-induced status epilepticus did not block seizures.[72] However, this may simply reflect that there are many potential foci after pilocarpine-induced status, and that neurogenesis does not explain seizures and epilepsy completely.

Thus, the hilar-CA3 region may contribute to seizures as one of many epileptic foci. After status, the development of ectopic hilar granule cells would occur, and this would be followed, hypothetically, by the formation of circuitry between the new cells and surviving CA3 neurons. This would ultimately lead to repetitive "interictal" types of burst discharges. It could directly foster a transition between status and subsequent seizures.

Notably, this "focus" may actually include other cells as well, such as the hilar mossy cells that survive repeated seizures.[88] These neurons are glutamatergic hilar neurons that discharge synchronously with pyramidal cells in slices of pilocarpine-treated rats. Although we cannot be sure at the present time what percentage of hilar mossy cells survive status and chronic seizures, and what subset participates in pyramidal cell burst discharges, it is clear that at least some do.[88]

The implications would be slight if in fact these cell types, area CA3 pyramidal cells, ectopic hilar granule cells, and hilar mossy cells, did not have such substantial projections to other excitatory neurons. However, the opposite is actually the case. CA3 projections include adjacent CA3 neurons, hilar neurons and CA1 pyramidal cells in the normal rat, and may include other cell types in the epileptic rat if sprouting of the CA3 axon occurs. CA3 also projects to the contralateral hippocampus. Ectopic hilar granule cells have, at least, projections to the inner molecular layer and CA3 pyramidal cells, and may also innervate hilar neurons because of the substantial collateralization of their axons in the hilus. These collaterals have numerous varicosities, indicating potential synapses. The inner molecular layer innervation may include granule cell dendrites, but it might also include processes of GABAergic neurons, analogous to the projections of sprouted mossy fibers from granule cells located in the granule cell layer. Mossy cells project to granule cells in the normal rat, both proximally and distally, ipsilateral and contralateral. They also innervate interneurons. Therefore, the initial focal discharges between relatively small populations of pyramidal cells, ectopic granule cells and mossy cells could have substantial impact on other neuronal cell types, and have potential to exit the hippocampus.

Figure 6 shows schematically how bursts within the CA3 region, ectopic granule cells and mossy cells could lead to seizure-like activity in the hippocampus. Empirical observations suggest CA3 bursts would occur first because simultaneous recordings showed that the first action potential of a spontaneous burst in a CA3 pyramidal cell always occurred before the first action potential of a simultaneously recorded ectopic hilar granule cell or mossy cell. Second, ectopic granule cells or mossy cells could excite the granule cells located in the granule cell layer because both project to the inner molecular layer. This would be particularly effective in depolarizing granule cells if the ectopic neurons and mossy cells were synchronized; asynchronous

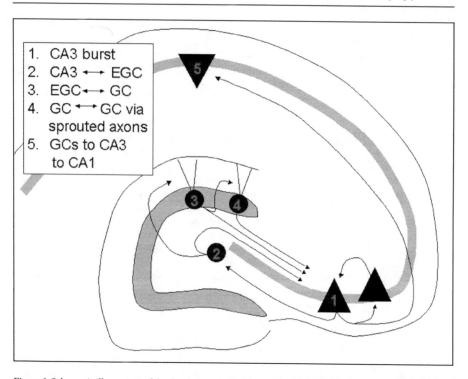

Figure 6. Schematic illustration of the development of widespread seizure activity from initial synchronized burst discharges among CA3 pyramidal cells and ectopic hilar granule cells. 1. CA3 burst. Initially, activity that is synchronized in area CA3 neurons develops after status epilepticus. The reasons may include increased recurrent excitatory activity due to loss of interneurons, as well as sprouting among residual pyramidal cells. 2. CA3 to hilus. Based on empirical findings in pilocarpine-treated rats, synchronized burst discharges develop among pyramidal cells, newly-born hilar ectopic granule cells (EGCs) and hilar mossy cells (MCs) several months after status epilepticus. This is likely to be due to the normal projection of pyramidal cells to hilar neurons, and the mossy fiber axon that develops in EGCs. 3. hilus to GC. Because EGCs have mossy fiber collaterals that contribute to the sprouted fiber plexus in the inner molecular layer[87], and MCs also project there, the bursts can potentially lead to activation of granule cells located in the granule cell layer. This would be limited by whatever interneurons are innervated by EGCs and MCs. 4. GC to GC. GC activation may be amplified because they are interconnected by sprouted fibers. However, this also would be limited by the extent that sprouted fibers activate GABAergic neurons. 5. GCs to CA3 to CA1. Strong excitatory activity that develops in the granule cell layer could potentially exit the hippocampus and trigger a limbic seizure by the trisynaptic circuit.

depolarizations may not reach threshold in a granule cell, because granule cells are quite hyperpolarized normally. However, synchronous release of glutamate, particularly on the proximal portion of the granule cell dendritic tree, where both the terminals of mossy cells and ectopic hilar granule cells are located, would be likely to depolarize a granule cell above its threshold. Action potentials in even a few granule cells would be likely to be amplified throughout the population because of recurrent excitatory connections within the sprouted network. Indeed, this activity may serve as a trigger for the sprouted network that normally appears to be silent. Of course, one would expect that at least some of this excitatory activity would be limited by concurrent activation of GABAergic neurons, but if interneurons are damaged or lost after seizures, as indeed appears to be the case,[16,18,50] GABAergic neurons might not limit the glutamatergic activity enough to stop its progression. The balance of excitation and inhibition may shift as neuroactive peptides in the hippocampus wax and wane with the circadian

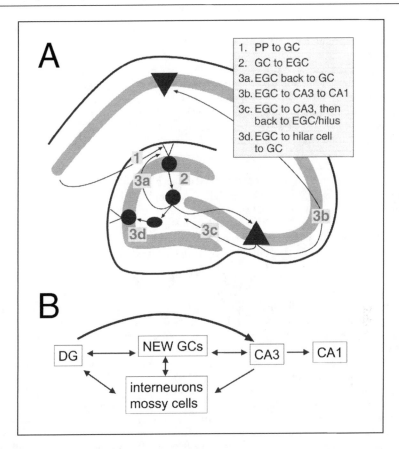

Figure 7. Effects of ectopic hilar granule cells on normal signal processing in the trisynaptic circuit. It is proposed that the development of ectopic hilar granule cells could contribute to cognitive deficits in epileptics by interfering with the normal trisynaptic circuit. A) The trisynaptic circuit is schematized as perforant path axons innervating distal granule cell dendrites (1), granule cell mossy fibers innervating proximal CA3 pyramidal cell dendrites (2), and CA3 pyramidal cell projections to apical dendrites of CA1 pyramidal cells (3). The accumulation of hilar ectopic granule cells with axons that innervate pyramidal cells, hilar neurons, and the inner molecular layer, suggests a multitude of complex pathways that would potentially interfere with normal transmission along the trisynaptic pathway, as shown diagrammatically in B.

rhythm.[10,15,23] Indeed, there may be several times of day when GABAergic inhibition would not balance the effect of glutamatergic neurons on the sprouted granule cell network. Thus, GABAergic inhibition could perhaps keep the network in control for the majority of the time, but at certain times inhibition might decrease, leading to periodic large discharges and possibly seizures.

Thus, the focal burst discharges in CA3-hilus could generate a larger degree of reverberatory activity, and this might eventually leave the hippocampus via area CA3 neuronal projections to CA1. Based on the findings thus far, the results support the hypothesis that new granule cells may not necessarily act to ameliorate seizures, but might even contribute to them.

These circuit considerations also may have implications for the cognitive deficits following seizures. Such deficits might in part be due to new circuits that develop between new hilar granule cells and the host brain. These new circuits would be likely to disrupt the trisynaptic circuit, and thus potentially interfere with normal learning and memory (Fig. 7A).

In Figure 7A, the trisynaptic circuit is illustrated schematically as: 1) the synapse of the perforant path onto granule cell outer dendrites; 2) the synapse of granule cell mossy fibers onto CA3 pyramidal proximal dendrites; and 3) the synapse of CA3 pyramidal Schaffer collaterals onto CA1 pyramidal apical dendrites. The activity along the trisynaptic pathway would clearly be more circuitous by the addition of hilar granule cells, particularly given the evidence that these new neurons have perforant path input, pyramidal cell input, and their own mossy fiber axons.[87] Thus, information from the perforant path that would normally pass through the trisynaptic circuit would be likely to pass through hilar granule cells. Information could become "stalled" in a circular nest of intermediary pathways within the CA3-dentate region (Fig. 7B).

How Do the New Cells Interact with the Host Brain?

This question actually can be addressed at a number of levels. One is the functional relationship of the new neurons and surrounding adult neurons or "host environment," once the new neurons have been born, developed, and matured. This is discussed above.

A second issue is how the new cells and host interact as the new cells are developing. In other words, do the mature cells influence those that are newly-born, and vice-versa?

Influence of the Host Environment on Newly-Born Cells

Clearly the host environment normally is quite influential on new neurons. Studies in culture have shown repeatedly that addition of specific factors to the culture medium can have a striking influence on neurodevelopment. In addition, transplantation of neurons into a new location often leads to the transformation of those neurons into those with characteristics of the new environment.

How might the characteristics of the epileptic "host" dentate gyrus influence cells that have recently divided in the subgranular zone? One effect may be a spatial influence, because in the epileptic dentate gyrus there are fewer cells in the hilar region than normal,[50] and this might provide an impetus to move into the hilus because there would be less competition for space there. Thus, hilar cell loss after seizures may facilitate the entry of new cells to the hilar region. In the normal adult hippocampus, the situation may be quite different, and indeed, newly-born granule cells are found in the hilus under very few conditions other than those following seizures. Furthermore, seizures without cell loss result in few new granule cells in the hilus (Scharfman et al, unpublished results).

Besides spatial cues, there is a likely chemical influence of numerous peptides, growth factors, and other substances in the host dentate gyrus (Fig. 8). These may provide chemoattractive/repulsive forces as well as supportive, proliferative and developmental influence on developing neurons of the subgranular zone. Indeed, one common question of past studies about the dentate gyrus has been the reason why so many substances that specifically influence development are present in adult dentate neurons. The answer may lie in the fact that ongoing neurogenesis occurs in the adult. Thus, some of the substances in adult neurons may aid growth and development of newly-born granule cells. These substances are even more likely to influence newly-born neurons after seizures because they appear to increase their expression after seizures. Thus, although many neuroactive substances in the adult dentate gyrus modulate adult neurotransmission, they could also have another function: to modulate neurogenesis.

The neurotrophin BDNF is one example. BDNF is normally synthesized in granule cells at much greater concentration than other neurons in the brain,[25,117] and until now the reason for this has been unclear. In addition, the expression of BDNF increases after seizures.[12,39] Why should granule cells possess high concentrations of BDNF? And why should this increase after seizures? It may be due to the need for BDNF in the environment of the subgranular zone, where newly-born cells arise. Indeed, BDNF is localized in granule cell axons, not their dendrites or soma, and the axons of course collateralize in the hilar region, where the subgranular zone exists. One might ask how BDNF is released so that it can affect newly-born cells, and this

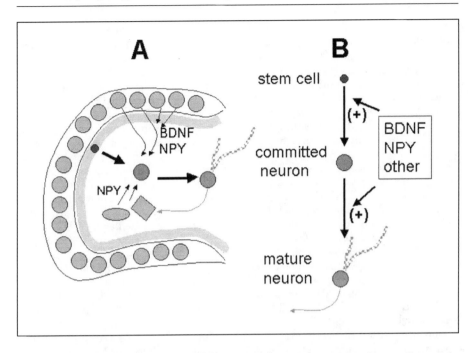

Figure 8. Interaction of newly-born granule cells and the host environment. A-B) A schematic illustrates the potential interactions between developing granule cells born after seizures and the adjacent adult (host) neurons. After seizures, new cells are born in the subgranular zone, and at a similar time, there is altered expression of various proteins in the surrounding adult granule cells and adult hilar neurons. The fact that these two phenomena occur at a similar time may be no coincidence, because the proteins may enhance growth, survival, and proliferation of newly-born cells. Examples include BDNF, zinc, neuropeptide Y and somatostatin. BDNF and zinc are normally present in adult granule cells, and BDNF expression increases in granule cells after seizures. NPY and somatostatin are present in a subset of dentate gyrus GABAergic neurons. After seizures, some of these cells increase their expression of NPY, and many somatostatin-containing neurons die. BDNF, NPY, and zinc appear to have a positive influence and somatostatin a negative influence on processes relating to cell proliferation, so the increase in BDNF, NPY and decrease in somatostatin may foster phenomena associated with neurogenesis. Interestingly, these same substances also have effects on adult synaptic transmission, making their overall influence on the epileptic dentate gyrus network difficult to predict. For further discussion, see text.

is not clear at the present time. Although it is likely that BDNF can be released using mechanisms common to peptide release at synapses,[25] whether it can be secreted in a non-vesicular manner or transported in some other way to specifically affect newly-born cells is not presently known.

Neuropeptide Y is another example. This peptide has been shown to promote neuroproliferation outside the hippocampus,[46] and it may do so in the dentate gyrus as well.[51] It is localized to a subset of hilar GABAergic neurons normally, and after seizures its expression increases in those neurons.[100] It also appears in granule cell axons of the hilus after seizures.[66] For many years the physiological role of NPY has been unclear, and recently it has been associated with depression of synaptic transmission.[109] These effects have been attributed to Y2 and Y5 receptors, mostly.[109] Yet there are also Y1 receptors in hippocampus, and specifically in the subgranular zone as well as the molecular layer.[109] What role might NPY play at its Y1 receptor? The answer may be related to neurogenesis, because NPY's proliferative effects appear to be mediated by this receptor.[46,51] The new data make it seem likely that NPY has effects on synaptic transmission as well as a role in neuroproliferation in the dentate gyrus.

Similar arguments may be possible with other peptides in interneurons of the dentate hilus besides NPY, such as somatostatin, which appears to influence proliferation in peripheral tissues and tumor growth.[57,78,83] Interestingly, somatostatin appears to influence proliferation negatively, and has been suggested as a potential therapy to treat cancer. Therefore, it may be no coincidence that, of all the peptidergic neurons in the dentate gyrus, the neurons which synthesize somatostatin are the ones that are most vulnerable to seizures.[80,108] As a result, neuroproliferation after seizures may be positively modulated, i.e., by the loss of somatostatin.

Another player could very well be zinc, long known to be a native constituent of granule cells both before and after seizures. It has often been asked why granule cells require zinc, and moreover, so much more zinc than other neurons. A potential purpose is to influence proliferation. The basis for this hypothesis comes from studies of the immune system, where zinc stimulates proliferation of T cells by effects on DNA synthesis[75] and the cell cycle.[28] Given that zinc is proliferative in the periphery, perhaps zinc also has a role in proliferation in the dentate gyrus. Zinc could potentially influence neuroproliferation, or zinc could play a role in glial changes, since some glia arise from the same lineage as some immune cells. Thus, analogous to NPY and BDNF, zinc may not only influence dentate gyrus physiology,[19,63,96,112] but also have effects on proliferation.

Finally, GABA is a potential substance that might influence newly-born granule cells of the dentate gyrus. It is normally present in hilar interneurons and, to a lesser extent, granule cells.[35,49,81] After seizures, GABA increases in both the interneurons and the granule cells[34] (see also Sperk et al, this volume). GABA has many actions that are trophic and influence development.[8,65]

Thus, many substances that are made in neurons of the dentate gyrus, substances previously associated with an influence only on neurotransmission of adult neurons, may have an important role in the development and synaptogenesis of newly-born granule cells. This role may explain the remarkable number of substances ordinarily synthesized by neurons in the dentate gyrus, and why they appear to increase expression after seizures.

An important element that is unknown presently is how the substances would be released into the extracellular milieu from the axons of granule cells and other hilar neurons so that they could influence newly-born cells; one would predict that this would be critical if they were to exert effects on immature neurons located in the hilus.

It is important to consider that some of the relevant factors may be released from astrocytes, which are known to influence neurogenesis, although the exact mechanisms are as yet unclear.[99] Glia also have important roles in synaptic transmission in hippocampus.[22,53,54] After seizures, these actions could become more potent as glia become increased in the hilar region. In addition, there is a transformation of many glia into reactive microglia, which may have additional functions besides those of normal glia.

Influence of New Cells on Host Neurons

One cannot rule out the potential influence of the developing neurons in the hilus or granule cell layer on the host hippocampus, even before they are matured completely. This is because immature granule cells have a few characteristics that may affect their neighbors. One is a high incidence of electrical coupling. A second is a chloride reversal potential that produces depolarizing responses to GABA rather than hyperpolarizations (see Staley, this volume). These two characteristics are interesting in light of the fact that granule cells and interneurons of the dentate gyrus increase synthesis of GAD and GABA after seizures. In the case of the granule cells, it is not altogether clear when GABA is released, but it may not be through conventional chemical neurotransmission, since the granule cells may not have the appropriate vesicular trans- porter.[58,101] Instead, GABA could be released by reverse transport (see Richerson and Wu, this volume, see also refs. 74,114). In any case, one would predict that release of GABA onto the immature hilar granule cells that are networked by gap junctions would lead to their synchro- nous depolarization. If this were to happen at a time when the synapses of the immature cells had formed on host neurons, such as pyramidal cells or granule cells, it might lead to synchronous activation and stimulate epileptiform activity. Indeed, it may be no coincidence that the maturation of the new granule cells in the hilus is similar in length to the so-called latent period, i.e., the time between status epilepticus and the first spontaneous seizure.

Is Neurogenesis Increased after Seizures in Man?

As described in part above, new data now exist that suggest that neurogenesis after seizures may play an important role in epilepsy. However, one of the areas of research that has been puzzling is that there has been minimal evidence for increased neurogenesis in human epileptics. Thus, Blumcke et al[14] showed that only the tissue from pediatric cases provided some evidence consistent with laboratory animals, that neurogenesis might increase after seizures. These data are surprising in light of the strong evidence in non-epileptics that neurogenesis increases throughout life.[33] One explanation could be that there is a difference between the stages of epileptogenesis. Although acute seizures or status epilepticus increase neurogenesis in a non-epileptic brain, after repetitive seizures or the chronic condition there may actually be a decreased rate of neurogenesis. Thus, if tissue is examined at a chronic stage of the disease, there may be little evidence of neurons that were recently-born. Another hypothesis is that the tissue examined thus far comes from a select population of pharmacologically refractory epileptics who may not be representative of all individuals. It is also possible that the patients examined by Blumcke et al[14] were not representative even of medically-intractable cases. Indeed, a different study of calbindin-immunoreactivity in human epileptics demonstrated neurons in the hilus that had the general morphology of granule cells.[98]

Yet even if the phenomenon of dentate neurogenesis is not as important to human epilepsy as it appears to be in the rodent, there are other aspects of these findings that may be quite relevant to the clinical condition. Indeed the possible abnormalities that arise from misplaced neurons in the dentate gyrus underscore that in perhaps many of the idiopathic epilepsies there are small areas of abnormal circuits. Each might be too small for current imaging techniques to recognize, hence the use of the term idiopathic. Thus, a large subset of the idiopathic epilepsies are actually epileptics with small migrational disorders. Whether the abnormal neurons were born prenatally or born in the adult brain, it points out the potential importance of understanding every step in development. Indeed, the importance of understanding the molecular basis of neuronal development has become a major emphasis in epilepsy research.

Summary

In this review, the focus has been on the remarkable changes in neurogenesis and function of adult rats after status epilepticus. A number of robust alterations of structure and function occur, independent of the plethora of changes already demonstrated and discussed concerning seizure-induced gene expression and mossy fiber sprouting. Our challenge will be to identify which changes are associated with epileptogenesis and which simply exemplify the remarkable plasticity of the epileptic brain. If this can be done, it may be possible to construct new strategies that are tailored to block epileptogenesis and epilepsy.

Acknowledgements

I thank Annamaria Vezzani and Devin Binder for their comments on the manuscript and Chris Hough for discussions about zinc. This work was supported by NS 37562, 38285, and the Human Frontiers Science Program.

References

1. Acsady L, Kamondi A, Sik A et al. GABAergic cells are the major postsynaptic targets of mossy fibers in the rat hippocampus. J Neurosci 1998; 18:3386-3403.
2. Altman J, Bayer SA. Migration and distribution of two populations of hippocampal granule cell precursors during perinatal and postnatal periods. J Comp Neurol 1990; 301:365-381.
3. Altman J, Bayer SA. Mosaic organization of the hippocampal neuroepithelium and the multiple germinal sources of dentate granule cells. J Comp Neurol 1990; 301:325-342.

4. Amaral DG. A Golgi study of cell types in the hilar region of the hippocampus in the rat. J Comp Neurol 1978; 182:851-914.
5. Azizi SA. Exploiting nonneuronal cells to rebuild the nervous system: from bone marrow to brain. The Neuroscientist 2000; 6:353-361.
6. Baimbridge KG, Miller JJ. Immunohistochemical localization of calcium binding proteins in hippocampus, cerebellum and olfactory bulb of the rat. Brain Res 1982; 245:223-229.
7. Baimbridge KG, Mody I, Miller JJ. Reduction of rat hippocampal calcium-binding protein following commissural, amygdala, septal, perforant path, and olfactory bulb kindling. Epilepsia 1985; 26:460-465.
8. Behar TN, Schaffner AE, Colton CA et al. GABA-induced chemokinesis and NGF-induced chemotaxis of embryonic spinal cord neurons. J Neurosci 1994; 14:29-38.
9. Bengzon J, Kokaia Z, Elmer E et al. Apoptosis and proliferation of dentate gyrus neurons after single and intermittent limbic seizures. Proc Natl Acad Sci USA 1997; 94:10432-10437.
10. Berchtold NC, Oliff HS, Isackson P et al. Hippocampal BDNF mRNA shows a diurnal regulation, primarily in the exon III transcript. Brain Res Mol Brain Res 1999; 71:11-22.
11. Bernabeu R, Sharp FR. NMDA and AMPA/kainate glutamate receptors modulate dentate neurogenesis and CA3 synapsin-I in normal and ischemic hippocampus. J Cereb Blood Flow Metab 2000; 20:1669-1680.
12. Binder DK, Croll SD, Gall CM et al. BDNF and epilepsy: too much of a good thing? Trends Neurosci 2001; 24:47-53.
13. Blackstad TW, Brink K, Hem J et al. Distribution of hippocampal mossy fibers in the rat. An experimental study with silver impregnation methods. J Comp Neurol 1970; 138:433-449.
14. Blumcke I, Schewe J-C, Normann S et al. Increase of nestin-immunoreactive neural precursor cells in the dentate gyrus of pediatric patients with early-onset temporal lobe epilepsy. Hippocampus 2001; 11:311-321.
15. Bova R, Micheli MR, Qualadrucci P et al. BDNF and trkB mRNAs oscillate in rat brain during the light-dark cycle. Brain Res Mol Brain Res 1998; 57:321-324.
16. Buckmaster PS, Dudek FE. Neuron loss, granule cell axon reorganization, and functional changes in the dentate gyrus of epileptic kainate-treated rats. J Comp Neurol 1997; 385:385-404.
17. Buckmaster PS, Dudek FE. In vivo intracellular analysis of granule cell axon reorganization in epileptic rats. J Neurophysiol 1999; 81:712-721.
18. Buckmaster PS, Jongen-Relo AL. Highly specific neuron loss preserves lateral inhibitory circuits in the dentate gyrus of kainate-induced epileptic rats. J Neurosci 1999; 19:9519-9529.
19. Buhl EH, Otis TS, Mody I. Zinc-induced collapse of augmented inhibition by GABA in a temporal lobe epilepsy model. Science 1996; 271:369-373.
20. Cameron HA, Hazel TG, McKay RDG. Regulation of neurogenesis by growth factors and neurotransmitters. J Neurobiol 1998; 36:287-306.
21. Cameron HA, McEwen BS, Gould E. Regulation of adult neurogenesis by excitatory input and NMDA receptor activation in the dentate gyrus. J Neurosci 1995; 15:4687-4692.
22. Castonguay A, Levesque S, Robitaille R. Glial cells as active partners in synaptic functions. Prog Brain Res 2001; 132:227-240.
23. Cirelli C, Tononi G. Gene expression in the brain across the sleep-waking cycle. Brain Res 2000; 885:303-321.
24. Claiborne BJ, Amaral DG, Cowan WM. A light and electron microscopic analysis of the mossy fibers of the rat dentate gyrus. J Comp Neurol 1986; 246:435-458.
25. Conner JM, Lauterborn JC, Yan Q et al. Distribution of brain-derived neurotrophic factor (BDNF) protein and mRNA in the normal adult rat CNS: evidence for anterograde axonal transport. J Neurosci 1997; 17:2295-2313.
26. Covolan L, Ribeiro LTC, Longo BM. Cell damage and neurogenesis in the dentate granule cell layer of adult rats after pilocarpine-or kainate-induced status epilepticus. Hippocampus 2000; 10:169-180.
27. D'Ambrosio R, Maris DO, Grady MS et al. Impaired K(+) homeostasis and altered electrophysiological properties of post-traumatic hippocampal glia. J Neurosci 1999; 19:8152-8162.
28. Dardenne M. Zinc and immune function. Eur J Clin Nutr 2002; 56 Suppl 3:S20-23.
29. Derrick BE, York AD, Martinez Jr JL. Increased granule cell neurogenesis in the adult dentate gyrus following mossy fiber stimulation sufficient to induce long-term potentiation. Brain Res 2000; 857:300-307.
30. Dragunow M, Yamada N, Bilkey DK et al. Induction of immediate-early gene proteins in dentate granule cells and somatostatin interneurons after hippocampal seizures. Brain Res Mol Brain Res 1992; 13:119-126.

31. Ekdahl CT, Mohapel P, Elmer E et al. Caspase inhibitors increase short-term survival of progenitor-cell progeny in the adult rat dentate gyrus following status epilepticus. Eur J Neurosci 2001; 14:937-945.
32. Enderlin S, Norman AW, Celio MR. Ontogeny of the calcium binding protein calbindin D-28k in the rat nervous system. Anat Embryol (Berl) 1987; 177:15-28.
33. Eriksson PS, Perfilieva E, Bjork-Eriksson T et al. Neurogenesis in the adult human hippocampus. Nat Med 1998; 4:1313-1317.
34. Esclapez M, Houser CR. Up-regulation of GAD65 and GAD67 in remaining hippocampal GABA neurons in a model of temporal lobe epilepsy. J Comp Neurol 1999; 412:488-505.
35. Freund TF, Buzsaki G. Interneurons of the hippocampus. Hippocampus 1996; 6:347-470.
36. Gaarskjaer FB. Organization of the mossy fiber system of the rat studied in extended hippocampi. I. Terminal area related to number of granule and pyramidal cells. J Comp Neurol 1978; 178:49-72.
37. Gaarskjaer FB. The hippocampal mossy fiber system of the rat studied with retrograde tracing techniques. Correlation between topographic organization and neurogenetic gradients. J Comp Neurol 1981; 203:717-735.
38. Gaarskjaer FB, Laurberg S. Ectopic granule cells of hilus fasciae dentatae projecting to the ipsilateral regio inferior of the rat hippocampus. Brain Res 1983; 274:11-16.
39. Gall CM. Seizure-induced changes in neurotrophin expression: implications for epilepsy. Exp Neurol 1993; 124:150-166.
40. Gass P, Herdegen T, Bravo R et al. Spatiotemporal induction of immediate early genese in the rat brain after limbic seizures: effects of NMDA receptor antagonist MK-801. Eur J Neurosci 1993; 5:933-943.
41. Goodman JH, Wasterlain CG, Massarweh WF et al. Calbindin-D28k immunoreactivity and selective vulnerability to ischemia in the dentate gyrus of the developing rat. Brain Res 1993; 606:309-314.
42. Gould E, McEwen BS, Tanapat P et al. Neurogenesis in the dentate gyrus of the adult tree shrew is regulated by psychosocial stress and NMDA receptor activation. J Neurosci 1997; 17:2492-2498.
43. Gould E, Tanapat P. Stress and hippocampal neurogenesis. Biol Psychiatry 1999; 46:1472-1479.
44. Gould E, Tanapat P, McEwen BS et al. Proliferation of granule cell precursors in the dentate gyrus of adult monkeys is diminished by stress. Proc Natl Acad Sci USA 1998; 95:3168-3171.
45. Gray WP, Sundstrom LE. Kainic acid increases the proliferation of granule cell progenitors in the dentate gyrus of the adult rat. Brain Res 1998; 790:52-59.
46. Hansel DE, Eipper BA, Ronnett GV. Neuropeptide Y functions as a neuroproliferative factor. Nature 2001; 410:940-944.
47. Hess DC, Hill WD, Martin-Studdard A et al. Bone marrow as a source of endothelial cells and NeuN-expressing cells after stroke. Stroke 2002; 33:1362-1368.
48. Holm IE, Geneser FA, Zimmer J et al. Immunocytochemical demonstration of the calcium-binding proteins calbindin-D 28k and parvalbumin in the subiculum, hippocampus and dentate area of the domestic pig. Prog Brain Res 1990; 83:85-97.
49. Houser CR, Esclapez M. Localization of mRNAs encoding two forms of glutamic acid decarboxylase in the rat hippocampal formation. Hippocampus 1994; 4:530-545.
50. Houser CR, Esclapez M. Vulnerability and plasticity of the GABA system in the pilocarpine model of spontaneous recurrent seizures. Epilepsy Res 1996; 26:207-218.
51. Howell OW, Scharfman HE, Beck-Sickinger AG et al. Neuropeptide Y is neuroproliferative for hippocampal stem cells and neuroblasts. J Neurochem 2003; 86:646-659.
52. Hsu M, Buzsaki G. Vulnerability of mossy fiber targets in the rat hippocampus to forebrain ischemia. J Neurosci 1993; 13:3964-3979.
53. Kang J, Jiang L, Goldman SA et al. Astrocyte-mediated potentiation of inhibitory synaptic transmission. Nat Neurosci 1998; 1:683-692.
54. Keyser DO, Pellmar TC. Synaptic transmission in the hippocampus: critical role for glial cells. Glia 1994; 10:237-243.
55. Kotti T, Riekkinen PJ, Sr., Miettinen R. Characterization of target cells for aberrant mossy fiber collaterals in the dentate gyrus of epileptic rat. Exp Neurol 1997; 146:323-330.
56. Kuhn HG, Dickinson-Anson H, Gage FH. Neurogenesis in the dentate gyrus of the adult rat: age-related decrease of neuronal progenitor proliferation. J Neurosci 1996; 16:2027-2033.
57. Kunert-Radek J, Stepien H, Radek A et al. Somatostatin suppression of meningioma cell proliferation in vitro. Acta Neurol Scand 1987; 75:434-436.
58. Lamas M, Gomez-Lira G, Gutierrez R. Vesicular GABA transporter mRNA expression in the dentate gyrus and in mossy fiber synaptosomes. Mol Brain Res 2001; 93:209-214.
59. LeGallaSalle G. Long-lasting and sequential increase of c-fos oncoprotein expression in kainic acid-induced status epilepticus. Neurosci Lett 1988; 88:127-130.

60. Liptakova S, Jacobi H, Sperber EF. Kindling increases dentate granule neurogenesis in immature rats. Epilepsia 1999; 40:13.
61. Liu J, Bernabeu R, Lu A et al. Neurogenesis and gliogenesis in the postischemic brain. The Neuroscientist 2000; 6:362-370.
62. Liu J, Solway K, Messing RO et al. Increased neurogenesis in the dentate gyrus after transient global ischemia in gerbils. J Neurosci 1998; 18:7768-7778.
63. MacDonald RL, Kapur J. Pharmacological properties of recombinant and hippocampal dentate granule cell GABAA receptors. Adv Neurol 1999; 79:979-990.
64. Madsen TM, Treschow A, Bengzon J et al. Increased neurogenesis in a model of electroconvulsive therapy. Biol Psychiatry 2000; 47:1043-1049.
65. Maric D, Liu QY, Maric I et al. GABA expression dominates neuronal lineage progression in the embryonic rat neocortex and facilitates neurite outgrowth via GABA$_A$ autoreceptor/Cl$^-$ channels. J Neurosci 2001; 21:2343-2360.
66. Marksteiner J, Ortler M, Bellmann R et al. Neuropeptide Y biosynthesis is markedly induced in mossy fibers during temporal lobe epilepsy of the rat. Neurosci Lett 1990; 112:143-148.
67. Marti-Subirana A, Soriano E, Garcia-Verdugo JM. Morphological aspects of the ectopic granule-like cellular populations in the albino rat hippocampal formation: a Golgi study. J Anat 1986; 144:31-47.
68. Mezey E, Chandross KJ, Harta G et al. Turning blood into brain: cells bearing neuronal antigens generated in vivo from bone marrow. Science 2000; 290:1779-1782.
69. Nakagawa E, Aimi Y, Yasuhara O et al. Enhancement of progenitor cell division in the dentate gyrus triggered by initial limbic seizures in rat models of epilepsy. Epilepsia 2000; 41:10-18.
70. Okazaki MM, Evenson DA, Nadler JV. Hippocampal mossy fiber sprouting and synapse formation after status epilepticus in rats: visualization after retrograde transport of biocytin. J Comp Neurol 1995; 352:515-534.
71. Parent JM, Janumpalli S, McNamara JO et al. Increased dentate granule cell neurogenesis following amygdala kindling in the adult rat. Neurosci Lett 1998; 247:9-12.
72. Parent JM, Tada E, Fike JR et al. Inhibition of dentate granule cell neurogenesis with brain irradiation does not prevent seizure-induced mossy fiber synaptic reorganization in the rat. J Neurosci 1999; 19:4508-4519.
73. Parent JM, Yu TW, Leibowitz RT et al. Dentate granule cell neurogenesis is increased by seizures and contributes to aberrant network reorganization in the adult rat hippocampus. J Neurosci 1997; 17:3727-3738.
74. Patrylo PR, Spencer DD, Williamson A. GABA uptake and heterotransport are impaired in the dentate gyrus of epileptic rats and humans with temporal lobe sclerosis. J Neurophysiol 2001; 85:1533-1542.
75. Prasad AS, Beck FW, Endre L et al. Zinc deficiency affects cell cycle and deoxythymidine kinase gene expression in HUT-78 cells. J Lab Clin Med 1996; 128:51-60.
76. Rakic P, Nowakowski RS. The time of origin of neurons in the hippocampal region of the rhesus monkey. J Comp Neurol 1981; 196:99-128.
77. Rami A, Brehier A, Thomasset M et al. The comparative immunocytochemical distribution of 28 kDa cholecalcin (CaBP) in the hippocampus of rat, guinea pig and hedgehog. Brain Res 1987; 422:149-153.
78. Reisine T. Somatostatin. Cell Mol Neurobiol 1995; 15:597-614.
79. Ribak CE, Tran PH, Spigelman I et al. Status epilepticus-induced hilar basal dendrites on rodent granule cells contribute to recurrent excitatory circuitry. J Comp Neurol 2000; 428:240-253.
80. Riekkinen PJ, Pitkanen A. Somatostatin and epilepsy. Metabolism 1990; 39:112-115.
81. Sandler R, Smith AD. Coexistence of GABA and glutamate in mossy fiber terminals of the primate hippocampus: an ultrastructural study. J Comp Neurol 1991; 303:177-192.
82. Sankar R, Shin D, Liu H et al. Granule cell neurogenesis after status epilepticus in the immature rat brain. Epilepsia 2000; 40 (S6):134-143.
83. Sara VR, Rutherford R, Smythe GA. The influence of maternal somatostatin administration on fetal brain cell proliferation and its relationship to serum growth hormone and brain trophin activity. Horm Metab Res 1979; 11:147-149.
84. Scharfman HE. Differentiation of rat dentate neurons by morphology and electrophysiology in hippocampal slices: granule cells, spiny hilar cells and aspiny 'fast-spiking' cells. Epilepsy Res Suppl 1992; 7:93-109.
85. Scharfman HE. Electrophysiological diversity of pyramidal-shaped neurons at the granule cell layer/hilus border of the rat dentate gyrus recorded in vitro. Hippocampus 1995; 5:287-305.
86. Scharfman HE. The role of nonprincipal cells in dentate gyrus excitability and its relevance to animal models of epilepsy and temporal lobe epilepsy. Adv Neurol 1999; 79:805-820.

87. Scharfman HE, Goodman JH, Sollas AL. Granule-like neurons at the hilar/CA3 border after status epilepticus and their synchrony with area CA3 pyramidal cells: functional implications of seizure-induced neurogenesis. J Neurosci 2000; 20:6144-6158.
88. Scharfman HE, Smith KL, Goodman JH et al. Survival of dentate hilar mossy cells after pilocarpine-induced seizures and their synchronized burst discharges with area CA3 pyramidal cells. Neuroscience 2001; 104:741-759.
89. Scharfman HE, Sollas AL, Goodman JH. Spontaneous recurrent seizures after pilocarpine-induced status epilepticus activate calbindin-immunoreactive hilar cells of the rat dentate gyrus. Neuroscience 2002; 111:71-81.
90. Schlessinger AR, Cowan WM, Gottlieb DI. An autoradiographic study of the time of origin and the pattern of granule cell migration in the dentate gyrus of the rat. J Comp Neurol 1975; 159:149-176.
91. Schwarzer C, Sperk G, Samanin R et al. Neuropeptides-immunoreactivity and their mRNA expression in kindling: functional implications for limbic epileptogenesis. Brain Res Brain Res Rev 1996; 22:27-50.
92. Scott BW, Wang S, Burnham WM et al. Kindling-induced neurogenesis in the dentate gyrus of the rat. Neurosci Lett 1998; 248:73-76.
93. Scott BW, Wojtowicz JM, McIntyre-Burnham W. Neurogenesis in the dentate gyrus of the rat following electroconvulsive shock seizures. Exp Neurol 2000; 165:231-236.
94. Sharp FR, Liu J, Bernabeu R. Neurogenesis following brain ischemia. Dev Brain Res 2002; 134:23-30.
95. Shetty AK, Turner DA. Hippocampal interneurons expressing glutamic acid decarboxylase and calcium-binding proteins decrease with aging in Fischer 344 rats. J Comp Neurol 1998; 394:252-269.
96. Shumate MD, Lin DD, Gibbs JW, 3rd et al. GABA$_A$ receptor function in epileptic human dentate granule cells: comparison to epileptic and control rat. Epilepsy Res 1998; 32:114-128.
97. Sloviter RS. Calcium-binding protein (calbindin-D28k) and parvalbumin immunocytochemistry: localization in the rat hippocampus with specific reference to the selective vulnerability of hippocampal neurons to seizure activity. J Comp Neurol 1989; 280:183-196.
98. Sloviter RS, Sollas AL, Barbaro NM et al. Calcium-binding protein (calbindin-D28K) and parvalbumin immunocytochemistry in the normal and epileptic human hippocampus. J Comp Neurol 1991; 308:381-396.
99. Song H, Stevens CF, Gage FH. Astroglia induce neurogenesis from adult neural stem cells. Nature 2002; 417:39-44.
100. Sperk G, Marksteiner J, Gruber B et al. Functional changes in neuropeptide Y-and somatostatin-containing neurons induced by limbic seizures in the rat. Neuroscience 1992; 50:831-846.
101. Sperk G, Schwarzer C, Hellman J. Membrane and vesicular GABA transporters after kainate-induced seizures. Soc Neurosci Abs 2001; 551.553.
102. Spigelman I, Yan XX, Obenaus A et al. Dentate granule cells form novel basal dendrites in a rat model of temporal lobe epilepsy. Neuroscience 1998; 86:109-120.
103. St John JL, Rosene DL, Luebke JI. Morphology and electrophysiology of dentate granule cells in the rhesus monkey: comparison with the rat. J Comp Neurol 1997; 387:136-147.
104. Staley KJ, Otis TS, Mody I. Membrane properties of dentate gyrus granule cells: comparison of sharp microelectrode and whole-cell recordings. J Neurophysiol 1992; 67:1346-1358.
105. Terada N, Hamazaki T, Oka M et al. Bone marrow cells adopt the phenotype of other cells by spontaneous cell fusion. Nature 2002; 416:542-545.
106. Tonder N, Kragh J, Bolwig T et al. Transient decrease in calbindin immunoreactivity of the rat fascia dentata granule cells after repeated electroconvulsive shocks. Hippocampus 1994; 4:79-83.
107. van Praag H, Schinder AF, Christie BR et al. Functional neurogenesis in the adult hippocampus. Nature 2002; 415:1030-1034.
108. Vezzani A, Schwarzer C, Lothman EW et al. Functional changes in somatostatin and neuropeptide Y containing neurons in the rat hippocampus in chronic models of limbic seizures. Epilepsy Res 1996; 26:267-279.
109. Vezzani A, Sperk G, Colmers WF. Neuropeptide Y: emerging evidence for a functional role in seizure modulation. Trends Neurosci 1999; 22:25-30.
110. Wenzel HJ, Robbins CA, Tsai LH et al. Abnormal morphological and functional organization of the hippocampus in a p35 mutant model of cortical dysplasia associated with spontaneous seizures. J Neurosci 2001; 21:983-998.
111. Wenzel HJ, Woolley CS, Robbins CA et al. Kainic acid-induced mossy fiber sprouting and synapse formation in the dentate gyrus of rats. Hippocampus 2000; 10:244-260.
112. Williamson A, Spencer D. Zinc reduces dentate granule cell hyperexcitability in epileptic humans. Neuroreport 1995; 6:1562-1564.

113. Williamson A, Spencer DD, Shepherd GM. Comparison between the membrane and synaptic properties of human and rodent dentate granule cells. Brain Res 1993; 622:194-202.
114. Williamson A, Telfeian AE, Spencer DD. Prolonged GABA responses in dentate granule cells in slices isolated from patients with temporal lobe sclerosis. J Neurophysiol 1995; 74:378-387.
115. Woldbye DPD, Greisen MH, Bolwig TG et al. Prolonged induction of c-fos in neuropeptide Y and somatstatin-immunoreactive neurons of the rat dentate gyrus after electroconvulsive stimulation. Brain Res 1996; 720:111-119.
116. Yackel JW, Puri PS. Neurons in the rat dentate gyrus granular layer substantially increase during juvenile and adult life. Science 1982; 216:890-892.
117. Yan Q, Rosenfeld RD, Matheson CR et al. Expression of brain-derived neurotrophic factor protein in the adult rat central nervous system. Neuroscience 1997; 78:431-448.

CHAPTER 15

Febrile Seizures and Mechanisms of Epileptogenesis:
Insights from an Animal Model

Roland A. Bender, Celine Dubé and Tallie Z. Baram

Abstract

Temporal lobe epilepsy (TLE) is the most prevalent type of human epilepsy, yet the causes for its development, and the processes involved, are not known. Most individuals with TLE do not have a family history, suggesting that this limbic epilepsy is a consequence of acquired rather than genetic causes. Among suspected etiologies, febrile seizures have frequently been cited. This is due to the fact that retrospective analyses of adults with TLE have demonstrated a high prevalence (20->60%) of a history of prolonged febrile seizures during early childhood, suggesting an etiological role for these seizures in the development of TLE. Specifically, neuronal damage induced by febrile seizures has been suggested as a mechanism for the development of mesial temporal sclerosis, the pathological hallmark of TLE. However, the statistical correlation between febrile seizures and TLE does not necessarily indicate a causal relationship. For example, preexisting (genetic or acquired) 'causes' that result independently in febrile seizures and in TLE would also result in tight statistical correlation. For obvious reasons, complex febrile seizures cannot be induced in the human, and studies of their mechanisms and of their consequences on brain molecules and circuits are severely limited. Therefore, an animal model was designed to study these seizures. The model reproduces the fundamental key elements of the human condition: the age specificity, the physiological temperatures seen in fevers of children, the length of the seizures and their lack of immediate morbidity. Neuroanatomical, molecular and functional methods have been used in this model to determine the consequences of prolonged febrile seizures on the survival and integrity of neurons, and on hyperexcitability in the hippocampal-limbic network. Experimental prolonged febrile seizures did not lead to death of any of the seizure-vulnerable populations in hippocampus, and the rate of neurogenesis was also unchanged. Neuronal function was altered sufficiently to promote synaptic reorganization of granule cells, and transient and long-term alterations in the expression of specific genes were observed. The contribution of these consequences of febrile seizures to the epileptogenic process is discussed.

Introduction: The Human Problem

Among the epilepsies, temporal lobe epilepsy is often intractable and is associated with significant morbidity in terms of cognitive and psychosocial dysfunction. The most common pathology identified in resected temporal lobe tissue from patients with intractable TLE is the constellation of mesial temporal lobe sclerosis.[6] This entity is characterized by selective neuronal loss, gliosis and synaptic reorganization in discrete regions of the hippocampal formation and related structures.[6,16,30,56] The neuroanatomical alterations of the mesial temporal lobe,

Recent Advances in Epilepsy Research, edited by Devin K. Binder and Helen E. Scharfman. ©2004 Eurekah.com and Kluwer Academic / Plenum Publishers.

and particularly the hippocampus, can be observed using sophisticated neuroimaging studies, including magnetic resonance imaging[15,17,21,33] permitting their recognition in vivo, in individual patients. The convergence of temporal lobe seizures, MRI changes and the pathological findings of mesial temporal sclerosis have been increasingly recognized as a distinct entity, mesial temporal lobe epilepsy, which is quite likely the most common of all epileptic syndromes in humans.[25]

Whereas the neuropathological features of mesial TLE, i.e., mesial temporal sclerosis, have been defined and extensively studied for decades, the relationship of the anatomical abnormalities to the seizures has remained controversial.[27,52,55,57,81] A significant body of evidence, including the presence of early features of mesial temporal sclerosis in young children, suggests that in some patients mesial temporal sclerosis precedes the TLE, and is thus not a consequence of the seizures.[18,31,55] This has been interpreted to suggest that the hippocampal injury is also the cause of the TLE. In contrast, progression of the hippocampal lesion on magnetic resonance imaging in individuals with TLE who were imaged repetitively, and a correlation of the hippocampal atrophy with the number of partial and generalized seizures have also been reported.[13,49,61,78] These observations are in support of the notion of induction of mesial temporal sclerosis by the seizures themselves. These conflicting views illustrate that an understanding of the causal relationship of neuronal loss in the hippocampal formation to temporal lobe seizures remains incomplete, and is further hampered by the difficulty inherent in human studies, i.e., their correlational nature.

A second striking correlation found in patients with TLE is the frequent history of childhood febrile seizures, and particularly prolonged ones.[72] Thus, whereas the overall frequency of febrile seizures of the general population in western countries is 2-5%,[42,74] retrospective analyses of populations with intractable TLE indicate a frequency of a febrile seizure history of 20->60%.[1,17,35,40,65] This remarkable statistical relationship has raised the hypothesis that febrile seizures—particularly complex ones (i.e., focal, prolonged or repetitive)—may produce hippocampal injury that evolves into mesial temporal sclerosis.[41,48,76,79] However, the high concordance of childhood febrile seizures in patients with mesial TLE is also consistent with a functional or structural 'predisposing factor' that leads independently to both conditions. Put differently, a genetically determined malformation or molecular dysfunction, or an early 'acquired' lesion or insult (e.g., pre- or perinatal injury or infection) may predate and actually cause both the hippocampal injury/mesial temporal sclerosis, as well as the complex febrile seizures.[8,17,18,31,52,73]

Given the high frequency of both febrile seizures and TLE, understanding the true impact of the former, and specifically their causal relationship to TLE, is of enormous clinical significance. However, studying the key questions relating to the acute and chronic effects of febrile seizures on neuronal integrity and function cannot be achieved in the human, for obvious reasons: Ethical considerations prevent the induction of febrile seizures in humans, and naturally occurring febrile seizures are typically sudden and unexpected, and rarely occur in circumstances where electrographic monitoring is possible. Furthermore, the evolution of molecular and fine structural alterations cannot be studied in the live human with currently available technology. Thus, studying febrile seizures and their consequences on the immature brain requires controlled and reproducible experiments which can only be achieved in an appropriate animal model. Here we describe such an immature rat model for prolonged febrile seizures, and discuss the contribution of data obtained using this model to the understanding of the consequences of complex febrile seizures on the developing hippocampal circuit.

The 'Optimal' Animal Model

Animal models used for the study of human conditions are by their nature only an approximation of the 'real', actual disorder. Therefore, care should be taken to define the key characteristics which are essential for any meaningful modeling of the human condition. In addition, the features and nature of a given model dictate the scope of questions that can be addressed using it. For example, substitution of hyperthermia for fever (it should be noted that hyperthermia, typically drug-induced, is a not uncommon cause of seizures in children[22,47,54]) does not permit determination of the mechanisms involved in fever generation. Thus, it is conceivable that separate models might be needed to address different questions related to the same human problem. In addition to reproducing certain key elements of the human condition, it is advantageous for a model to be relatively simple and inexpensive, and much should be known about the relevant brain structures. These considerations have led to the choice of rodents over primates. In the following paragraphs we discuss several other characteristics of a meaningful animal model for prolonged febrile seizures.

Age Specificity

Febrile seizures are seen almost exclusively in infants and young children, specifically between 6 months and ~ 5 years of age, with peak incidence at ~18 months.[42] An appropriate rat model for febrile seizures should therefore employ rats which are in a developmental age equivalent to the seizure-sensitive period in humans. However, data comparing rodent and human brain development are rare. In addition, different brain regions develop at variable rates and chronological ages, in terms of neurogenesis, migration, connectivity and function, and these processes are not necessarily parallel in human and rat. These facts are important considerations in defining the 'appropriate age' of an animal model of febrile seizures.

Early, detailed studies correlated rat and human brain development based on neuronal birth dates, myelination and saltatory growth stages, and suggested that the 5-7 day old rat may be "equivalent" to the human full-term newborn.[24,39] More selective comparative neuroanatomical studies have focused on maturational milestones in discrete limbic regions, specifically the hippocampus and the prefrontal cortex.[7,43] Overall, these converging and complementary studies suggest that in the rat the first postnatal week may be comparable to the third trimester gestational period of the human fetus, and the second postnatal week to the first year of human life. For the hippocampal formation, specifically, we have recently summarized the information on comparative developmental milestones in human and rodent (see Table 1).[7] For example, comparison of the maturation of synaptic communication indicates that the maturational state of the hippocampus of a 8 day old rat[4,64] is roughly equivalent to the maturational state of the human hippocampus at 7 months of age (see refs. 7, 71 for analysis of hippocampal neurogenesis, connectivity and maturation of select synapses). Thus, based on anatomical data, rat hippocampal development during the second postnatal week seems to correspond best to the developmental stage at which human infants and young children are most susceptible to febrile seizures.

The ontogenetic profile of physiological responses of the developing rat brain to hyperthermia supports the end of the second postnatal week as the 'rat equivalent' of the human susceptibility period for febrile seizures: The threshold temperatures required to generate experimental febrile seizures are lowest during postnatal days 10-13, and rise rapidly thereafter[44] (Eghbal-Ahmadi & Baram, unpublished observations). Thus, because febrile seizures are a developmental phenomenon confined to infants and young children, and because limbic neuronal circuits (and particularly the hippocampal formation) are those suspected of involvement in and vulnerability to febrile seizures, models of febrile seizures should employ developing animals in which the stage of development of these limbic structures corresponds to the state of maturation of the human infant and young child, i.e., the middle of the second postnatal week in rat, and the end of the second postnatal week in mouse.

Temperature

Febrile seizures are defined as those occurring in children with fever (rectal temperature of at least 38.4°C) but without evidence of intracranial infection.[60] In addition, it has been suggested that in most normal children, threshold temperatures for febrile seizures exceed ~ 41°C.[50] Hence, animal models for febrile seizures should involve temperatures which are relevant to the human condition, and are observed in children with fever and should not rely on extreme temperatures (e.g., ref. 59). The ability to tightly regulate the temperature is an advantage of a

Table 1. Selected milestones in hippocampal development: Human and rat

Category	Event	Human	Rat
General	Maximal growth velocity	2-3 postnatal months	8-12 postnatal days
	Hippocampal volume approximates that of adult	10 postnatal months	11-30 postnatal days
	Hippocampal-dependent learning/ memory function	4-5 postnatal years	15-16 postnatal days
Neuronal Formation	Birth of pyramidal cells	1st half of gestation	2nd half of gestation
	Onset:	before 15th week	~15th day
	End:	24th week	~19th day
	Birth of dentate gyrus (DG) granule cells	~70% prenatal (majority by 34th week)	~85% postnatal (majority by 1st postnatal month)
	Onset:	13-14th weeks	~18th prenatal day
	End:	throughout life	throughout life
Differentiation, Synaptogenesis	'Thorny excrescences' on proximal CA3 pyramidal cell dendrites.	postnatal years	1st postnatal month
	Onset:	3-7 months	~9th day
	Maturation:	3-5 years	21st day
	Pedunculate spines: distal CA3 pyramidal cell dendrites	postnatal years	1st postnatal month
	Onset:	birth	7th day
	Maturation:	3-5 years	21st day
	Peak synapse overshoot (DG)	1-2 postnatal years	9-14 postnatal days
	Synapse density reaches adult levels: DG molecular layer	7-10 postnatal months	21st postnatal day
	Period of maximal mossy cell differentiation (DG)	7-30 postnatal months (adult-like by 5 years)	7-14 postnatal days (adult-like by 14th day)
Afferent input	Entorhinal cortex toDG	Prenatal: 19-20th weeks	17th prenatal day
	Supramammillary afferents reach juxtagranular and CA2 pyramidal cell layers	Prenatal: ~20th week	Presumed 1st postnatal week

Note: Gestation lasts 270-280 and 21 days in human and rat, respectively.
Modified from ref. 7, with permission.

hyperthermia-based model compared to fever-inducing agents. In addition, the majority of established pyrogens do not induce substantive fever in the immature rat[32,51] (Hatalski & Baram, unpublished observations).

The relationship of brain and core temperatures should be carefully considered. In animal models, core temperature is routinely monitored, rather than direct measurements of brain temperatures. It should be noted that the relationship between the two may not be consistent throughout the range of temperatures induced to provoke a hyperthermic seizure.[75] This may result in core temperatures that do not reflect actual brain temperatures. Therefore, calibration and standardization of parameters that influence the relationship of core and brain temperatures: the rate of heating, and volume, direction and diffusion of the heat, should carefully be standardized. Ideally, chronic measurement of brain temperature should be employed, but this is not feasible in the immature rat.

Ascertainment and Localization of the Seizures

The behaviors induced by hyperthermia may resemble those seen during seizures, but automatisms, stiffening and other motor phenomena may result from nonconvulsive discharges in brainstem, basal ganglia or other subcortical regions. Therefore, electrophysiological correlation of such observed behaviors should be obtained from behaving animals. In infant rats, hyperthermia induces stereotyped seizure behaviors[27,44] consisting of tonic body flexion accompanied by biting and chewing ('facial myoclonus'). These are typical for seizures of limbic origin.[9,46] Electrographic correlates should therefore be sought also in limbic regions, particularly in amygdala and hippocampus. In addition to pinpointing the likely source of the seizures, electrophysiological recording also from limbic rather than only from cortical regions is justified by the normal sequence of maturation in these regions: cortical maturation is incomplete during the second postnatal week in the rat, resulting typically in poorly organized and low-voltage cortical EEG activity.[9,69] It should be noted that because EEG correlations of genuine febrile seizures in humans are exceedingly rare,[59] they are not helpful in guiding placement of electrodes in experimental animals.

Absence of Immediate Morbidity and Mortality

Febrile seizures, whether single, short and nonfocal (simple) or recurrent, more prolonged or focal (complex) are typically not associated with immediate morbidity or mortality. Therefore, an appropriate animal model of febrile seizures should demonstrate similar low morbidity and mortality. In addition, a fundamental question related to hyperthermic seizures in the young human is whether they result in long-term effects, i.e., loss of hippocampal neurons and/or alteration of hippocampal circuitry leading to epilepsy. Therefore, an optimal model should be suitable for long-term survival without the confounding effects of major stress, burns, infection or general moribund states. An optimal model should use a defined, benign mechanism for increasing brain and core temperatures, which is suitable for repeated exposures. In summary, benign outcome not only reproduces the human situation, but permits meaningful prospective long-term studies of the long-term consequences of these seizures on epileptogenesis and neuronal function in general.

Hyperthermic Controls

The goals of setting up models of febrile seizures are to study the mechanisms or the outcomes of these seizures. However, by definition, each model involves subjecting the brain to hyperthermia, to simulate fever. Therefore, the effects of the hyperthermia per se must be distinguished from those of the associated seizures. The potential effects of hyperthermia may not be inconsequential. Hyperthermia induces a significant number of genes (e.g., heat shock proteins), and both deleterious and protective effects on neuronal function and integrity. For example, hyperthermia enhances seizure severity and the neuronal loss induced by kainic acid,[53] and when extreme, results in neuronal injury by itself.[38] Therefore, experiments using febrile seizure models should compare three sets of animals: normothermic controls, hyperthermic controls, in which hyperthermia was induced, but seizures were prevented[14,19,27,76] and the experimental group which has experienced both hyperthermia and seizures.

The Immature Rat Model

Based on the criteria outlined above, a rat model for prolonged febrile seizures was developed:[27,28,78] In this model, 10-11 day-old Sprague-Dawley rats are subjected to hyperthermia which raises body and brain temperatures gradually via a regulated stream of moderately heated air (using a hair-dryer, low-intermediate settings, air temperature ~43°C). The air stream is

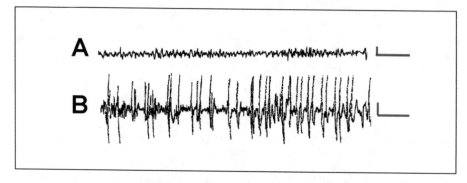

Figure 1. Electrophysiological characteristics of hyperthermia-induced seizures in immature rats. Records were performed via bipolar hippocampal electrodes in freely-moving 11-day-old rats. A) Baseline tracing of hippocampal activity during normothermia, showing a nonrhythmic pattern in the theta range. B) The hyperthermia procedure provoked hippocampal electrographic seizures, manifest as trains of spike-waves. Calibration: vertical, 50 mV; horizontal 1 sec.

directed ~ 30 cm above the rats, which are placed (1-2 at a time) on a towel in a 3 liter glass jar. Core temperatures are measured prior to initiating the hyperthermia, then every two minutes as well as at the onset of hyperthermia-induced seizures. These core temperatures have been extensively correlated with brain temperatures (Eghbal-Ahmadi and Baram, unpublished observations). In over 400 animals, we have found that raising core and brain temperatures to an average 40.88°C resulted in behavioral seizures in over 98%.

As mentioned above, the behavioral seizures are stereotyped, consisting of arrest of the heat-induced hyperkinesis, body flexion and biting of an extremity, occasionally followed by clonus. The epileptic nature of these seizures was confirmed by electrophysiological recording from the hippocampi of behaving pups, using bipolar electrodes. As shown in (Fig. 1), hyperthermia induces a change in hippocampal activity from a nonrhythmic pattern in the theta range (Fig. 1A) to the onset of rhythmic epileptiform spikes (Fig. 1B), which correlate with the onset of behavioral seizures. In hyperthermic controls, given the rapid- and short-acting barbiturate pentobarbital prior to the procedure, the behavioral as well as the electrophysiological seizures are blocked.

Animals are maintained hyperthermic (39-41.5°C) for 30 minutes, which is designed to generate seizures lasting about 20 minutes. This reproduces the human condition of prolonged, or complex febrile seizures (defined as longer than 15 minutes, and comprising only ~10% of all febrile seizures.[12] It is these longer seizures which have been statistically implicated in the development of TLE.[5,12,76] The 20-minute duration also avoids the onset of status epilepticus (defined as continuous seizures for 30 minutes), which may carry distinct implications for outcome (see refs. 2, 23). Following the hyperthermia, animals are moved to a cool surface to regain normal body and brain temperatures, then returned to their mothers for rehydration. It should be noted that weighing the animals before and after the procedure indicates little evidence of dehydration (< 3% change in body weight). Furthermore, animals regain normal activity rapidly after the procedure, and mortality has been <1%.

In summary, the immature rat model described above reproduces key features of prolonged febrile seizures, those that have been statistically correlated with the development of epilepsy and/or the presence of mesial temporal sclerosis. Using this model, we have started addressing the question of whether the relationship of these prolonged febrile seizures to neuronal loss and hyperexcitability is causal. In other words, the experiments described below query whether prolonged experimental febrile seizures cause epilepsy. Furthermore, if these seizures do induce hippocampal hyperexcitability, what are the underlying mechanisms?

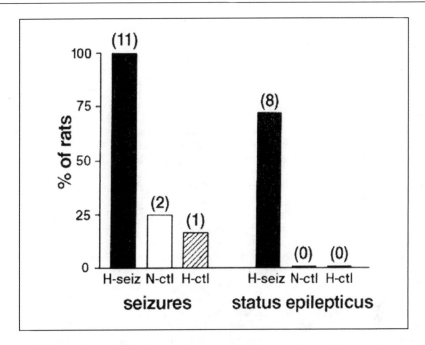

Figure 2. Differential induction of seizures and of status epilepticus in adult rats by low-dose kainic acid as a function of prolonged hyperthermic seizures early in life. Kainic acid led to seizures in all adult rats that had experienced prolonged hyperthermic seizures on postnatal days 10-11 (H-seiz; n=11). The majority of these (n=8) developed status epilepticus. In contrast, only 2 out of 8 normothermic (N-ctl) and 1 out of 6 hyperthermic control rats developed brief seizures, none of them status epilepticus (reproduced from ref. 27, with permission).

Do Prolonged Experimental Febrile Seizures Increase Seizure Susceptibility?

To determine whether the susceptibility to seizures is altered after prolonged febrile seizures, rats were allowed to mature (three months), and then underwent extensive hippocampal-electrophysiology and behavioral seizure monitoring.[27] Both the hippocampal tracings and the behavioral measures to date have failed to demonstrate the occurrence of spontaneous seizures. However, when challenged with a sub-convulsant dose of the AMPA/ kainate-type glutamate receptor agonist kainic acid, adult animals which had sustained developmental febrile seizures were far more sensitive than age-matched controls to the development of further seizures. In essence, a dose that failed to provoke seizures in normothermic and hyperthermic littermate controls led to severe seizures in all adult animals which had sustained prolonged experimental febrile seizures early in life (Fig. 2), demonstrating a ~four-fold increased sensitivity to kainic acid. This increased susceptibility to limbic convulsants was confirmed in vitro:[27] Spontaneous epileptiform discharges were not observed in hippocampal-entorhinal cortex slices derived from either control or experimental groups. However, Schaffer collateral stimulation induced prolonged, self-sustaining, status-epilepticus-like discharges exclusively in slices from experimental rats. These data indicate that experimental prolonged febrile seizures do not cause spontaneous limbic seizures during adulthood. However, they induce persistent enhancement of hippocampal excitability that may facilitate the emergence of subsequent seizures in response to even a mild (and perhaps not clearly demonstrable in the human situation) trigger later in life.

Do Prolonged Experimental Febrile Seizures Cause Neuronal Death and/or Synaptic Reorganization?

Neuronal loss and resulting changes in hippocampal circuitry (e.g., mossy fiber sprouting) in specific hippocampal subfields are characteristic of mesial temporal sclerosis in patients with TLE (reviewed in refs. 6, 45). The loss of seizure-sensitive neuronal populations can critically alter the balance of excitation and inhibition in the hippocampus, which may lead to long-term hyperexcitability and a reduced seizure threshold later in life, as indeed found in this model of prolonged febrile seizures. Therefore, we studied the short- and long-term effects of experimental febrile seizures on neuronal survival and synaptic connectivity.

Acute neuronal death was studied using the in situ end labeling (ISEL) technique for visualizing apoptotic cell death, as well as using the Gallyas silver stain method ("dark" neuron[36]) for visualizing neuronal injury. ISEL demonstrated no evidence for acute neuronal death in the hippocampus when studied 1, 4, 8.5, 24 or 48 hours after the seizures. However, the seizures did impact neuronal structure: the Gallyas method demonstrated dark, argyrophilic neurons starting within 24 hours and lasting as long as two weeks after the seizures. Whereas the precise mechanisms which render neurons argyrophilic are not known, selective uptake of the silver stain is considered to arise from alterations in proteins constituting the cytoskeleton. These changes have often been suggested to signify cell death. However, such 'dark' neurons can also be generated by subjecting the brain to postmortem trauma, indicating that this process is independent from the process of cell death.[37] Indeed, neuronal counts carried out in the central nucleus of the amygdala, where ~30% of neurons became silver-stained after experimental febrile seizures, demonstrated no loss of cells.[78] These findings suggest that the onset of the avidity to silver may not necessarily mean neuronal death: the changes or injury which render a cell argyrophilic may be reversible and not lead to cell loss.

That experimental prolonged febrile seizures do not cause neuronal cell death was further confirmed in a long-term study,[11] in which neuronal densities in the hippocampal formations of seizure-experiencing animals and age-matched controls were analyzed three months after the seizures. No difference was evident in the neuronal numbers of specific, seizure-sensitive hippocampal cell populations of these experimental groups. However, the density of the mossy fibers, the axons of granule cells, in granule cell and molecular layers was significantly increased in seizure-experiencing compared to control rats 3 months after the seizures. These findings indicate that despite the absence of seizure-induced neuronal loss, reorganization of the hippocampal circuit, evident by mossy fiber sprouting, did occur.

Do Prolonged Experimental Febrile Seizures Alter the Rate of Granule Cell Neurogenesis?

Altered neurogenesis of dentate gyrus granule cells, promoting aberrant, excitatory connectivity in the hippocampus, has recently been proposed as an additional mechanism by which seizures can modulate the hippocampal network.[63] Seizure-induced neurogenesis may be particularly disruptive during hippocampal development, since neurogenesis in the dentate gyrus peaks during the first and second postnatal weeks.[3,70] Therefore, we examined the influence of experimental prolonged febrile seizures on granule cell proliferation: Rats experiencing experimental febrile seizures and age-matched controls were injected with BrdU 3, 7 or 28 days after the seizures, and numbers of BrdU-labeled cells were determined 48 hrs later. No differences were found between seizure-experiencing and control animals at any of the time-points studied.[11] Thus, although granule cell neurogenesis in the immature hippocampus may be influenced by seizures during development,[11,58,66] prolonged febrile seizures had no significant effect on this process. This might be due to the relatively short duration of these seizures, or to other, as yet unresolved model-specific factors.

Molecular Plasticity after Experimental Prolonged Febrile Seizures

Electrophysiological analyses in acute hippocampal slices from seizure-experiencing rats revealed a surprising result: Despite the increased network hyperexcitability in the hippocampus, the inhibitory perisomatic drive onto CA1 pyramidal cells was increased, rather than decreased.[19] Conversion of enhanced, GABA-mediated hyperpolarization into neuronal depolarization may be mediated by activation of the hyperpolarization-activated, cyclic nucleotide-gated (I_h) current.[26,62] This led to the hypothesis that the I_h-current was altered after prolonged febrile seizures.[20,77] Indeed, whole-cell patch clamp recordings from CA1 pyramidal cells demonstrated that the biophysical properties of the I_h-current were altered by the experimental febrile seizures.[20] In slices from seizure-experiencing rats, the I_h-current was activated and deactivated much more slowly, and its half-maximal activation (V_{50}) was shifted towards a more depolarized membrane potential. Both changes opposed the increased presynaptic hyperpolarizing input, and could convert it to a depolarizing overshoot and action potential burst firing.[20,77] These changes persisted for at least 3 months.

In teasing out the mechanisms which might mediate these changes of the I_h-current, it was found that, unlike typical short-term modulation of the properties of the current, they did not depend on alteration of cellular cyclic nucleotides. This led to the notion that this long-lasting alteration of the I_h-current might derive from transcriptional regulation of the molecules which constitute the h-channels: The I_h-current is generated by a specific type of channels, the hyperpolarization-activated cyclic nucleotide-gated cation channels (HCNs). Recently, four different genes encoding HCN isoforms (HCN1-4) have been discovered, each isoform forming channels with significantly differing physiological properties (reviewed in ref. 67). Three of these isoforms (HCN1, HCN2, HCN4) are expressed in CA1 pyramidal cells of the immature rat during the age when febrile seizures can be provoked.[10] The relative abundance of each of these isoforms in a given cell has been shown to be critical for the physiological properties of the channels, which, in turn, govern the overall HCN properties of the cell.[34,68] In CA1 pyramidal cells, the HCN1 isoform, forming fast activating and deactivating channels with limited conductance, seems to be dominant under normal conditions.[34,68] However, recent results indicate that prolonged experimental febrile seizures—occurring during a period of rapid evolution of the HCN isoform expression pattern[10]—influence the mRNA and protein expression of these channel molecules. The expression of HCN1 mRNA was significantly decreased and the expression of HCN2 mRNA significantly increased in seizure-sustaining animals by one week later.[14] Both of these changes increase the relative abundance of HCN2 compared to HCN1 in a given neuron, favoring the formation of slower kinetics (and potentially larger-conductance) HCN2-homomeric channels, with altered biophysical properties. This alteration in the molecular make-up of the HCNs would promote neuronal activity-dependent depolarization and enhance the excitability of the hippocampal circuit.

Summary

What Has the Immature Rat Model of Prolonged Febrile Seizures Taught Us so Far?

The original working hypotheses driving these studies of experimental febrile seizures suggested that these seizures would either kill vulnerable neurons, or lead only to transient injury, without long-term effects on the hippocampal circuit. As evident from the data above, both hypotheses were refuted. The scenario emerging from the experimental data indicates that the process of epileptogenesis in the immature rat—the transformation of a 'normal' limbic network to a pro-epileptic one—is far more intricate and subtle than a simple composite of direct or compensatory changes in response to cell death.

Indeed, the data clearly indicate that experimental prolonged febrile seizures do not result in death of neurons in amygdala and hippocampus. Vulnerable populations, such as the mossy cells or specific interneuronal subtypes, were specifically labeled and counted, and no loss or reduction in their numbers were found.[11] The preservation of neuronal numbers was not due to the birth of new neurons, since BrdU analyses demonstrated that the rate of neurogenesis was not altered.

However, prolonged experimental febrile seizures were not "benign". In the aftermath of the seizures, the hippocampus was far more susceptible to minor excitatory input (electrical current in the slice, kainic acid in vivo) compared with a hippocampus not previously involved in febrile seizures. These striking and long-lasting changes rendered the animal more likely to generate seizures. Thus, pro-epileptogenic changes may occur without the requirement for neuronal death.

Insight into the mechanisms contributing to the hyperexcitability resulting from prolonged experimental febrile seizures was derived from electrophysiological and molecular analyses, which, to date, have provided specific clues. Early (within hours of the seizures) regulation of calcium entry is modified, due to transient reduction of GluR2 expression and creation of calcium-permeable AMPA channels.[29] By several days after the seizures, striking changes in ion channels, specifically in the molecular make-up—and hence the kinetics and voltage-dependence—of the HCNs, emerge and persist long-term. These contribute significantly to conversion of augmented GABA-induced hyperpolarization to activity-dependent hyperexcitation. Many questions remain: How do the seizures lead to down-regulation of GluR2? What other critical alterations result from altered calcium entry? What are the mechanisms governing HCN expression in a spatially and temporally constrained pattern?

These and related questions are the focus of ongoing studies. These studies are carried out in the hope that they will lead to further clues and to the discovery of the specific molecular targets, the key determinants, of this seizure-induced long-term hyperexcitability. It is the discovery of such specific molecular targets which could lead to the design of compounds which will specifically prevent the consequences of the seizures, and thus perhaps prevent these pro-epileptogenic effects of prolonged febrile seizures.

Acknowledgment

Authors work supported by NIH NINDS 35439 (TZB) and Epilepsy Foundation and Milken family awards (RAB, CD).

References

1. Abou-Khalil B, Andermann E, Andermann F et al. Temporal lobe epilepsy after prolonged febrile convulsions: excellent outcome after surgical treatment. Epilepsia 1993; 34:878-883.
2. Alldredge BK, Lowenstein DL. Status epilepticus: new concepts. Curr Opin Neurol 1999; 12:183-190.
3. Altman J, Bayer SA. Migration and distribution of two populations of hippocampal granule cell precursors during the perinatal and postnatal periods. J Comp Neurol 1990; 301:365-381.
4. Amaral DG, Dent JA. Development of the mossy fibers of the dentate gyrus: I. A light and electron microscopic study of the mossy fibers and their expansions. J Comp Neurol 1981; 195:51-86.
5. Annegers JF, Hauser WA, Shirts SB et al. Factors prognostic of unprovoked seizures after febrile convulsions. N Engl J Med 1987; 316:493-498.
6. Armstrong DD. The neuropathology of temporal lobe epilepsy. J Neuropath Exp Neurol 1993; 52:433-443.
7. Avishai-Eliner S, Brunson KL, Sandman CA et al. Stressed out or in (utero)? Trends Neurosci 2002; 25:518-524.
8. Baram TZ. Mechanisms and outcome of febrile seizures: What have we learned from basic science approaches, and what needs studying? In: Baram TZ, Shinnar S, eds. Febrile seizures. San Diego, CA: Academic Press, 2002:325-328.
9. Baram TZ, Hirsch E, Snead III OC et al. Corticotropin-releasing hormone-induced seizures in infant rats originate in the amygdala. Ann Neurol 1992; 31:488-494.

10. Bender RA, Brewster A, Santoro B et al. Differential and age-dependent expression of hyperpolarization-activated, cyclic nucleotide-gated cation channel isoforms 1-4 suggests evolving roles in the developing rat hippocampus. Neuroscience 2001; 106:689-698.
11. Bender RA, Dubé C, Gonzalez-Vega R et al. Mossy fiber plasticity and enhanced hippocampal excitability, without hippocampal cell loss or altered neurogenesis, in an animal model of prolonged febrile seizures. Hippocampus 2002; 13:399-412.
12. Berg AT, Shinnar S. Complex febrile seizures. Epilepsia 1996; 37:126-133.
13. Bower SP, Kilpatrick CJ, Vogrin SJ et al. Degree of hippocampal atrophy is not related to a history of febrile seizures in patients with proved hippocampal sclerosis. J Neurol Neurosurg Psychiatry 2000; 69:733-8.
14. Brewster A, Bender RA, Chen Y et al. Developmental febrile seizures modulate hippocampal gene expression of hyperpolarization-activated channels in an isoform and cell-specific manner. J Neurosci 2002; 22:4591-4599.
15. Briellmann RS, Kalnins RM, Berkovic SF et al. Hippocampal pathology in refractory temporal lobe epilepsy. T2-weighted signal change reflects dentate gliosis. Neurology 2002; 58:265-271.
16. Bruton CJ. The neuropathology of temporal lobe epilepsy (Maudsley Monographs, No 31). New York, NY: Oxford University Press, 1988.
17. Cendes F, Andermann F, Dubeau F et al. Early childhood prolonged febrile convulsions, atrophy and sclerosis of mesial structures and temporal lobe epilepsy. An MRI volumetric study. Neurology 1993; 43:1083-1087.
18. Cendes F, Cook MJ, Watson C et al. Frequency and characteristics of dual pathology in patients with lesional epilepsy. Neurology 1995; 45:2058-2064.
19. Chen K, Baram TZ, Soltesz I et al. Febrile seizures in the developing brain result in persistent modification of neuronal excitability in limbic circuits. Nat Med 1999; 5:888-894.
20. Chen K, Aradi I, Thon N et al. Persistently modified h-channels after complex febrile seizures convert the seizure-induced enhancement of inhibition to hyperexcitability. Nat Med 2001; 7:331-337.
21. Cook MJ, Fish DR, Shorvon SD et al. Hippocampal volumetric and morphometric studies in frontal and temporal lobe epilepsy. Brain 1992; 115:1001-1015.
22. Cooper AJ, Egleston C. Accidental ingestion of Ecstasy by a toddler: unusual cause for convulsion in a febrile child. J Accid Emerg Med 1997; 14:183-184.
23. Coulter DA, DeLorenzo RJ. Basic mechanisms of status epilepticus. Adv Neurol 1999; 79:725-733.
24. Dobbing J, Sands J. Quantitative growth and development of human brain. Arch Dis Child 1973; 48:757-767.
25. Engel Jr J, Williamson PD, Wieser HG. Mesial temporal lobe epilepsy. In: Engel Jr J, Pedley TA, eds. Epilepsy: A comprehensive textbook. Philadelphia, PA: Lippincott-Raven Publishers, 1997:2417-2426.
26. DiFrancesco D. Pacemaker mechanisms in cardiac tissue. Annu Rev Physiol 1993; 55:455-472.
27. Dubé C, Chen K, Eghbal-Ahmadi M et al. Prolonged febrile seizures in the immature rat model enhance hippocampal excitability long term. Ann Neurol 2000; 47:336-344.
28. Dubé C. Do prolonged febrile seizures in an immature rat model cause epilepsy? In: Baram TZ, Shinnar S, eds. Febrile seizures. San Diego, CA: Academic Press, 2002:215-229.
29. Eghbal-Ahmadi M, Yin H, Stafstrom CE et al. Altered expression of specific AMPA type glutamate receptor subunits after prolonged experimental febrile seizures in CA3 of immature rat hippocampus. Soc Neurosci Abstr 2001; 31:684.6.
30. Falconer MA, Serafetinides EA, Corsellis JAN. Etiology and pathogenesis of temporal lobe epilepsy. Arch Neurol 1964; 10:233-248.
31. Fernandez G, Effenberger O, Vinz B et al. Hippocampal malformation as a cause of familial febrile convulsions and subsequent hippocampal sclerosis. Neurology 1998; 50:909-917.
32. Fewell JE, Wong VH. Interleukin-1beta-induced fever does not alter the ability of 5- to 6-day-old rat pups to autoresuscitate from hypoxia-induced apnoea. Exp Physiol 2002; 87:17-24.
33. Fish DR, Spencer SS. Clinical correlations: MRI and EEG. Magn Reson Imaging 1995; 13:1113-1117.
34. Franz O, Liss B, Neu A et al. Single-cell mRNA expression of HCN1 correlates with a fast gating phenotype of hyperpolarization-activated cyclic nucleotide-gated ion channels (Ih) in central neurons. Eur J Neurosci 2000; 12:2685-2693.
35. French JA, Williamson PD, Thadani VM et al. Characteristics of medial temporal lobe epilepsy: I. Results of history and physical examination. Ann Neurol 1993; 34:774-780.
36. Gallyas F, Guldner FH, Zoltay G et al. Golgi-like demonstration of "dark" neurons with an argyrophil III method for experimental neuropathology. Acta Neuropathol (Berlin) 1990; 79:620-628.

37. Gallyas F, Zoltay G, Horvath Z et al. Light microscopic response of neuronal somata, dendrites, axons to postmortem concussive head injury. Acta Neuropathol (Berlin) 1992; 83:499-503.
38. Germano IM, Zhang YF, Sperber EF et al. Neuronal migration disorders increase seizure susceptibility to febrile seizures. Epilepsia 1996; 37:902-910.
39. Gottlieb A, Keydar I, Epstein HT. Rodent brain growth stages: an analytical review. Biol Neonate 1977; 32:166-176.
40. Hamati-Haddad A, Abou-Khalil B. Epilepsy: diagnosis and localization in patients with antecedent childhood febrile convulsions. Neurology 1998; 50:917-922.
41. Harvey AS, Grattan-Smith JD, Desmond PM et al. Febrile seizures and hippocampal sclerosis: frequent and related findings in intractable temporal lobe epilepsy of childhood. Pediatr Neurol 1995; 12:201-206.
42. Hauser WA. The prevalence and incidence of convulsive disorders in children. Epilepsia 1994; 35(Suppl 2) S1-S6.
43. Herschkowitz N, Kagan J, Zilles K. Neurobiological bases of behavioral development in the first year. Neuropediatrics 1997; 28:296-306.
44. Hjeresen DL, Diaz J. Ontogeny of susceptibility to experimental febrile seizures in rats. Dev Psychobiol 1988; 21:261-275.
45. Houser CR. Neuronal loss and synaptic reorganization in temporal lobe epilepsy. Adv Neurol 1999; 79:743-761.
46. Ikonomidou-Turski C, Cavalheiro EA, Turski WA et al. Convulsant action of morphine, [D-Ala^2, D-Leu^5] -enkephalin and naloxone in the rat amygdala: electroencephalographic, morphological and behavioral sequelae. Neuroscience 1987; 20:671-686.
47. Ioos C, Fohlen M, Villeneuve N et al. Hot water epilepsy: A benign and unrecognized form. J Child Neurol 2000; 15:125-128.
48. Jackson GD, McIntosh AM, Briellmann RS et al. Hippocampal sclerosis studied in identical twins. Neurology 1998; 51:78-84.
49. Kälviäinen R, Salmonperä T, Partanen K et al. Recurrent seizures may cause hippocampal damage in temporal lobe epilepsy. Neurology 1998; 50:1377-1382.
50. Knudsen FU. Febrile seizures—treatment and outcome. Brain Dev 1996; 18:438-449.
51. Lagerspetz KY, Vaatainen T. Bacterial endotoxin and infection cause behavioural hypothermia in infant mice. Comp Biochem Physiol A 1987; 88:519-521.
52. Lewis DV. Febrile convulsions and mesial temporal lobe sclerosis. Curr Opin Neurol 1999; 12:197-201.
53. Liu Z, Gatt A, Mikati M et al. Effect of temperature on kainic acid-induced seizures. Brain Res 1993; 631:51-58.
54. Mani KS, Mani AJ, Ramesh CK et al. Hot-water epilepsy—a peculiar type of reflex epilepsy: clinical and EEG features in 108 cases. Trans Am Neurol Assoc 1974; 99:224-226.
55. Mathern GW, Babb TL, Vickrey BG et al. The clinical-pathogenic mechanisms of hippocampal neuron loss and surgical outcomes in temporal lobe epilepsy. Brain 1995; 118:105-118.
56. Mathern GW, Babb TL, Armstrong DL. Hippocampal sclerosis. In: Engel Jr J, Pedley TA, eds. Epilepsy: A comprehensive textbook. Philadelphia, PA: Lippincott-Raven Publishers, 1997:133-155.
57. Mathern GW, Pretorius JK, Leite JP et al. Hippocampal neuropathology in children with severe epilepsy. In: Nehlig A, Motte J, Moshé SL, Plouin P, eds. Childhood epilepsies and brain development. London, England: John Libbey & Co., 1999:171-185.
58. McCabe BK, Silveira DC, Cilio MR et al. Reduced neurogenesis after neonatal seizures. J Neurosci 2001; 21:2094-2103.
59. Morimoto T, Nagao H, Sano N et al. Electroencephalographic study of rat hyperthermic seizures. Epilepsy 1991; 32:289-293.
60. National Insitutes of Health Febrile seizures: Consensus development conference summary. Bethesda, MD: National Institutes of Health,1980:3(2).
61. O'Brien TJ, So EL, Meyer FB et al. Progressive hippocampal atrophy in chronic intractable temporal lobe epilepsy. Ann Neurol 1999; 45:526-529.
62. Pape HC. Queer current and pacemaker: the hyperpolarization-activated cation current in neurons. Annu Rev Physiol 1996; 58:299-327.
63. Parent JM, Yu TW, Leibowitz RT et al. Dentate granule cell neurogenesis is increased by seizures and contributes to aberrant network reorganization in the adult rat hippocampus. J Neurosci 1997; 17:3727-3738.
64. Ribak CE, Seress L, Amaral DG. The development, ultrastructure and synaptic connections of the mossy cells of the dentate gyrus. J Neurocytol 1985; 14:835-857.

65. Rocca WA, Sharbrough FW, Hauser WA et al. Risk factors for complex partial seizures: a population-based case-control study. Ann Neurol 1987; 21:22-31.
66. Sankar R, Shin D, Liu H et al. Granule cell neurogenesis after status epilepticus in the immature rat brain. Epilepsia Suppl 2001; 7:53-56.
67. Santoro B, Tibbs GR. The HCN gene family: molecular basis of the hyperpolarization-activated pacemaker channels. Ann NY Acad Sci 1999; 868:741-764.
68. Santoro B, Chen S, Lüthi A et al. Molecular and functional heterogeneity of hyperpolarization-activated pacemaker channels in the mouse CNS. J Neurosci 2000; 20:5264-5275.
69. Schickerova R, Mares P, Trojan S. Correlation between electrocorticographic and motor phenomena induced by pentamethylenetetrazol during ontogenesis in rats. Exp Neurol 1984; 84:153-164.
70. Schlessinger AR, Cowan WM, Gottlieb ID. An autoradiographic study of the time of origin and the pattern of granule cell migration in the dentate gyrus of the rat. J Comp Neurol 1975; 159:149-176.
71. Seress L, Mrzljak L. Postnatal development of mossy cells in the human dentate gyrus: a light microscopic Golgi study. Hippocampus 1992; 2:127-142.
72. Shinnar S. Prolonged febrile seizures and mesial temporal lobe sclerosis. Ann Neurol 1998; 43:411-412.
73. Shinnar S. Human Data: What do we know about febrile seizures and what further information is needed? In: Baram TZ, Shinnar S, eds. Febrile seizures. San Diego, CA: Academic Press, 2002:317-324.
74. Stafstrom CE. The incidence and prevalence of febrile seizures. In: Baram TZ, Shinnar S, eds. Febrile seizures. San Diego, CA: Academic Press, 2002; 1-25.
75. Sundgren-Andersson AK, Ostlund P, Bartfai T. Simultaneous measurement of brain and core temperature in the rat during fever, hyperthermia, hypothermia and sleep. Neuroimmunomodulation 1998; 5:241-247.
76. Theodore WH, Bhatia S, Hatta J et al. Hippocampal atrophy, epilepsy duration and febrile seizures in patients with partial seizures. Neurology 1999; 52:132-136.
77. Thon N, Chen K, Aradi I et al. Physiology of limbic hyperexcitability after experimental complex febrile seizures: interactions of seizure-induced alterations at multiple levels of neuronal organization. In: Baram TZ, Shinnar S, eds. Febrile seizures. San Diego, CA: Academic Press, 2002:203-213.
78. Toth Z, Yan XX, Haftoglou S et al. Seizure-induced neuronal injury: vulnerability to febrile seizures in an immature rat model. J Neurosci 1998; 18:4285-4294.
79. VanLandingham KE, Heinz ER, Cavazos JE et al. Magnetic resonance imaging evidence of hippocampal injury after prolonged focal febrile convulsions. Ann Neurol 1998; 43:413-426.
80. Van Paesschen W, Duncan JS, Stevens JM et al. Longitudinal quantitative hippocampal magnetic resonance imaging study of adults with newly diagnosed partial seizures: one-year follow-up results. Epilepsia 1998; 39:633-639.
81. Zimmerman HM. The histopathology of convulsive disorders in children. J Pediatr 1940; 13:359-390.

CHAPTER 16

The Tetanus Toxin Model of Chronic Epilepsy

Timothy A. Benke and John Swann

Introduction

In experimental models of epilepsy, single and recurrent seizures are often used in an attempt to determine the effects of the seizures themselves on mammalian brain function. These models attempt to emulate as many features as possible of their human disease counterparts without many of the confounding factors such as underlying disease processes and medication effects. Numerous models have been used in the past to address different questions. Nevertheless, the basic questions are often the same:

1. Do seizures cause long-term damage?
2. Do seizures predispose to chronic epilepsy (epileptogenesis), that is long-term spontaneous repetitive seizures?
3. Are these results developmentally regulated?
4. Are the underlying mechanisms of epileptogenesis and brain damage related?

In pursuing these questions, the goal is to determine how seizures exert their effects and to minimize any side effects from the methods employed to induce the seizures themselves. This requires a detailed characterization of the methods used to induce seizures.

In this chapter, we will review the literature regarding the tetanus toxin model of chronic epilepsy with regard to its mechanisms of action, clinical comparisons, how it is experimentally implemented and the results obtained thus far. These results will be compared to other models of chronic epilepsy in order to make generalizations about the effects of repetitive seizures in adult and early life. At this time, it appears that repetitive seizures cause long-term changes in learning ability and may cause a predisposition to chronic seizures at all ages. In younger animals, both features of learning impairment and epilepsy are not typically associated with cell loss as they are in adult animals. At all ages, some form of synaptic reorganization has been demonstrated to occur.

Mechanisms of Tetanus Toxin

Clinical Syndrome

Tetanus is a neurological disease manifested by trismus (lockjaw) and severe generalized muscular spasms. The syndrome is caused by the neurotoxin produced by the spore-forming anaerobic bacterium *Clostridium tetani* in a contaminated wound. Clostridial spores are worldwide, and ubiquitous, found in soil, dust, street dirt and human and animal feces. Spores are resistant to heat and disinfection. The bacterium itself is simply a wound contaminant, causing neither tissue destruction nor an inflammatory response.

Symptoms, caused by an exotoxin, may begin locally in areas contiguous to a wound with local muscle spasms. The incubation period is from 2 days to months and frequently without an obvious wound. Cephalic tetanus involves cranial nerves in association with head and neck wounds. Onset is gradual, typically with dysphagia and trismus (50%), occurring over 1 to 7

Recent Advances in Epilepsy Research, edited by Devin K. Binder and Helen E. Scharfman.
©2004 Eurekah.com and Kluwer Academic / Plenum Publishers.

days, and can progress to severe, generalized, extensor muscle spasms (opisthotonus), which frequently are triggered and aggravated by any external stimulus. Opisthotonic spasms are often mistaken for seizures. Severe spasms persist for 1 week or more and subside in a period of weeks in those who survive (90%). Involvement of the autonomic nervous system results in cardiovascular instability, with labile hypertension and tachyarrhythmias. As little as 150 μg of tetanus toxin is lethal to humans. Neonatal tetanus, a common cause of neonatal mortality in developing countries, yet also known to occur in association with the popular practice of "home birth" in developed countries, arises from contamination of the umbilical stump. Mortality in neonates ranges from 30-80%. Approximately 10-30% of survivors of neonatal tetanus exhibit long-term sequelae such as mental retardation and spasticity, which is thought to be due to hypoxic-ischemic injury associated with severe spasms. However, the possibility of permanent damage to spinal cord anterior horn cells has not been ruled out.[1-3]

Structural Details and Molecular Mechanism of Action

Tetanus toxin shares many features with the other clostridial neurotoxins, however certain structural features confer unique mechanisms of action. Tetanus toxin (TT), molecular weight 150 kDa, is released following bacterial cell wall lysis. The inactive protein is activated by a specific protease resulting in a heavy chain (100 kDa) and light chain (50 kDa) that remain associated by non-covalent interactions and a disulfide bridge, whose integrity is essential for neurotoxicity. The heavy chain C-terminus appears to modulate binding to ganglioside-containing neuronal membranes and a reportedly specific, though as yet not fully characterized, TT receptor. Both low-affinity (nM) and high-affinity (sub-nM) binding sites have been identified. The light chain N-terminus is a zinc metalloprotease. These unique structural features account for the 5 steps thought to describe neurointoxication:

1. binding,
2. internalization,
3. transport,
4. membrane translocation and
5. proteolytic cleavage of substrate (Fig.1).[4]

Step (3) appears unique to TT compared to the other clostridial neuronal toxins and likely accounts not only for its unique clinical syndrome but also for potential novel features associated with its use as a model for experimental epilepsy.

Initial neuronal-specific binding (Step 1) is thought to be mediated by the TT receptor located at axonal endings that may terminate in a wound, for instance. Such axonal endings could be motor or sensory axons. Following subsequent rapid internalization into an endocytic vesicle (Step 2), TT is vesicularly transported within the axon to the soma. In the case of anterior horn cell neurons, it is then transported to dendrites where, it is thought, it is then released in an exocytic fashion and subsequently taken up again endocytically by the terminals of spinal inhibitory interneurons. Following acidification of the internal milieu of the vesicle, the light chain translocates (Step 4) across the vesicle membrane into the neuronal cytosol. There the metalloprotease specifically cleaves (Step 5) a site

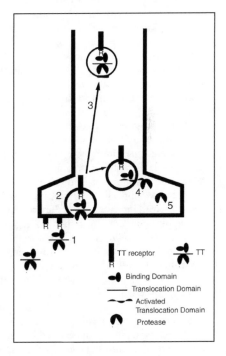

Figure 1. Sequence of toxin intoxication. Adapted from reference 4.

on synaptobrevin,[5-7] a vesicle protein whose function is required for vesicle fusion with membranes. Synaptobrevin is associated with neurotransmitter-containing synaptic vesicles.[8] Hence, in the case of glycinergic and GABAergic interneurons, by inhibiting the release of glycine or GABA onto anterior horn cell motor neurons, TT exerts its main clinical effects of a spastic paralysis. Since this is the main clinical effect of TT, one could assume that its effects on inhibitory transmission are primary and specific.

These processes have been confirmed experimentally. Experimental application of TT has been shown to prevent the release of GABA in spinal cord cultures,[9] hippocampal slices[10,11] and cortical membrane preparations.[12] Injection of TT into spinal cord and cerebellum blocks synaptic inhibition.[13,14] Application of TT to spinal cord cultures results in epileptiform discharges associated with a presynaptic blockade of inhibitory synaptic transmission while no effect is found on post-synaptic responsiveness to GABA or glycine.[9]

However, both clinical and experimental evidence suggests that the toxin's effects are not solely due to its interference with inhibitory synaptic transmission. Clinically, autonomic dysfunction suggests that following uptake by sensory neurons, TT disrupts cholinergic neurotransmission. Experimentally, TT prevents not only release of GABA[11] but also glutamate, aspartate and met-enkephalin release at similar concentrations.[12] EEG activity is acutely suppressed following intrahippocampal injection of TT,[15] suggesting an effect on excitatory transmission. In hippocampal cultures, TT preferentially binds to presynaptic glutamatergic terminals where it is taken up by synaptic vesicle endocytosis. Once in vesicles there, the characterized sequence of toxin action follows, resulting in cytosolic cleavage of synaptobrevin.[16]

The actions of TT may not be entirely due to a presynaptic blockade of vesicular fusion. Since synaptobrevin is suggestively located on vesicles associated with insertion of membrane receptors,[17,18] TT could also prevent their insertion in postsynaptic membranes. Application of TT to *Torpedo* organ results in not only a reduction in miniature endplate potential frequency, but also miniature endplate potential size[7]. Indeed, intracellular application of TT can prevent post-synaptic insertion of both GABA[19] and glutamate[17,20] receptors in hippocampal neurons. However, responses to exogenous GABA are not altered in hippocampus following intrahippocampal TT injection, though these studies investigated only somatic GABA receptors on pyramidal neurons.[21] Thus, the ultimate effects of TT depend on transport within neurons and their associated neuronal networks as well as the half-life of the toxin.

Trans-synaptic transfer appears limited to "single-jumps". That is, following initial uptake and transport, TT moves across a single synapse to then exert its effect. This appears to be the case for horseradish peroxidase-conjugated heavy chain TT fragment when injected intramuscularly. In this case, it was found to be present in motor neurons and to eventually accumulate following trans-synaptic transfer in inhibitory terminals but did not continue beyond from there.[22] That is, TT did not then move to inhibitory somata and beyond. Similar results were found at cholinergic terminals in spinal preparations.[23] In both of these studies, some TT was observed extrasynaptically and found to persist for several days at the sites of synaptic transfer, leaving the possibility for further intoxication of these and other local neurons as possible. Following injection of TT into the CA3 region of hippocampus, transport would be expected to at least encompass the projection of axon terminals and dendrites present in the CA3 region (Fig. 2). This would include mossy fiber terminals from dentate granule neurons, association fiber terminals from CA3 pyramidal neurons and commissural axons from the contralateral CA3. Dentate granule neurons follow a laminar projection, while CA3 associational and Schaffer collateral axons to CA1, project diffusely to both dorsal and ventral hippocampus (ipsilateral).

Theoretically, following transport to CA3 cell bodies, TT could also be further transported via Schaffer commissural axons projecting to the contralateral CA1 and CA3 associational regions.[24] Following intrahippocampal injection of the conjugated heavy chain TT fragment, transport was noted into the contralateral hippocampal CA3 region but not in either ventral hippocampus or in cortex.[25] However, it is not clear if the presence of the light chain may facilitate multi-synaptic transfer. This could take place at any infected terminal. Indeed, upon

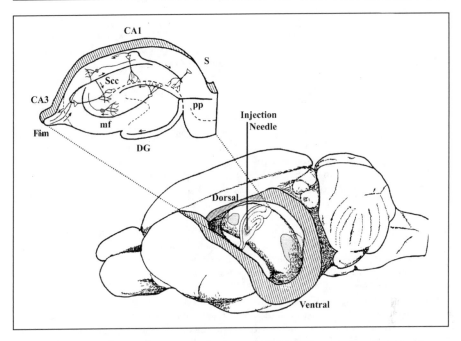

Figure 2. Spread of tetanus toxin. Following initial injection of TT into dorsal hippocampus (double cross hatch), likely pathways of transport (cross hatch) are shown. Inset shows pathways in an idealized transverse hippocampal slice. Pyramidal neurons are denoted by triangles, granule neurons by open circles, and GABAergic interneurons by filled circles. DG=dentate gyrus, S=subiculum, mf=mossy fiber pathway, pp=perforant pathway, Scc=Schaffer collateral commissural pathway, Fim=fimbria pathway for commissural pathway, arrows=usual direction of transmission. Adapted from reference 24.

intrahippocampal injection of whole tetanus toxin in young animals, multiple epileptic foci are established.[26] Whether or not this is a property of toxin transfer or the establishment of epilepsy in developing animals is unclear. As will be discussed, since epileptic foci appear limited to either bilateral hippocampi[26] or to the site of injection into cortex[27] (but see ref. 28) in older animals, this may be due to the development of epilepsy in young animals and not to transfer of the toxin to distant cortical regions. Also, since the theoretical actions of TT include both presynaptic as well as postsynaptic mechanisms, the contribution of each in areas into which TT is transported is unclear. That is, the presence of TT in dendrites could lead to down-regulation of post-synaptic receptors. As will be discussed, electrophysiological evidence for this is equivocal.

Discrepancies exist about the half-life of TT. As mentioned, the clinical syndrome can persist from weeks to months. Recovery at neuromuscular junctions occurs only following axonal sprouting and the creation of new terminal endplates,[29] suggesting that intoxicated terminals are permanently damaged. In vitro studies have suggested that, following rapid uptake[30] the toxin half-life is from 3[31] to 6[32] days, while in vivo less than 1% persisted up to 9 days.[33] Labeling studies have noted the presence of HRP-conjugated TT for several days.[22,23] However, blockade of synaptic transmission in spinal cord cultures persists for over one month.[32] Depression of GABAergic inhibitory transmission in hippocampus following intrahippocampal injection persists for several weeks; however, this has been interpreted as a change in the properties of inhibitory interneurons secondary to epileptic activity since GABA release in this system is only impaired for 2 weeks.[34] Persistent blockade of transmission likely occurs with fewer than 10 TT molecules[31] and theoretically occurs with only one TT molecule[4] due to its enzymatic nature. The long duration of poisoning is likely due to its compartmentalization and a lack of cytosolic proteases capable of destroying TT.[30,34] Thus, interpretation of results using TT in experimental models of epilepsy must seek to separate those results that could be due to a persistent action of TT from those due to the seizures themselves.

Experimental Implementation

Goals of Model Development

Intra-cortical injection of tetanus toxin as a means of inducing seizures was first demonstrated over a century ago.[35] The technique re-emerged and was developed in the 1960s and 1970s as an experimental tool for studying epilepsy. Its development became directed by the need for an animal model that exhibited a semi-chronic epileptic syndrome in order to explore the effect of epileptic discharges on cortical function. That is, the goal was to develop an animal model with spontaneous daily seizures for several weeks that then abated so that the effect of chronic seizures versus the acute seizures themselves on cortical function could be studied. Such functions included mood, memory, recurrent seizures and the mechanisms involved in their generation. Thus, given the persistence of the mechanism of tetanus toxin, it seems an ideal agent. However, as is often argued, it often becomes difficult to determine which of the effects observed in this model are due to the toxin or the seizures themselves. As it arises, this will be mentioned.

Laboratory personnel are typically immunized with tetanus toxoid and anti-titers are verified prior to studies. In general, the method requires an appropriate humane animal anesthetic followed by stereotactic injection of the toxin into the brain region of choice under aseptic conditions (see Fig. 2). Injection sites are accessed by means of a burr hole. Typically only a single injection of a very small volume (nanoliters) is given. However, in some instances, multiple or bilateral injections are performed and in some cases this is done with the semi-permanent implantation of a cannula.[28,36] Following wound closure and care, animals are returned to their cages. The dose of toxin is often noted in either amount in nanograms or in terms of mouse LD_{50}. In older studies, a mouse LD_{50} was defined as the amount required to kill half of a group of mice injected intramuscularly within 7 days (10^4 mouse LD_{50}s are equivalent to 1 rat LD_{50}).[15] Typically the equivalent of a few mouse LD_{50}s are used; 1 ng is approximately 3 mouse LD_{50}s.[28] Currently, an MD_{100}, defined as the minimal effective dose (0.25 ng) required to elicit hind limb paralysis in all animals following intramuscular injections in adult mice, has been used to assess potency.[26]

Chronic Epilepsy in Adult Animals

Intra-Hippocampal Injections of TT

Overview

The usage and consequences of intra-hippocampal (typically into CA3 region) TT injections in adult (21 days and older) rats have been reviewed.[11,37,38] Studies have employed and, where noted, included bilateral and unilateral injections of TT into either dorsal or ventral hippocampus. The findings are updated and summarized here with respect to a behavioral and seizure description, correlation with EEG findings, duration and severity of epilepsy and in vitro electrophysiological findings. This is followed by a summary of the pathological and biochemical findings. Long-term behavioral consequences to animals are discussed. The suggested mechanistic schemes for these findings are reviewed.

Seizures, EEG Correlates and Behavior

Injected animals begin to have seizures 1-2 days following the injection. Seizures in adult rats typically involve a behavioral arrest followed by focal or generalized myoclonus or facial twitching. Occasionally they begin in sleep. Head rearing and rearing on hind feet follows this. Occasionally the animals have so-called "wild running fits". The ictal stage lasts 25 to 75 seconds and is followed by a post-ictal stage of unresponsiveness lasting up to 4 minutes followed by resumption of normal activities.[15] Intracortical recordings obtained within hours following injection show that both evoked excitation and inhibition is impaired near the site of injection.[39] (This is also seen in vitro following acute application of very high doses of TT to hippocampal slices.[40]) One to two days later, the EEG shows epileptiform activity arising from the injected hippocampus initially and later bilaterally. Electrographic seizures appear to begin in either hippocampus with bursts of 3-20 Hz spike-wave activity that subsequently generalize. The non-injected hippocampus often develops larger discharges during the ictus and may discharge independently following an ictus.[41] Following the electrographic seizure, relative electrical silence is noted coinciding with the time the animal is unresponsive post-ictally. Interictal spike frequency increases as the number of seizures increases. Animals have 4-5 seizures per hour for 4 weeks, however this varies around roughly a 4-day cycle and most seizures occur during the afternoon. One week after injection, some of the hippocampal discharges occur without any overt behavioral seizure.[41] After 5-7 weeks, the EEG returns to normal and seizures desist.[15,37] Rarely (<10%), seizures or EEG abnormalities may recur or persist.[11] In a radial arm maze-learning task, rats learned more slowly than controls long after seizures had abated.[42] Rats developed life-long aggressiveness and hyperactivity soon after TT injection.[15]

Further in Vivo and in Vitro Electrophysiological Characterization

While initial studies into the electrophysiological characterization of the TT model involved in vivo hippocampal recordings,[42] the use of the hippocampal in vitro slice preparation has allowed a more thorough characterization of the nature and origin of the epileptic discharges and their persistence following TT injection. The TT model is unusual in this regard since most models of epilepsy do not retain epileptic discharges in vitro.[43] Three major findings are apparent. First, hippocampal CA3 pyramidal neurons show permanently decreased synaptic excitability by mossy fiber (from ipsilateral dentate granule neurons) and commissural (from contralateral CA3 pyramidal neurons) inputs 4-6 months following bilateral hippocampal TT injections. This is apparently not due to diminished intrinsic excitability since antidromic population spikes are normal.[42] Similar findings were found at 16 weeks following unilateral TT injections.[21] Interestingly, given that these animals had impaired learning, these same animals did not have impairment of long-term potentiation (LTP) in the pathway of commissural inputs to CA3 neurons.[42] LTP is thought to be an in vitro correlate of learning and memory.[44] Conversely, permanently enhanced excitability of dentate gyrus granule neurons (determined by larger, single orthodromic population spikes) and CA1 pyramidal neurons (determined by the presence of multiple orthodromic population spikes) is present ipsilaterally and contralaterally to unilateral TT injections.[21] Second, spontaneous epileptiform activity arising from the CA3 region is temporarily present in vitro at similar times to the observed epileptic syndrome, that is, up to 7 weeks following bilateral TT injection[43] and up to, but not beyond, 16 weeks following unilateral TT injection.[21] This epileptic activity is present bilaterally even following unilateral TT injection[45] and may be more consistently present contralaterally to TT injection.[34] Third, synaptic inhibition in CA3 and dentate gyrus is temporarily abolished in the hippocampus ipsilateral to the TT injection site for up to 8 weeks following injection.[21,46] Synaptic inhibition contralateral to TT injection, while not abolished, is apparently diminished somewhat, and it was found that the stimuli required to generate monosynaptic fast IPSPs were twice that found compared to control slices. This suggested that interneurons were either less excitable (intrinsically or extrinsically) and/or that they were less efficient in GABA release.[34] Such network alterations may explain the permanent changes in excitability noted in CA1 and dentate gyrus. Concurrent studies by these authors suggested that GABA release contralateral to TT injection was normal since they observed that GABA release ipsilateral to the injection was transiently impaired only in the first two weeks following injection.[34]

Regardless, inhibition is apparently only impaired up to 8 weeks following TT injection, while epileptic activity persists beyond this but it too diminishes finally 16 weeks following TT injection.[21]

Neuropathological Correlates

The TT model of epilepsy in adult rats is unique in that it is not associated with apparently significant pyramidal neuron loss compared to other models.[15] Remote from the injection site, there is bilateral 10% cell loss in CA1 following unilateral injection of TT into dorsal hippocampus and 30% cell loss in CA1 following unilateral injection of TT into ventral hippocampus[47] (and reviewed in[38]). Consistent with some degree of cell loss, microglial activation is noted in these regions.[47] Much larger doses of TT than are normally used do apparently cause pyramidal neuron loss at the site of injection.[48] However, morphologically it has been briefly reported that there is dendritic spine loss on CA3 pyramidal neurons near the site of injection noted 4 weeks following TT injection.[49] CA1 pyramidal neurons ipsilateral to the site of injection demonstrate a higher incidence of dye coupling, presumably via gap junctions along with simplified apical and basal dendritic arbors.[50] Aberrant mossy fibers that have apparently sprouted enter the outer and inner molecular layers of the dentate gyrus and are noted bilaterally 4 weeks following unilateral TT injection.[51] Expression of mRNA encoding specific glutamate receptor isoforms is transiently elevated bilaterally in a specific pattern (GluR1 flip in dentate gyrus and GluR2 flip in all hippocampal sub-fields) at 4 weeks following unilateral TT injections.[52] While presumably this could underlie increased excitability, the significance of this is speculative given the unknown function of these specific glutamate receptor isoforms. Disparate conclusions have been obtained in an attempt to address whether increased excitability is mediated by a loss of inhibitory interneurons. In a study utilizing interneuron-specific expression of glutamic acid decarboxylase (GAD), inhibitory interneurons are apparently unaffected and may exhibit a transient up-regulation of GAD.[53] Others have observed a selective, bilateral loss of somatostatin-immunoreactive interneurons in the hilus of the dentate gyrus noted at 8 weeks following unilateral TT injection.[54]

Effects of Anti-Convulsant Drugs

Two studies have looked at the effect of anti-convulsant drugs on the development of seizures in the adult TT model. The first study[28] was different from others in that it involved multiple unilateral intrahippocampal injections of TT in cats. The resulting seizures were so severe, that if control was not attempted with phenobarbital, the animals succumbed to status epilepticus. This study, despite its limited size, suggested that if seizures were initially controllable with phenobarbital, this led to remission of seizures. However, in those animals that had remission of seizures, repetitive seizures could be re-induced with a single injection of TT that required life-long phenobarbital for control for the resulting chronic epileptic syndrome. Importantly, it has not been determined whether or not treatment with anti-convulsants resulted in normal performance in learning tasks or minimization of any of the neuropathological changes found.

In a subsequent study,[55] adult rats were pretreated with either intrahippocampal TT or kainate, with or without suppression of seizures by phenobarbital, followed by a treatment with intrahippocampal TT. Animals pretreated with kainate were more likely to develop chronic spontaneous repetitive seizures, regardless of seizure suppression by phenobarbital, and in correlation with damage to the CA3 region of the hippocampus. All animals that received only TT treatment (with and without phenobarbital treatment) recovered from their epileptic syndrome. Treatment with phenobarbital during the pretreatment with either kainate or TT did result in a less severe epileptic syndrome during the subsequent administration of TT. This study did not investigate the presence of subclinical seizures nor did it ascertain performance in learning tasks.

Intra-Cortical Injections of TT

Compared to intra-hippocampal injections of TT, injections of TT into adult rat neocortex result in a more severe and prolonged epileptic syndrome. Seizures, abnormal EEG and in vitro epileptiform activity are present for greater than 7 months following injection. These abnormalities are noted locally and diffusely, presumably due to transport of TT.[27] Similar findings have been found following injections of TT into cat motor cortex where it has been noted that no significant cell loss is present.[36] Similar to rat hippocampus, injection of TT into neocortex results in a transient depression of GABA release only ipsilateral to the injection.[56] Also, neocortical injection of TT is associated with a focal up-regulation of GAD mRNA, as well as mRNA for type II calcium/calmodulin-dependent protein kinase, GluR2[57] and brain-derived neurotrophic factor (BDNF) which may play a role in aberrant neuronal sprouting.[58] While injection of TT into other brain regions has been attempted, these have been only briefly described.[15]

Discussion

The conclusion of many studies has suggested that the epileptic syndrome that follows TT injection is due to impaired inhibition. However, investigators have argued that impaired inhibition is a consequence of the seizures themselves and not directly due to the toxin itself. This argument is based on the observations that the measured half-life of the toxin is short, release of GABA is impaired only locally and in a transient fashion that mirrors the toxin half-life, the contralateral hippocampus which has only mildly impaired inhibition has more dramatic epileptic discharges in vitro, epileptic discharges persist after the recovery of inhibition and, when seizures are controlled with anti-convulsant drugs, the syndrome is prevented. It is still unclear if there is a central nervous system counterpart to the recovery from TT in the peripheral nervous system that requires axonal sprouting and the creation of new terminal endplates.[29] The mechanism(s) of denervation (loss of CA3 dendritic spines) are equally unclear. It is thought that focal loss of CA3 dendritic spines could underlie impaired CA3 excitation and performance in learning tasks. This has not been studied thoroughly, however. Since it is known that TT inhibits insertion of postsynaptic glutamate receptors,[17,18] and that blockade of this process can lead to spine retraction,[59] a direct effect is theoretically possible. Perhaps by studying dendritic spines in other hippocampal regions such as CA1 or dentate gyrus this could be clarified. Despite these reservations, the observations found in young animals suggest that it is the impact of recurrent seizures that induces the epileptic syndrome.

Chronic Epilepsy in Young Animals

Intra-Hippocampal Injections of TT

Overview

The use of intra-hippocampal injections of TT in young animals (9-11 days old) has been performed in our laboratory. These have always involved single, unilateral injections of low-dose TT into the dorsal CA3 region. The findings are summarized here with respect to a behavioral and seizure description, correlation with EEG findings, duration and severity of epilepsy and in vitro electrophysiological findings. This is followed by a summary of the neuropathological findings. Long-term behavioral consequences to animals are discussed. The suggested mechanistic schemes for these findings coupled with that found in adult animals are presented.

Seizures, EEG Correlates and Behavior

Suppression of background activity was observed in EEG recordings in the first 12-24 hours following TT injection. Multifocal spikes from either hippocampi or neocortex developed in the next 24 hours. Behavioral seizures, described as "wild-running" or "wet-dog shakes", began 24-72 hours following TT injection, were correlated with EEG recordings, and were found to arise either focally in the injected hippocampus or bilaterally. At their peak, approximately 1-3 days following TT injection, as many as 4-5 wild-running seizures per hour were observed. Subsequently most seizures were sub-clinical in nature after 4 weeks. While only 20% of animals might be observed to have behavioral seizures in adulthood, when these animals were reexamined in adulthood (up to 24 weeks following TT injection) with EEG, nearly all had abnormal EEGs consisting of either brief bursts of fast generalized spike-wave activity or brief bursts of multifocal interictal spikes, typically without behavioral correlate.[26,60] When tested as adults, TT-injected animals learned the Morris water maze more slowly compared to control animals, findings which were independent of EEG abnormalities and any impairment in motor function.[25]

Further in Vitro Electrophysiological Characterization

Most hippocampal slices obtained from rats injected with TT in infancy exhibited epileptiform bursting from both the ipsilateral and contralateral CA3 region; this was independent of the time following (up to 10 weeks) TT injection.[26,61] Such bursting was not present in CA1, however some slices exhibited unusual rhythmic (2-8 Hz) IPSPs. Membrane properties of CA3 pyramidal neurons were similar to control animals. Blockade of IPSPs in TT-injected rats induced rhythmic epileptiform bursting in both contralateral and ipsilateral hippocampi compared to rhythmic single spikes in control animals suggesting excitation was enhanced in TT-injected animals.[61]

Neuropathological Correlates

Injection of TT in infancy shares many findings with that in adulthood with a few quantitative differences. No total cell loss was noted in any hippocampal region,[26] however there was found to be a dispersion of stratum pyramidale that was more prominent in the CA1 region and could not be explained by gliosis.[25] Aberrant mossy fiber sprouting was also found bilaterally following TT injection.[60] Dendritic spine loss along apical and basal dendrites of pyramidal neurons in CA3 has been quantified and found to exist bilaterally while axons from the same CA3 neurons appear normal with regards to length and branching pattern.[62]

Effects of Anti-Convulsant Drugs

The effect of the anti-convulsant carbamazepine in infant rats following unilateral TT injection in infancy was markedly different from that found in adult animals.[63] While able to significantly limit the number of observed seizures, sub-clinical seizures persisted, though still fewer than that found in control animals. Upon withdrawal of carbamazepine, behavioral seizures recurred and subsequently abated with a similar, though delayed, time course comparable to animals not treated with carbamazepine. This suggests that the seizures themselves appear to limit their recurrence during development. Longer-term evaluation of these animals has not yet been performed to determine if treated animals had normalization of their EEGs, as was found in the study on adult animals. Again, as in adult animals, it has not been determined whether or not treatment with anti-convulsants resulted in normal performance in learning tasks or minimization of any of the neuropathological changes found.

Discussion

The persistent epileptic state in rats following unilateral hippocampal injection of TT in infancy suggests two important conclusions. First, since the epileptic state persists well beyond any conceivable persistence of TT action (at 24 weeks, assuming a half-life of 7 days, only one molecule of TT would remain), it is the effect of persistent seizures that causes the chronic epileptic state. Second, the developing brain is more sensitive to early-life seizures to cause the chronic epileptic state. An important distinction with the seizures in adult rats is noted, in that, recurrent seizures in adult rats do not necessarily predispose to chronic epilepsy unless, as recent reports have suggested,[55] recurrent seizures induced by TT are coupled with other agents or treatments (kainate,[55] hypoxia[64]) that cause hippocampal neuronal loss. Next, it is important to determine how recurrent seizures induced by TT in both adult and infant rats lead to learning impairment. While it appears that it is the effect of chronic seizures, one could also argue that it is the damage caused by TT. Mossy fiber sprouting is a common abnormality found in most animal models of epilepsy and is not likely to be a consequence of TT itself. This occurs even in the infant model, and its presence bilaterally following unilateral TT injection suggests it is secondary to the seizures and not TT. Sprouting may also explain the increase in neuropil found in CA1 regions. Once the TT model is compared to other models of epilepsy, a common denominator emerges in that in different models, early-life seizures lead to not only chronic epilepsy and similar neuropathological changes but also learning impairment.

Comparisons to Other Models of Chronic Epilepsy

While clear distinctions exist in other models of chronic epilepsy between infant and adult animals, these share many of the features of the TT model. For example, following kainate-induced status epilepticus in adult animals, there are spontaneous repetitive seizures, CA3 cell loss, mossy fiber sprouting into CA3 and dentate gyrus, sprouting into CA1 stratum pyramidale and stratum radiatum, and impaired learning in memory tasks.[65-67] However, if animals younger than 14-21 days old are treated, the animals are apparently unaffected, even following repetitive treatments.[68] Newer studies now suggest that while kainate treatment in animals younger than 14 days has neither neuropathological sequelae nor induction of spontaneous repetitive seizures, it is found to be associated with diminished LTP, learning task performance, kindling rate and increased inhibition in the dentate gyrus.[69] Furthermore, kainate insult in infancy and again later in adulthood results in more prominent memory impairment than a single insult at either time.[70] Similar findings are noted with lithium-pilocarpine-induced status epilepticus.[71-73] However, as suggested by kainate, repetitive episodes of status epilepticus in infancy caused by lithium-pilocarpine are not benign. Following 3 episodes of status epilepticus in this model, rats develop recurrent electrographic seizures (most of which were subclinical), increased CA1 excitability, and diminished learning, all without significant neuropathological changes such as cell loss (however mossy fiber sprouting was not assessed).[74-76] This is very similar to the findings of the chronic epileptic syndrome induced by TT in infancy and suggests that repetitive early-life seizures are a common mechanism capable of creating a chronic epileptic state that includes memory impairment without cell loss. Indeed, it is not just the seizures themselves, it is the molecular pathways they trigger, as has been demonstrated by the co-application of an NMDA receptor antagonist (MK-801) with lithium-pilocarpine in adult animals. In this case, status epilepticus was not prevented but all of the sequelae were.[77]

The effects of early-life seizures are certainly model-dependent. In the fluorothyl-model, early-life seizures are associated with impaired learning and mossy fiber sprouting but not spontaneous repetitive seizures.[78,79] In the hyperthermia model of febrile seizures, a single prolonged seizure results in life-long susceptibility to convulsants, enhanced inhibition in CA1, enhanced in vitro kindling, and enhancement of the membrane conductance I_h, with subsequent hyperexcitability in CA1.[80-82] Following a prolonged hypoxia-induced seizure, experimental animals develop acute seizures and a chronic reduced seizure threshold to convulsants, but neither mossy fiber sprouting nor cognitive disturbances. This model has now been shown to be associated with increased expression of calcium-permeable glutamate receptors, however the exact role of this in the pathogenesis of this model is unclear.[83] Status epilepticus induced by kainate in infant rats does not result in mossy fiber sprouting but does result in impaired LTP, impaired kindling and impaired learning associated with enhanced inhibition in the dentate gyrus.[69]

Thus, while each of these models has distinct mechanisms in the induction of seizures, their common outcome suggests that it is directed by the repetitive experience of the seizures themselves. Since there are likely many mechanisms that could lead to spontaneous repetitive seizures and/or learning impairment (similar to the clinical experience that there are many causes of epilepsy), seizures are the likely trigger of any or all of these pathways that could be favored by one treatment or another.

References

1. Tetanus. In: Pickering LK, ed. Red Book: Report of the Committee on Infectious Diseases. Elk Grove: American Academy of Pediatrics, 2000:563-568.
2. Volpe JJ. Bacterial and fungal intracranial infections. In: Neurology of the Newborn. Philadelphia: W.B. Saunders, 2001:774-810.
3. Anaerobic infections. In Johnson KB, Oski FA, eds. Oski's Essential Pediatrics. Philadelphia: Lippincott-Raven, 1997:209-212.
4. Rossetto O, Seveso M, Caccin P et al. Tetanus and botulinum neurotoxins: turning bad guys into good by research. Toxicon 2001; 39:27-41.
5. Hohne-Zell B, Ecker A, Weller U et al. Synaptobrevin cleavage by the tetanus toxin light chain is linked to the inhibition of exocytosis in chromaffin cells. FEBS Letters 1994; 355:131-134.
6. Schiavo G, Benfenati F, Poulain B et al. Tetanus and botulinum-B neurotoxins block neurotransmitter release by proteolytic cleavage of synaptobrevin. Nature 1992; 359:832-834.
7. Herreros J, Miralles FX, Solsona C et al. Tetanus toxin inhibits spontaneous quantal release and cleaves VAMP/synaptobrevin. Brain Res 1995; 699:165-170.
8. Rothman JE. Mechanisms of intracellular protein transport. Nature 1994; 372:55-63.
9. Bergey GK, MacDonald RL, Habig WH et al. Tetanus toxin: convulsant action on mouse spinal neurons in culture. J Neurosci 1983; 3:2310-2323.
10. Collingridge GL, Davies J. The in vitro inhibition of GABA release by tetanus toxin. Neuropharmacol 1982; 21:851-855.
11. Mellanby J. Tetanus toxin as a tool for investigating the consequences of excessive neuronal excitation. In: DasGupta BR, ed. Botulinum and Tetanus Neurotoxins. New York: Plenum, 1993:291-297.
12. McMahon HT, Foran P, Dolly JO et al. Tetanus toxin and botulinum toxins type A and B inhibit glutamate, gamma-aminobutyric acid, aspartate, and met-enkephalin release from synaptosomes. Clues to the locus of action. J Biol Chem 1992; 267:21338-21343.
13. Curtis DR, de Groat WC. Tetanus toxin and spinal inhibition. Brain Res 1968; 10:208-212.
14. Curtis DR, Felix D, Game CJA et al. Tetanus toxin and the synaptic release of GABA. Brain Res 1973; 51:358-362.
15. Mellanby J, George G, Robinson A et al. Epileptiform syndrome in rats produced by injecting tetanus toxin into the hippocampus. J Neurol Neurosurg Psychiatry 1977; 40:404-414.
16. Matteoli M, Verderio C, Rossetto O et al. Synaptic vesicle endocytosis mediates the entry of tetanus neurotoxin into hippocampal neurons. Proc Natl Acad Sci USA 1996; 93:13310-13315.
17. Lledo PM, Zhang X, Sudhof TC et al. Postsynaptic membrane fusion and long-term potentiation. Science 1998; 279:399-403.
18. Lin JW, Sheng M. NSF and AMPA receptors get physical. Neuron 1998; 21:267-270.
19. Ouardouz M, Sastry BR. Mechanisms underlying LTP of inhibitory synaptic transmission in the deep cerebellar nuclei. J Neurophysiol 2000; 84:1414-1421.
20. Lu W-Y, Man H-Y, Ju W et al. Activation of synaptic NMDA receptors induces membrane insertion of new AMPA receptors and LTP in cultured hippocampal neurons. Neuron 2001; 29:243-254.
21. Whittington MA, Jefferys JGR. Epileptic activity outlasts disinhibition after intrahippocampal tetanus toxin in the rat. J Physiol 1994; 481:593-604.
22. Fishman PS, Savitt JM. Transsynaptic transfer of retrogradely transported tetanus protein-peroxidase conjugates. Exp Neurol 1989; 106:197-203.
23. Schwab ME, Suka K, Thoenen H. Selective retrograde transsynaptic transfer of a protein, tetanus toxin, subsequent to its retrograde axonal transport. J Cell Biol 1979; 82:798-810.
24. Amaral DG, Witter MP. The three-dimensional organization of the hippocampal formation: a review of anatomical data. Neuroscience 1989; 31:571-591.
25. Lee CL, Hannay J, Hrachovy R et al. Spatial learning deficits without hippocampal neuronal loss in a model of early-onset epilepsy. Neuroscience 2001; 107:71-84.
26. Lee CL, Hrachovy RA, Smith KL et al. Tetanus toxin-induced seizures in infant rats and their effects on hippocampal excitability in adulthood. Brain Res 1995; 677:97-109.
27. Brener K, Amitai Y, Jefferys JGR et al. Chronic epileptic foci in neocortex: In vivo and in vitro effects of tetanus toxin. Eur J Neurosci 1991; 3:47-54.
28. Darcey TM, Williamson PD. Chronic/semichronic limbic epilepsy produced by microinjection of tetanus toxin in cat hippocampus. Epilepsia 1992; 33:402-419.
29. Duchen LW, Tonge DA. The effects of tetanus toxin on neuromuscular transmission and on the morphology of motor end-plates in slow and fast skeletal muscle of the mouse. J Physiol 1973; 228:157-172.
30. Critchley DR, Nelson PG, Habig WH et al. Fate of tetanus toxin bound to the surface of primary neurons in culture: evidence for rapid internalization. J Cell Biol 1985; 100:1499-1507.

31. Erdal E, Bartels F, Binscheck T et al. Processing of tetanus and botulinum A neurotoxin in isolated chromaffin cells. Naunyn Schmiedebergs Arch Pharmacol 1995; 351:67-78.
32. Habig WH, Bigalke H, Bergey GK et al. Tetanus toxin in dissociated spinal cord cultures: long-term characterization of form and action. J Neurochem 1986; 47:930-937.
33. Mellanby JH. Elimination of ^{125}I from rat brain after injection of small doses of ^{125}I-labelled tetanus toxin into the hippocampus. Neurosci Lett Suppl 1989; Suppl 36:S55
34. Empson RM, Jefferys JGR. Synaptic inhibition in primary and secondary chronic epileptic foci induced by intrahippocampal tetanus toxin in the rat. J Physiol 1993; 465:595-614.
35. Roux E, Borrel A. Tetanos cerebral et immunite contre le tetanos. Ann Inst Pasteur 1898; 4:225-239.
36. Louis ED, Williamson PD, Darcey TM. Chronic focal epilepsy induced by microinjection of tetanus toxin into the cat motor cortex. Electroencephalogy Clin Neurophysiol 1990; 75:548-557.
37. Mellanby J, Hawkins C, Mellanby H et al. Tetanus toxin as a tool for studying epilepsy. J Physiol (Paris) 1984; 79:207-215.
38. Jefferys JGR, Whittington MA. Review of the role of inhibitory neurons in chronic epileptic foci induced by intracerebral tetanus toxin. Epilepsy Res 1996; 26:59-66.
39. Sundstrom LE, Mellanby JH. Tetanus toxin blocks inhibition of granule cells in the dentate gyrus of the urethane-anaesthetized rat. Neuroscience 1990; 38:621-627.
40. Calabresi P, Beneditti M, Mercuri NB et al. Selective depression of synaptic transmission by tetanus toxin: a comparative study on hippocampal and neostriatal slices. Neuroscience 1989; 30:663-670.
41. Mellanby J, Strawbridge P, Collingridge GL et al. Behavioural correlates of an experimental hippocampal epileptiform syndrome in rats. J Neurol Neurosurg Psychiatry 1981; 44:1084-1093.
42. Brace HM, Jefferys JGR, Mellanby J. Long-term changes in hippocampal physiology and learning ability of rats after intrahippocampal tetanus toxin. J Physiol (Lond) 1985; 368:343-357.
43. Jefferys JGR. Chronic epileptic foci in vitro in hippocampal slices from rats with tetanus toxin epileptic syndrome. J Neurophysiol 1989; 62:458-468.
44. Bliss T, Collingridge G. A synaptic model of memory: long-term potentiation in the hippocampus. Nature 1993; 361:31-39.
45. Jefferys JGR, Empson RM. Development of chronic secondary epileptic foci following intrahippocampal injection of tetanus toxin in the rat. Exp Physiol 1990; 75:733-736.
46. Jordan SJ, Jefferys JGR. Sustained and selective block of IPSPs in brain slices from rats made epileptic by intrahippocampal tetanus toxin. Epilepsy Res 1992; 11:119-129.
47. Shaw JAG, Perry VH, Mellanby J. Tetanus toxin-induced seizures cause microglial activation in rat hippocampus. Neurosci Lett 1990; 120:66-69.
48. Bagetta G, Nistico G, Bowery NG. Prevention by the NMDA receptor antagonist, MK801 of neuronal loss produced by tetanus toxin in the rat hippocampus. Br J Pharmac 1990; 101:776-780.
49. Hughes JT, Mellanby J. Experimental epilepsy induced by tetanus toxin injected into the rat hippocampus: a Golgi study. Neuropath Appl Neurobiol 1985; 11:73
50. Colling SB, Man WD-C, Draguhn A et al. Dendritic shrinkage and dye-coupling between rat hippocampal CA1 pyramidal cells in the tetanus toxin model of epilepsy. Brain Res 1996; 741:38-43.
51. Mitchell J, Gatherer M, Sundstrom LE. Aberrant Timm-stained fibres in the dentate gyrus following tetanus toxin-induced seizures in the rat. Neuropath Appl Neurobiol 1996; 22:129-135.
52. Rosa MLNM, Jefferys JGR, Sanders MW et al. Expression of mRNAs encoding flip isoforms of GluR1 and GluR2 glutamate receptors is increased in rat hippocampus in epilepsy induced by tetanus toxin. Epilepsy Res 2002; 36:243-251.
53. Najlerahim A, Williams SF, Pearson RCA et al. Increased expression of GAD mRNA during the chronic epileptic syndrome due to intrahippocampal tetanus toxin. Exp Brain Res 1992; 90:332-342.
54. Mitchell J, Gatherer M, Sundstrom LE. Loss of hilar somatostatin neurons following tetanus toxin-induced seizures. Acta Neuropathologica 1995; 89:425-430.
55. Mellanby J, Milward AJ. Do fits really beget fits? The effect of previous epileptic activity on the subsequent induction of the tetanus toxin model of limbic epilepsy in the rat. Neurobiol Dis 2001; 8:679-691.
56. Empson RM, Amitai Y, Jefferys JGR et al. Injection of tetanus toxin into the neocortex elicits persistent epileptiform activity but only transient impairment of GABA release. Neuroscience 1993; 57:235-239.
57. Liang F, Jones EG. Differential and time-dependent changes in gene expression for type II calcium/calmodulin-dependent protein kinase, 67 kDa glutamic acid decarboxylase, and glutamate receptor subunits in tetanus toxin-induced focal epilepsy. J Neurosci 1997; 17:2168-2180.
58. Liang F, Jones EG. Reciprocal up- and down-regulation of BDNF mRNA in tetanus toxin-induced epileptic focus and inhibitory surround in cerebral cortex. Cerebral Cortex 1998; 8:481-491.

59. Bear MF, Rittenhouse CD. Molecular basis for induction of ocular dominance plasticity. J Neurobiol 1999; 41:83-91.
60. Anderson AE, Hrachovy RA, Antalffy BA et al. A chronic focal epilepsy with mossy fiber sprouting follows recurrent seizures induced by intrahippocampal tetanus toxin injection in infant rats. Neuroscience 1999; 92:73-82.
61. Smith KL, Lee CL, Swann JW. Local circuit abnormalities in chronically epileptic rats after intrahippocampal tetanus toxin injection in infancy. J Neurophysiol 1998; 79:106-116.
62. Jiang M, Lee CL, Smith KL et al. Spine loss and other persistent alterations of hippocampal pyramidal cell dendrites in a model of early-onset epilepsy. J Neurosci 1998; 18:8356-8368.
63. Rashid S, Lee I-G, Anderson AE et al. Insights into the tetanus toxin model of early-onset epilepsy from long-term video monitoring during anticonvulsant therapy. Brain Res 1999; 118:221-225.
64. Milward AJ, Meldrum BS, Mellanby JH. Forebrain ischaemia with CA1 cell loss impairs epileptogenesis in the tetanus toxin limbic seizure model. Brain 1999; 122:1009-1016.
65. Stafstrom CE, Thompson JL, Holmes GL. Kainic acid seizures in the developing brain: status epilepticus and spontaneous recurrent seizures. Brain Res Dev Brain Res 1992; 21:227-236.
66. Esclapez M, Hirsch JC, Ben-Ari Y et al. Newly formed excitatory pathways provide a substrate for hyperexcitability in experimental temporal lobe epilepsy. J Comp Neurol 1999; 498:449-460.
67. Yang Y, Tandon P, Liu Z et al. Synaptic reorganization following kainic acid-induced seizures during development. Dev Brain Res 1998; 107:169-177.
68. Sarkisian MR, Tandon P, Liu Z et al. Multiple kainic acid seizures in the immature and adult brain: ictal manifestations and long-term effects on learning and memory. Epilepsia 1997; 38:1157-1166.
69. Lynch M, Sayin U, Bownds J et al. Long-term consequences of early postnatal seizures on hippocampal learning and plasticity. Eur J Neurosci 2000; 12:2252-2264.
70. Koh S, Storey TW, Santos TC et al. Early-life seizures in rats increase susceptibility to seizure-induced brain injury in adulthood. Neurology 1999; 53:915-921.
71. Sankar R, Shin DH, Liu H et al. Patterns of status epilepticus-induced neuronal injury during development and long-term consequences. J Neurosci 1998; 18:8382-8393.
72. Sankar R, Shin D, Mazarati AM et al. Epileptogenesis after status epilepticus reflects age- and model-dependent plasticity. Ann Neurol 2000; 48:580-589.
73. Dube C, Marescaux C, Nehlig A. A metabolic and neuropathological approach to the understanding of plastic changes that occur in the immature and adult rat brain during lithium-pilocarpine-induced epileptogenesis. Epilepsia 2000; 41:S36-S43.
74. Priel MR, dos Santos NF, Cavalheiro EA. Developmental aspects of the pilocarpine model of epilepsy. Epilepsy Res 1996; 26:115-121.
75. dos Santos NF, Arida RM, Filho EM et al. Epileptogenesis in immature rats following recurrent status epilepticus. Brain Res Rev 2000; 32:269-276.
76. Santos NF, Marques RH, Correia L et al. Multiple pilocarpine-induced status epilepticus in developing rats: a long-term behavioral and electrophysiological study. Epilepsia 2000; 41:S57-S63.
77. Rice AC, Floyd CL, Lyeth BG et al. Status epilepticus causes long-term NMDA receptor-dependent behavioral changes and cognitive deficits. Epilepsia 1998; 39:1148-1157.
78. Holmes GL, Gairsa J-L, Chevassus-Au-Louis N et al. Consequences of neonatal seizures in the rat: morphological and behavioral effects. Ann Neurol 1998; 44:845-857.
79. Huang L, Cilio MR, Silveira DC et al. Long-term effects of neonatal seizures: a behavioral, electrophysiological and histological study. Brain Res Dev Brain Res 1999; 118:99-107.
80. Chen K, Baram TZ, Soltesz I. Febrile seizures in the developing brain result in persistent modification of neuronal excitability in limbic circuits. Nat Med 1999; 5:888-894.
81. Dube C, Chen K, Eghbal-Ahmadi M et al. Prolonged febrile seizures in the immature rat model enhance hippocampal excitability long term. Ann Neurol 2000; 47:336-344.
82. Chan K, Aradi I, Thon N et al. Persistently modified h-channels after complex febrile seizures convert the seizure-induced enhancement of inhibition to hyperexcitability. Nat Med 2001; 7:331-337.
83. Sanchez RM, Koh S, Rio C et al. Decreased glutamate receptor 2 expression and enhanced epileptogenesis in immature rat hippocampus after perinatal hypoxia-induced seizures. J Neurosci 2001; 21:8154-8163.

CHAPTER 17

Brain Stimulation As a Therapy for Epilepsy

Jeffrey H. Goodman

Abstract

The failure of current antiepileptic therapies to adequately treat a significant number of epileptic patients highlights the need for the development of new treatments for the disorder. A new strategy that is currently being developed is to deliver electrical stimulation directly to the brain to decrease or prevent seizure activity. Clinical evidence that electrical stimulation could interfere with seizure activity was initially reported in the 1930's. However, many of these early studies consisted of case reports or were poorly controlled. In addition, there were a number of studies that failed to observe any beneficial effect of brain stimulation on seizures. More recently, deep brain stimulation has been used successfully to treat patients with movement disorders and vagus nerve stimulation has been shown to effectively decrease seizure activity in a select population of epilepsy patients. These advances have led to a reexamination of the potential therapeutic benefits of deep brain stimulation for the treatment of epilepsy. There is now experimental and clinical evidence that direct electrical stimulation of the brain can prevent or decrease seizure activity. However, several fundamental questions remain to be resolved. They include where in the brain the stimulus should be delivered and what type of stimulation would be most effective. One goal of this research is to combine the beneficial aspects of electrical stimulation with seizure detection technology in an implantable responsive stimulator. The device will detect the onset of a seizure and deliver an electrical stimulus that will safely block seizure activity without interfering with normal brain function.

Introduction

Current pharmacologic therapies for epilepsy fail to effectively treat 25% of the patients with the disorder.[1] The remaining patients who derive some benefit from these treatments are at risk of developing a myriad of side effects as a consequence of long-term drug administration. As a result, there is a need for the development of new effective therapies with a low potential for side effects. One candidate is the use of direct electrical stimulation of the brain to stop or prevent seizures.

There have been numerous attempts to use electrical stimulation of the brain to block seizure activity but the reality is that while brain stimulation as a therapy for epilepsy holds great promise, it is still in its infancy. Evidence from a number of animal and clinical studies suggest that direct electrical stimulation of the brain can block or decrease the severity of seizures.[2-4] However, there are also a number of contradictory studies that suggest deep brain stimulation does not decrease seizure activity.[5-7] The problem with many of the clinical studies is that they were not properly controlled, so it is difficult to determine the exact meaning of their results. One controlled clinical study that examined the effect of centromedian thalamic stimulation in patients with intractable epilepsy did not observe a significant decrease in the number of seizures.[8] The only controlled clinical study that has reported a significant decrease in seizure frequency after electrical stimulation was the study that demonstrated the efficacy of vagus nerve stimulation (VNS).[9] Intermittent electrical stimulation applied to the left vagus

Recent Advances in Epilepsy Research, edited by Devin K. Binder and Helen E. Scharfman.
©2004 Eurekah.com and Kluwer Academic / Plenum Publishers.

nerve provides effective seizure control in a select population of epilepsy patients. The mechanism by which VNS decreases seizure activity is unknown.[10]

Given the limited success of VNS, it is tempting to hypothesize that the delivery of an electrical stimulus directly to the CNS, instead of a peripheral nerve, will result in a more effective decrease in seizure activity. Deep brain stimulation (DBS) of the basal ganglia and thalamic structures is an approved and effective therapy for movement disorders. The success obtained with DBS as a treatment of movement disorders has led several investigators to examine the potential anticonvulsant effect of electrical stimulation delivered to these same areas. There are several clinical reports of decreased seizure activity after high-frequency stimulation of the anterior thalamus,[11] the subthalamic nucleus,[12] and the hippocampus.[13] The results from these clinical studies and recent studies using animal seizure models strongly suggest that DBS has the potential to become a treatment for those patients with epilepsy who are currently unresponsive to pharmacologic therapy or are not candidates for surgical intervention.

However, before DBS can be considered as a therapy for epilepsy several fundamental questions need to be answered. First, there is the overriding concern of safety. The effective stimulus should not cause tissue damage, interfere with normal brain function, and most importantly, lead to seizures. Secondly, where in the brain should the stimulus be delivered? Should the stimulus be delivered directly at an identified seizure focus or should it be delivered at a site outside the focus that has inhibitory projections to the focus? Finally, what type of stimulus will be most effective? Will one type of stimulation be effective against many forms of epilepsy or will the stimulus have to be customized to meet the needs of the individual patient?

Once an effective stimulation paradigm is developed, a method for delivering the stimulus to the patient will also have to be developed. The stimulators presently used for VNS, and those used for movement disorders, have been defined as blind or unintelligent since they deliver current at a predetermined intensity on a continual basis, independent of the patient's physiological state.[14] An intelligent stimulator would be able to detect a seizure and deliver a stimulus at the optimal time to prevent or stop a seizure. By decreasing the time that current is delivered to the patient, issues of safety and the potential for side effects are minimized. Recent advances in seizure prediction technology raise the possibility that in the near future an intelligent stimulator will be available as a treatment option for patients with epilepsy.[14,15]

Vagus Nerve Stimulation

Vagus nerve stimulation is the only approved therapy for the treatment of epilepsy that uses electrical stimulation. The efficacy of VNS has been rated to be equal to that of a new anticonvulsant agent for patients with partial seizures but less effective than surgical resection.[10] The mechanism underlying the anticonvulsant effect of VNS is unknown. Several pathways have been hypothesized as routes by which activation of the vagus nerve could lead to activation of central structures with the ultimate effect of decreasing seizure activity.

The vagus nerve is a mixed nerve, which in addition to carrying parasympathetic fibers also carries sensory afferents into the CNS. The cell bodies for these afferents are located in the nodose ganglion. It has been hypothesized that activity in the vagus can enter the CNS by traveling from the nodose ganglion to the nucleus tractus solitarius (NTS). One of the key structures innervated by the NTS is the parabrachial nucleus. Neurons in this structure project to a number of thalamic nuclei thereby providing a mechanism for activation of thalamocortical circuits with the potential for modulation of seizure activity.[16] The finding that electrical stimulation of the NTS interfered with the development of convulsive seizures in amygdaloid kindled cats supports an anticonvulsant role for the NTS.[17] Results from animal studies suggest that the NTS also projects to the locus ceruleus, a brainstem structure with widespread projections to cortical and limbic structures. The locus ceruleus has also been implicated in mediating the effects of VNS, because when lesions were placed in the locus ceruleus, the anticonvulsant effect of VNS was prevented.[18] Therefore, the anticonvulsant effect of NTS activation and VNS may be mediated through activation of the locus ceruleus.

Independent of the mechanism of action by which VNS decreases seizure activity, VNS is very effective in a small population of patients. However, the number of patients that do not respond to this therapy is significant enough that the need for new therapies still exists. One disadvantage of VNS is that it is impossible to predict before surgery which patients will benefit and which will not.[10] In addition, VNS is not free of side effects. Besides the potential complications from surgery, some patients complain of hoarseness, voice changes and dyspnea.[10] When VNS is ineffective, additional surgery is required to remove the implanted electrodes.

DBS As a Therapy for Epilepsy

The recent success obtained with deep brain stimulation for movement disorders and vagus nerve stimulation for seizures has led to a renewed research effort to develop direct stimulation of the brain as a treatment for epilepsy. However, it is unclear where the stimulus should be delivered to obtain the optimal anticonvulsant effect. Two approaches are being used to assess the efficacy of DBS for epilepsy. The first is to examine the effect of stimulation of areas away from the focus that may be able to disinhibit areas of the brain that regulate seizure susceptibility. The second approach is to deliver the stimulus directly at an identifiable focus.

Activation of Seizure-Gating Networks—Animal Studies

The results from a number of experimental animal studies have identified several areas in the brain that, when stimulated, act to regulate seizure susceptibility by raising seizure threshold or by interfering with seizure propagation.[19] These areas are referred to as seizure-gating networks.[19] It is interesting that many of the sites that decrease seizure activity are the same areas that have been used in the treatment of movement disorders. A decrease in experimentally-induced seizure activity has been reported to occur after stimulation of the basal ganglia,[20] cerebellum,[21,22] anterior thalamus,[23] subthalamic nucleus,[24] mammillary bodies,[25] locus ceruleus,[26,27] substantia nigra (SN),[28-31] and disinhibition of the superior colliculus.[32]

Of the sites listed above, the SN, the superior colliculus, and the subthalamic nucleus have received the most attention. The pars reticulata of the SN is believed to be a central player in a network that includes the striatum and the superior colliculus.[19] Manipulations that inhibit the SN tend to be anticonvulsant.[31,33] The SN sends inhibitory projections to the superior colliculus.[34,35] Therefore, any manipulation that inhibits the release of GABA from the SN should disinhibit the superior colliculus.

The anticonvulsant effect of electrical stimulation of the striatum and the subthalamic nucleus may also be mediated through the superior colliculus.[32,36] The striatum sends inhibitory projections to the SN. Electrical stimulation of the striatum inhibits the SN, which disinhibits the superior colliculus.[32] The subthalamic nucleus sends excitatory fibers to the pars reticulata of the SN.[37] Stimulation of the subthalamic nucleus is believed to inhibit its output.[32] Inhibition of the output of the subthalamic nucleus results in an inhibition of the SN and a disinhibition of the superior colliculus.[32,38] The mechanism by which activation of the superior colliculus is anticonvulsant remains to be determined.

The locus ceruleus sends widespread noradrenergic projections throughout the cortex and limbic system. Evidence from a number of studies suggests that activation of these ascending monoamine pathways can modulate seizure activity in several different experimental models of epilepsy. Indirect activation of the locus ceruleus has been proposed as a possible mechanism for the anticonvulsant effect of VNS.[18] Direct stimulation of the locus ceruleus suppressed penicillin-induced epileptiform activity in the rat[26,27] and locus ceruleus lesions result in an increase in epileptiform activity in the rat.[39] Studies that examined kindling acquisition revealed a decrease in the time to kindle when norepinephrine was depleted[40] and the alpha 2-adrenoreceptor agonist clonidine suppresses amygdala-kindled seizures.[41] The increased seizure susceptibility of the genetically epilepsy-prone rat has been partially attributed to a genetic deficiency in norepinephrine.[42] More recently it has been reported that transgenic mice that lack the ability to make norepinephrine have an increased susceptibility to seizures and that an intact noradrenergic system is required for the ketogenic diet to inhibit flurothyl-induced seizures in mice.[43,44] The results from these studies strongly suggest that activation of the locus ceruleus could be beneficial to epileptic patients.

The effects of anterior thalamic and cerebellar stimulation on experimentally-induced seizures have also been examined. High-frequency stimulation of the anterior thalamic nucleus raised the threshold for clonic seizures in rats injected with pentylenetetrazol (PTZ).[23] However, the effect of cerebellar stimulation is less clear. Several studies failed to observe any change in seizure activity after cerebellar stimulation.[5,6] Maiti and Snider[21] observed a decrease in electrically-induced seizure activity in the amygdala and hippocampus of cats and monkeys after midline cerebellar stimulation and cerebellar stimulation has been reported to decrease the number and duration of paroxysmal events induced by penicillin in the cat.[22]

Activation of Seizure-Gating Networks—Clinical Studies

Clinical studies of DBS and epilepsy have not yielded the positive results that have been obtained in animal models. The results from many of the early clinical studies are contradictory because some find a decrease in seizure activity while others fail to observe any effect of DBS. In addition, only a limited number of these clinical studies were properly controlled. The controlled study by Fisher et al[8] that examined the effect of centromedian thalamic stimulation on intractable seizures failed to detect a change in seizure frequency. Nevertheless, recent clinical studies have examined the effect of stimulation of structures that have been shown to decrease seizure activity in animal models. There are reports of decreased seizure activity after stimulation of the caudate,[45,46] the cerebellum,[45] the anterior thalamus,[11] the subthalamic nucleus,[12] the central median thalamic nucleus,[47] and the hippocampus.[13] Robert Fisher at Stanford is currently heading a clinical trial to investigate the effect of anterior thalamic stimulation on refractory partial seizures.[14]

Focal Stimulation

An alternative approach to using DBS to disinhibit seizure-gating networks is to apply the stimulus to block seizures directly at an identified focus. Lesser et al[48] were able to block stimulation-induced seizure afterdischarges by applying a brief burst of high-frequency stimulation. Velasco et al[13] observed a decrease in seizure activity in patients with intractable temporal lobe seizures after application of direct stimulation to the hippocampus. More recently, direct cortical stimulation has been used to decrease interictal spiking from the cortex of a patient with mesial temporal epilepsy.[49]

The effect of focal stimulation on seizure activity has also been examined in the kindling model of epilepsy. There are several reports that showed that electrical stimulation, delivered at a kindled focus, raised seizure threshold[50,51] and decreased seizure frequency.[52-55]

Using the in vitro hippocampal slice preparation, several laboratories have demonstrated that direct stimulation[56-59] and the application of electric fields[60-62] can disrupt experimentally-induced epileptiform activity. Low-frequency stimulation effectively blocked ictal activity in the hippocampal-entorhinal slice[63] and in the amygdala-perirhinal slice.[64]

Selection of Stimulation Parameters

Many of the experimental and clinical studies that have examined the effect of DBS on seizure activity have reported decreased seizure activity after stimulation. However, there are also studies that failed to detect a change in seizure activity and some studies even observed an increase in seizure activity after stimulation. A possible explanation for the variability of reported effects of electrical stimulation on seizures may be related to the type of stimulus used. Parameters such as stimulus frequency, intensity, duration and waveform may all be critical as to whether a given stimulus will successfully interfere with seizure activity.

Stimulus frequency seems to be a key parameter for disinhibition of seizure-gating networks. Many of the studies that report a decrease in seizure activity after stimulation of the basal ganglia and thalamic nuclei used a high-frequency stimulus. Mirski et al[23] observed an elevation in PTZ-induced clonic seizure threshold after 100 Hz stimulation of the anterior thalamus, while 8 Hz stimulation was proconvulsant. Why high-frequency stimulation was effective is unclear. It has been suggested that high-frequency stimulation has a lesion-like effect, causing the stimulated nucleus to stop functioning due to depolarization block, neural jamming (disruption of the network caused by stimulation-induced neuronal impulses) or preferential activation of inhibitory neurons.[65]

Low-frequency stimulation appears to be effective for focal stimulation in the kindling model[50-55] and in the in vitro slice preparation.[58,63,64] The low-frequency stimulus used in these studies is reminiscent of stimulations that lead to long-term depression (LTD). Durand and Bikson[66] have suggested "…low-frequency stimulation possesses the greatest potential for clinical benefit since the effect of the stimulation can last well beyond the duration of the pulse." High-frequency stimulation would increase synaptic efficacy, which could be epileptogenic.[66] Consistent with this hypothesis is the observation that the increase in kindled seizure threshold that occurred after low-frequency stimulation lasted for days.[50] It is interesting that high-frequency stimulation of the NTS, a site distant from the kindled focus, interfered with kindling acquisition.[17] Together, these results suggest that low-frequency stimulation may be more effective at the focus and high-frequency stimulation may be more effective in disinhibiting seizure-gating networks. However, there are clinical findings that don't support this hypothesis. Lesser et al,[48] blocked cortical afterdischarge activity induced during brain mapping stimulation with short bursts of high-frequency stimulation, and high-frequency focal hippocampal stimulation was effective in patients with mesial temporal sclerosis.[13]

It is unclear whether the choice of stimulus waveform is important. Several of the studies that reported decreases in kindled seizure threshold and seizure frequency used preemptive sine wave stimulation.[51-53,55] But low-frequency square wave stimulation also appears to be effective.[50,54]

Safety

The two primary safety issues associated with DBS are the requirement of chronic implantation of intracranial electrodes and the possibility that the repeated presentation of electric current could induce tissue damage. There are conflicting reports as to whether DBS causes tissue damage. The consequence of the long-term electrode implantation and the passing of current through these electrodes was examined in cat brains.[67] The only histological change observed in the tissue was gliosis around the electrode tract, but no additional damage was associated with the passage of current or the chronic presence of the foreign material in the brain. No tissue damage was found in the brains from patients with Parkinson's disease examined at autopsy, which received up to 70 months of DBS.[68] However, there is a case report of thalamic damage in a patient who received DBS for Parkinson's disease.[69] There is also the potential for tissue damage due to high current density at the electrode tip with low-frequency sine wave stimulation.

There are reports that DBS can cause a decline in cognitive function accompanied by behavioral changes.[70,71] Since many of the structures stimulated for movement disorders are also being stimulated for epilepsy, it is likely that changes in cognitive function and alterations in behavior after stimulation will also be a concern in patients with epilepsy.

The Future Is Now: Seizure Prediction Combined with Pre-Emptive Stimulation

Any advancement that improves our ability to predict when a clinical seizure will occur will enhance our ability to intervene and potentially prevent the seizure from occurring. New advances in seizure detection technology have provided evidence that seizures develop over time and that the onset of seizure activity may actually begin hours rather than seconds before the appearance of a clinical seizure. This evidence came from studies that incorporated chaos theory in their analysis of EEGs from epilepsy patients.[72-76] This was consistent with reports from patients who report symptoms hours before the seizure.[77] In an important retrospective study by Litt and colleagues[1] three different methods were used to detect changes in the EEG that might predict seizure onset in a group of patients with mesial temporal lobe epilepsy. The three methods used were: 1) to measure the total EEG energy in 5-min epochs; 2) detection of focal subclinical seizures; 3) measurement of the accumulated energy for one hour, 50 min before a seizure and 10 min after the seizure compared to one hour baseline measurements. Data were collected from depth electrodes and subtemporal strip electrodes. EEG energy was determined by squaring the voltage at each time point in the EEG and summing those values over a predetermined period of time. Using these methods, they were able to detect a series of subclinical electrographic seizures that tended to increase in frequency around clinical seizures. They found increases in EEG energy occurred as early as 7 hours before the unequivocal onset of seizure activity. Increases in energy were not detected at electrodes distant from the focus. These results suggest that once the appropriate algorithms are developed a responsive intelligent stimulator can be designed. An intelligent stimulator would detect the early onset of seizure activity and be able to deliver a preemptive blocking stimulus before a clinical seizure occurs. One of the benefits of an intelligent stimulator over the stimulators currently used for DBS and VNS is that an intelligent stimulator would deliver a stimulus only when needed, rather than continuously delivering current independent of the patient's physiological state.[14] This would decrease the likelihood that the stimulus will cause tissue damage or interfere with normal brain function. An external responsive neurostimulator is currently being tested in patients with intractable epilepsy.[78]

Conclusion

Direct electrical stimulation of the brain has the potential to become a safe new therapy for epilepsy. Several issues remain to be resolved. These include a determination of the optimal sites and types of stimuli that will be most effective. It is anticipated that once this occurs, the development of intelligent stimulators that combine seizure detection with the ability to deliver therapeutic stimulation will be possible.

References

1. Litt B, Esteller J, Eschauz M et al. Epileptic seizures may begin hours in advance of clinical onset: a report of five patients. Neuron 2001; 30:51-64.
2. Walker AE. An oscillographic study of the cerebello-cerebral relationship. J Neurophysiol 1938; 1:16-23.
3. Cooper IS, Amin I, Gilman S. The effect of chronic cerebellar stimulation upon epilepsy in man. Trans Am Neurol Assoc 1973; 98:192-196.
4. Cooper IS, Upton AR. Effects of cerebellar stimulation on epilepsy, the EEG and cerebral palsy in man. Electroencephalogr Clin Neurophysiol 1978; S34:349-354.
5. Myers RR, Burchiel KJ, Stockard JJ et al. Effects of acute and chronic paleocerebellar stimulation on experimental models of epilepsy in the cat: studies with enflurane, pentylenetetrazol, penicillin and chlorolose. Epilepsia 1975; 16:257-267.
6. Lockard JS, Ojemann GA, Condon WC et al. Cerebellar stimulation in alumina-gel monkey model: inverse relationship between clinical seizures and EEG interictal bursts. Epilepsia 1979; 20:223-234.
7. Wright GD, McLellan DL, Brice JG. A double-blind trial of chronic cerebellar stimulation in twelve patients with severe epilepsy. J Neurol Neurosurg Psychiatry 1984; 47:769-774.
8. Fisher RS, Uematsu S, Krauss GL et al. Placebo-controlled pilot study of thalamic stimulation in treatment of intractable seizure. Epilepsia 1992; 33:841-851.
9. The Vagus Nerve Stimulation Group. A randomized controlled trial of chronic vagus nerve stimulation for treatment of medically-intractable seizures. Neurology 1995; 45:224-230.
10. Schmidt D. Vagus nerve stimulation for the treatment of epilepsy. Epilepsy and Behavior 2001; 2:S1-S5.
11. Hodaie M, Wennberg RA, Dostrovsky JO et al. Chronic anterior thalamus stimulation for intractable epilepsy. Epilepsia 2002; 43:603-608.

12. Benabid AL, Minotti L, Koudsie A et al. Antiepileptic effect of high-frequency stimulation of the subthalamic nucleus (Corpus Luysi) in a case of medically intractable epilepsy caused by focal dysplasia: a 30-month follow-up: Technical Case Report. Neurosurgery 2002; 50:1385-1392.
13. Velasco M, Velasco F, Velasco AL et al. Subacute electrical stimulation of the hippocampus blocks intractable temporal lobe seizures and paroxysmal EEG activities. Epilepsia 2000; 41:158-169.
14. Litt B, Baltuch G. Brain stimulation for epilepsy. Epilepsy and Behavior 2001; 2:S61-S67.
15. Litt B, Lehnertz K. Seizure prediction and the preseizure period. Curr Opin Neurol 2002; 15:173-177.
16. Krout KE, Loewy AD. Parabrachial nucleus projections to midline and intralaminar thalamic nuclei of the rat. J Comp Neurol 2000; 428:475-494.
17. Magdaleno-Madrigal VM, Valdez-Cruz A, Martinez-Vargas D et al. Effect of electrical stimulation of the nucleus of the solitary tract on the development of electrical amygdaloid kindling in the cat. Epilepsia 2002; 43:964-969.
18. Krahl SE, Clark KB, Smith DC et al. Locus coeruleus lesions suppress the seizure-attenuating effects of vagus nerve stimulation. Epilepsia 1998; 39:709-714.
19. Gale K. Subcortical structures and pathways involved in convulsive seizure generation. J Clin Neurophysiol 1992; 9:264-277.
20. Deransart C, Riban V, Le-Pham BT et al. Evidence for the involvement of the pallidum in the modulation of seizures in a genetic model of absence epilepsy in the rat. Neurosci Lett 1999; 265:131-134.
21. Maiti A, Snider R. Cerebellar control of basal forebrain seizures: amygdala and hippocampus. Epilepsia 1975; 6:521-533.
22. Hablitz JJ, Rea G. Cerebellar nuclear stimulation in generalized penicillin epilepsy. Brain Res Bull 1976; 1:599-601.
23. Mirski MA, Rossell LA, Terry JB et al. Anticonvulsant effect of anterior thalamic high frequency electrical stimulation in the rat. Epilepsy Res 1997; 28:89-100.
24. Vercueil L, Benazzouz A, Deransart C et al. High-frequency stimulation of the subthalamic nucleus suppresses absence seizures in the rat: comparison with neurotoxic lesions. Epilepsy Res 1998; 31:39-46.
25. Mirski M, Fisher R. Electrical stimulation of the mammillary nuclei increases seizure threshold to pentylenetetrazol in rats. Epilepsia 1994; 35:1309-1316.
26. Neuman RS. Suppression of penicillin-induced focal epileptiform activity by locus ceruleus stimulation: mediation by an alpha-1-adrenoreceptor. Epilepsia 1986; 27:359-366.
27. Ferraro G, Sardo P, Sabatino M et al. Locus coeruleus noradrenaline system and focal penicillin hippocampal epilepsy: neurophysiological study. Epilepsy Res 1994; 19:215-220.
28. Grutta VL, Sabatino M. Substantia nigra-mediated anticonvulsant action: a possible role of a dopaminergic component. Brain Res 1990; 515:87-93.
29. Sabatino M, Gravante G, Ferraro G et al. Striatonigral suppression of focal epilepsy. Neurosci Lett 1989; 98:285-290.
30. Sabatino M, Ferraro G, Vella N et al. Nigral influence on focal epilepsy. Neurophysiol Clin 1990; 20:189-201.
31. Velisek L, Veliskova J, Moshe SL. Electrical stimulation of substantia nigra pars reticulata is anticonvulsant in adult and young male rats. Exp Neurol 2002; 173:145-152.
32. Bressand K, Dematteis M, Gao DM et al. Superior colliculus firing changes after lesion or electrical stimulation of the subthalamic nucleus in the rat. Brain Res 2002; 943:93-100.
33. Deransart C, Le-Pham BT, Hirsch E et al. Inhibition of the substantia nigra suppresses absences and clonic seizures in audiogenic rats, but not tonic seizures: evidence for seizure specificity of the nigral control. Neuroscience 2001; 105:203-211.
34. DiChiarra G, Poceddu ML, Morelli M et al. Evidence for a GABAergic projection from the substantia nigra to the ventromedial thalamus and to the superior colliculus of the rat. Brain Res 1979; 272:368-372.
35. Garant DS, Gale K. Substantia nigra-mediated anticonvulsant actions: role of nigral output pathways. Exp Neurol 1987; 97:143-159.
36. Chevlier G, Thierry AM, Shibazaki T et al. Evidence for a GABAergic inhibitory nigrotectal pathway in the rat. Neurosci Lett 1981; 21:67-70.
37. Kita H, Kitai ST. Efferent projections of the subthalamic nucleus in the rat: light and electron microscope analysis with the PHA-L method. J Comp Neurol 1987; 260:435-452.
38. Dybdal D, Gale K. Postural and anticonvulsant effects of inhibition of the rat subthalamic nucleus. J Neurosci 2000; 20:6728-6733.
39. Sullivan HC, Osorio I. Aggravation of penicillin-induced epilepsy in rats with locus ceruleus lesions. Epilepsia 1991; 32:591-596.

40. Corcoran ME. Characteristics of accelerated kindling after depletion of noradrenaline in the rat. Neuropharmacology 1988; 27:1081-1084.
41. Shouse M, Langer J, Bier M et al. The alpha 2-adrenoreceptor agonist clonidine suppresses seizures, whereas the alpha 2-adrenoreceptor antagonist idazoxan promotes seizures in amygdala-kindled kittens: a comparison of amygdala and pontine microinfusion effects. Epilepsia 1996; 37:709-717.
42. Jobe PC, Dailey JW, Reigel CE. Noradrenergic and serotonergic determinants of seizure susceptibility and severity in genetically epilepsy-prone rats. Life Sci 1986; 39:775-782.
43. Szot P, Weinshenker D, White SS et al. Norepinephrine-deficient mice have increased susceptibility to seizure-inducing stimuli. J Neurosci 1999; 19:10985-10992.
44. Szot P, Weinshenker D, Rho JM et al. Norepinephrine is required for the anticonvulsant effect of the ketogenic diet. Brain Res Dev Brain Res 2001; 129:211-214.
45. Sramka M, Chkhenkeli SA. Clinical experience in intraoperative determination of brain inhibitory structures and application of implanted neurostimulators in epilepsy. Stereotact Funct Neurosurg 1990; 54-55:56-59.
46. Chkhenkeli SA, Chkhenkeli IS. Effects of therapeutic stimulation of nucleus caudatus on epileptic electrical activity of brain in patients with intractable epilepsy. Stereotact Funct Neurosurg 1997; 69:221-224.
47. Velasco F, Velasco M, Jimenez F et al. Stimulation of the central median thalamic nucleus for epilepsy. Stereotact Funct Neurosurg 2001; 77:228-232.
48. Lesser RP, Kim SH, Beyderman DL et al. Brief bursts of pulse stimulation terminate afterdischarges caused by cortical stimulation. Neurology 1999; 53:2073-2081.
49. Yamamoto J, Ikeda A, Satow T et al. Low-frequency electric cortical stimulation has an inhibitory effect on epileptic focus in mesial temporal lobe epilepsy. Epilepsia 2002; 43:491-495.
50. Ullal GR, Ninchoji T, Uemura K. Low-frequency stimulation induces an increase in afterdischarge thresholds in hippocampal and amygdaloid kindling. Epilepsy Res 1989; 3:232-235.
51. McIntyre DC, Gilby K, Carrington CA. Effect of low-frequency stimulation on amygdala-kindled afterdischarge thresholds and seizure profile in fast and slow kindling rat strains. Epilepsia 2002; 43(S7):12.
52. Gaito J. The effect of variable duration one hertz interference on kindling. Can J Neurol Sci 1980; 7:59-64.
53. Gaito J, Nobrega JN, Gaito ST. Interference effect of 3 Hz brain stimulation on kindling behavior induced by 60 Hz stimulation. Epilepsia 1980; 21:73-84.
54. Velisek L, Veliskova J, Stanton PK. Low-frequency stimulation of the kindling focus delays basolateral amygdala kindling in immature rats. Neurosci Lett 2002; 326:61-63.
55. Goodman JH, Berger RE, Scharfman HE et al. Low-frequency sine wave stimulation decreases seizure frequency in amygdala-kindled rats. Epilepsia 2002; 43(S7):10.
56. Durand DM, Warren EN. Desynchronization of epileptiform activity by extracellular current pulses in rat hippocampal slices. J Physiol 1994; 480:527-537.
57. Warren RJ, Durand DM. Effects of applied currents on spontaneous epileptiform activity induced by low calcium in the rat hippocampus. Brain Res 1998; 806:186-195.
58. Jerger K, Schiff SJ. Periodic pacing an in vitro epileptic focus. J Neurophysiol 1995; 73:876-879.
59. Gluckman BJ, Neel EJ, Netoff TI et al. Electric field suppression of epileptiform activity in hippocampal slices. J Neurophysiol 1996; 76:4202-4205.
60. Gluckman BJ, Nguyen H, Weinstein SL et al. Adaptive electric field control of epileptic seizures. J Neurosci 2001; 21:590-600.
61. Ghai RS, Bikson M, Durand DM. Effects of applied electric fields on low-calcium epileptiform activity in the CA1 region of rat hippocampal slices. J Neurophysiol 2000; 84:274-280.
62. Bikson M, Lian J, Hahn PJ et al. Suppression of epileptiform activity by high frequency sinusoidal fields in rat hippocampal slices. J Physiol 2001; 531:181-191.
63. Barbarosie M, Avoli M. CA3-driven hippocampal-entorhinal loop controls rather than sustains in vitro limbic seizures. J Neurosci 1997; 17:9308-9314.
64. Kano T, D'Antuono M, d Guzman P et al. Low-frequency stimulation of the amygdala inhibits ictogenesis in the perirhinal cortex. Epilepsia 2002; 43(S7):129.
65. Benazzouz A, Hallet M. Mechanism of action of deep brain stimulation. Neurology 2000; 55(12 Suppl 6):S13-16.
66. Durand DM, Bikson M. Suppression and control of epileptiform activity by electrical stimulation: a review. Proc IEEE 2001; 89:1065-1082.
67. Stock G, Strum V, Schmitt HP et al. The influence of chronic deep brain stimulation on excitability and morphology of the stimulated tissue. Acta Neurochir (Wien) 1979; 47:123-129.
68. Haberler C, Alesch F, Mazal PR et al. No tissue damage by chronic deep brain stimulation in Parkinson's disease. Ann Neurol 2000; 48:372-376.

69. Henderson JM, O'Sullivan DJ, Pell M et al. Lesion of thalamic centromedian-parafascicular complex after chronic deep brain stimulation. Neurology 2001; 56:1576-1579.
70. Saint-Cyr JA, Trepanier LL, Kumar R et al. Neuropsychological consequences of chronic bilateral stimulation of the subthalamic nucleus in Parkinson's disease. Brain 2000; 123:2091-2108.
71. Dujardin K, Defebvre L, Krystkowiak P et al. Influence of chronic bilateral stimulation of the subthalamic nucleus on cognitive function in Parkinson's disease. J Neurol 2001; 248:603-611.
72. Iasemidis LD, Sackellares JC, Zaveri HP et al. Phase space topography and the Lyapunov exponent of electrocorticograms in partial seizures. Brain Topogr 1990; 2:187-201.
73. Martinerie J, Adam C, Quyen MLV et al. Epileptic seizures can be anticipated by nonlinear analysis. Nat Med 1998; 4:1173-1176.
74. Lehnertz K, Elger C. Can epileptic seizures be predicted? Evidence from nonlinear time series analysis of brain electrical activity. Phys Rev Lett 1998; 80:5019-5022.
75. Osorio I, Frei M, Wilkinson S. Real-time automated detection and quantitative analysis of seizure and short-term prediction of clinical onset. Epilepsia 1998; 39:615-627.
76. Jerger KK, Netoff TI, Francis JT et al. Early seizure detection. J Clin Neurophysiol 2001; 18:259-268.
77. Sperling M, O'Conner M. Auras and subclinical seizures: characteristics and prognostic significance. Ann Neurol 1990; 28:320-328.
78. Bergey GK, Britton JW, Cascino GD et al. Implementation of an external responsive neurostimulator system (eRNS) in patients with intractable epilepsy undergoing intracranial seizure monitoring. Epilepsia 2002; 43(S7):191.